Applied Mathematical Sciences
Volume 54

Applied Mathematical Sciences

Wolfgang Wasow

Linear Turning Point Theory

With 19 Illustrations

Springer-Verlag
New York Berlin Heidelberg Tokyo

Wolfgang Wasow
University of Wisconsin–Madison
Mathematics Department
Madison, Wisconsin 53706
U.S.A.

Editors

F. John
Courant Institute of
  Mathematical Sciences
New York University
New York, NY 10012
U.S.A.

J. E. Marsden
Department of
  Mathematics
University of California
Berkeley, CA 94720
U.S.A.

L. Sirovich
Division of
  Applied Mathematics
Brown University
Providence, RI 02912
U.S.A.

AMS Subject Classifications: 65L99, 34E99, 34A30, 34A50

Library of Congress Cataloging in Publication Data

Wasow, Wolfgang Richard,
  Linear turning point theory.

  (Applied mathematical sciences; vol. 54)
  Bibliography: p.
  1. Differential equations—Asymptotic theory.
I. Title. II. Series: Applied mathematical sciences
(Springer-Verlag New York Inc.); v. 54.
QA1.A647   vol. 54   510 s   84-20243
[QA372]   [513.3'5]
ISBN 0-387-96046-5

Typeset by Computype, Inc., St. Paul, Minnesota.
Printed and bound by R. R. Donnelley & Sons, Harrisonburg, Virginia.
Printed in the United States of America.

9 8 7 6 5 4 3 2 1

ISBN 0-387-96046-5 Springer-Verlag New York Berlin Heidelberg Tokyo
ISBN 3-540-96046-5 Springer-Verlag Berlin Heidelberg New York Tokyo

Preface

My book "Asymptotic Expansions for Ordinary Differential Equations" published in 1965 is out of print. In the almost 20 years since then, the subject has grown so much in breadth and in depth that an account of the present state of knowledge of all the topics discussed there could not be fitted into one volume without resorting to an excessively terse style of writing.

Instead of undertaking such a task, I have concentrated, in this exposition, on the aspects of the asymptotic theory with which I have been particularly concerned during those 20 years, which is the nature and structure of turning points. As in Chapter VIII of my previous book, only linear analytic differential equations are considered, but the inclusion of important new ideas and results, as well as the development of the necessary background material have made this an exposition of book length.

The formal theory of linear analytic differential equations without a parameter near singularities with respect to the independent variable has, in recent years, been greatly deepened by bringing to it methods of modern algebra and topology. It is very probable that many of these ideas could also be applied to the problems concerning singularities with respect to a parameter, and I hope that this will be done in the near future. It is less likely, however, that the analytic, as opposed to the formal, aspects of turning point theory will greatly benefit from such an algebraization. Also, mathematically trained readers who are mostly interested in applications may prefer the more elementary, narrative style in which my previous book was written, and which I have accepted for the present one as well.

The book is primarily concerned with general mathematical methods, but its contents may help applied mathematicians to understand better the many special cases, modifications and additions that can be found in the extensive literature.

The presupposed mathematical background includes the elements of advanced calculus, complex variable theory, linear algebra, and ordinary differential equations, but no knowledge that is not normally acquired by an undergraduate student of mathematics whose main mathematical interest is Analysis.

I am grateful to the Mathematics Department of the University of Wisconsin–Madison, from which I retired four years ago, for allowing me to draw freely on its secretarial services in preparing the manuscript of this book, and special thanks are due to its capable and patient staff of technical typists.

Wolfgang Wasow

Contents

Chapter VIII   Connection Problems

Chapter IX   Fedoryuk's Global Theory of Second-Order Equations

Chapter X   Doubly Asymptotic Expansions

Chapter XI   A Singularly Perturbed Turning Point Problem

Chapter XII Appendix: Some Linear Algebra for Holomorphic Matrices

CHAPTER I

Historical Introduction

1.1. Early Asymptotic Theory Without Turning Points

Turning point theory is a branch of the asymptotic theory of ordinary differential equations that depend in a singular manner on a parameter. "Turning points" are certain exceptional points in that theory. Their precise definition in a general framework is, in itself, a nontrivial matter which will be discussed later on. While turning points are exceptional, their analysis is essential to a full understanding of the asymptotic nature of the solutions of such differential equations. This is a general situation in Mathematics. In an analogous way, analytic functions can only be understood through a study of their singularities; the solutions of an ordinary differential equation depend decisively on the location and type of their critical points, etc. Also, the mathematical formulation of many problems of Physics and Engineering involves turning point problems, and this has been the principal motivation for most of the early investigations.

In spite of the importance of turning points, the first 100 years of the asymptotic theory of differential equations having singularities in a parameter avoided the consideration of such points. A very brief description of this earlier work is, however, necessary for the understanding of later developments.

That theory has been traced back to the year 1817, when the Italian astronomer F. Carlini published a long paper on the calculation of planetary orbits. This contribution would probably be completely forgotten by now if Jacobi had not reissued it in German translation in 1850 [7]. The surprisingly accurate asymptotic approximations derived there are based on a rather confused analysis, which has been substantially clarified by Schlissel in Ref. 70.

Except for Carlini's early isolated effort, the starting point of the whole asymptotic theory of differential equations is the year 1837, when Liouville [46] and G. Green [17] each published papers in which differential equations of the form

$$\epsilon^2 \frac{d^2u}{dx^2} + \varphi u = 0, \qquad (1.1\text{-}1)$$

as well as somewhat more general ones, were analyzed for small values of the real variable ϵ. Here, φ denotes a real, sufficiently smooth function of the real variable x. Both authors observed that the functions

$$\hat{u}(x,\epsilon) = \left[\varphi(x)\right]^{-1/4}\exp\left\{ \pm \frac{i}{\epsilon} \int^x \left[\varphi(t)\right]^{1/2} dt\right\} \qquad (1.1\text{-}2)$$

"almost" solve the differential equation in the sense that

$$\epsilon^2 \frac{d^2\hat{u}(x,\epsilon)}{dx^2} + \varphi(x)\hat{u}(x,\epsilon) = O(\epsilon\hat{u}(x,\epsilon)) \qquad \text{as} \quad \epsilon \to 0. \qquad (1.1\text{-}3)$$

There are several natural plausibility arguments that lead to the functions $\hat{u}$ as likely candidates for approximate solutions of the differential equation. Liouville's procedure for proving this conjecture, which is the germ of more general modern methods, can be paraphrased as follows:

Construct a linear transformation

$$u = a(x)v \qquad (1.1\text{-}4)$$

of the dependent variable and a transformation

$$\xi = \xi(x), \qquad (1.1\text{-}5)$$

such that the new differential equation differs from some special, already solvable differential equation by terms that tend to zero with ϵ. Liouville's choice for that special "comparison equation" was

$$\epsilon^2 \frac{d^2w}{d\xi^2} + w = 0. \qquad (1.1\text{-}6)$$

One verifies directly that the transformed differential equation is

$$a(\xi')^2\ddot{v} + (a\xi'' + 2a'\xi')\dot{v} + (a'' + \epsilon^{-2}\varphi a)v = 0, \qquad (1.1\text{-}7)$$

where the prime and the dot indicate differentiation with respect to x and to ξ, respectively. If

$$a\xi'' + 2a'\xi' = 0, \qquad (\xi')^2 = \varphi, \qquad (1.1\text{-}8)$$

and, therefore,

$$a = \varphi^{-1/4}, \qquad \xi = \int^x \varphi(t)^{1/2} dt, \qquad (1.1\text{-}9)$$

the transformed differential equation becomes

$$\epsilon^2 \ddot{v} + v + \epsilon^2 \varphi^{-3/4}(\varphi^{-1/4})'' v = 0. \tag{1.1-10}$$

Green accepts without proof the plausible conjecture that, at least for $\varphi(x) \neq 0$, equation (1.1-10) has solutions which are close to those of (1.1-6) when ϵ is small. In view of (1.1-4) and (1.1-5) this means, indeed, that (1.1-2) represents approximations to solutions of equation (1.1-1). The conclusion that a function which "almost" solves a differential equation is "almost" a solution is prevalent among applied scientists. In view of the many hypotheses and complications inherent in the application of mathematics to the material world, I consider it a justifiable position that this problem should be left for the mathematicians to worry about.

Liouville, who came to the asymptotic study of equation (1.1-1) in the framework of a purely mathematical theory, namely, the "Sturm–Liouville Theory" of generalized Fourier series, did proceed to prove carefully that the differential equation possesses, corresponding to given finite intervals in which $\varphi(x)$ does not vanish, true solutions u such that

$$u(x,\epsilon) - \hat{u}(x,\epsilon) = O\left\{\epsilon \exp\left[\pm \frac{i}{\epsilon}\int^x \varphi(t)^{1/2}\,dt\right]\right\} \qquad \text{as} \quad \epsilon \to 0+.$$

Incidentally, Liouville's article [46] is still a joy to read and quite up to our present standards of precise mathematical thinking, but it does not belong in this Introduction.

The assumption that x ranges over an interval where $\varphi(x) \neq 0$ is essential. Otherwise, the function $\hat{u}$ is not even everywhere defined. The zeros of φ are the "turning points" of the present problem. It is surprising that for 100 years none of the many mathematicians who have contributed to the development of the asymptotic theory took a close look at the behavior of the solutions near such turning points. The beginning of a serious interest in these questions came from physicists who were led to them through the mathematical study of natural phenomena.

1.2. Total Reflection and Turning Points

The earliest substantial publication that would now be said to deal with turning points is a paper that appeared in 1915 [14]. There, R. Gans, a theoretical physicist, studied the propagation of light in a nonhomogeneous medium as an application of Maxwell's equations and was led to equation (1.1-1). It turned out that the phenomenon of total reflection, when studied from the viewpoint of physical—as opposed to geometric—optics, required the solution of that differential equation for small ϵ in an interval where $\varphi(x)$ changes sign, and this is where the new difficulties come in. Gans assumes that φ has a zero at $x = 0$ which is of first order, so that one may write

$$\varphi(x) = x\psi(x), \qquad \psi(0) \neq 0. \qquad (1.2\text{-}1)$$

In intervals where $x \neq 0$, the approximation (1.1-2) can be used. To fix the ideas, assume that $\psi(0) > 0$ and that the lower end point of the integral in (1.1-2) is at zero. Then, for small $\epsilon > 0$, that formula represents two oscillating functions for $x > 0$, while the corresponding two functions for $x < 0$ have, respectively, growing and decaying exponential character, as $\epsilon \to 0$. Unfortunately, it is, in general, not true that the solutions approximated by one of the two functions in (1.1-2) on one side of the origin are also approximated on the other side by one of those functions. Rather, the continuation of the solution is approximately represented there by certain linear combinations of the two functions (1.1-2), and the determination of the coefficients of these linear combinations—they are functions of ϵ—is a difficult "connection problem."

Gans solved that problem by introducing the differential equation

$$\epsilon^2 u_0'' + x\psi(0)u_0 = 0, \qquad (1.2\text{-}2)$$

whose solutions he expected to be close to those of the full equation (1.1-1) as long as x was sufficiently small. Now, equation (1.2-2) can be solved explicitly in terms of Bessel functions of order $\frac{1}{3}$, if the new variable

$$\tau = \tfrac{2}{3}\big(\psi(0)\epsilon^{-1}x^3\big)^{1/2} \qquad (1.2\text{-}3)$$

is introduced (see [1], Chapter 10). This is the first systematic use of what is called a "stretching" transformation in the asymptotic theory. The term describes the fact that, as ϵ shrinks, (1.2-3) maps intervals on the x-axis into increasingly larger τ-intervals. An essentially equivalent, slightly simpler stretching transformation is

$$t = \big(\psi(0)\epsilon^{-2}\big)^{1/3}x, \qquad (1.2\text{-}4)$$

which takes (1.2-2) into the so-called Airy equation,

$$\frac{d^2u_0}{dt^2} + tu_0 = 0, \qquad (1.2\text{-}5)$$

whose solutions can be represented for all t by integrals or by power series.

If Gans' conclusions are accepted, there are now available three approximately known fundamental systems for the differential equation (1.1-1): (i) One that is approximated by the expressions (1.1-2) in intervals strictly to the left of $x = 0$; (ii) one approximated by the same expressions on the right of $x = 0$; (iii) one approximated by expressions involving Bessel functions of order $\frac{1}{3}$ of the variable τ in (1.2-3). The solutions of type (i) or (ii) are often called "outer" solutions. The third type is then an "inner" solution. If it can be shown that, taken together, the three intervals in which these pairs of expressions are close to true solutions do cover, for all sufficiently small $\epsilon > 0$, a full interval $|x| \leqslant x_0$, then a solution given by initial values at $-x_0$, say, can be approximated in the whole interval by "matching" it first with the appropriate linear combination of the inner solutions, and then with the right-hand outer solutions. In this way the "connection problem" across the turning point $x = 0$ can be solved.

Gans carried the formal part of this argument out, carefully and correctly and, I believe, fully aware that he was giving convincing plausibility arguments, not complete proofs. The terms "comparison equation," "stretching," "matching," "outer approximation," "inner approximation," "turning point," "connection problem" are not his, but it is remarkable that these concepts, under the names I have used, still dominate the modern asymptotic theory.

1.3. Hydrodynamic Stability and Turning Points

Proceeding chronologically, the next important stimulus to turning point theory came from physical problems concerning the stability of small disturbances in nonturbulent viscous plane flows. This question leads to a differential equation of the following type:

$$\epsilon^2 u^{(4)} + x\psi(x,\epsilon)u'' + b(x,\epsilon)u = 0. \tag{1.3-1}$$

(The notation used here brings out the mathematical structure but ignores several parameters important for the physical interpretation.) Equation (1.3-1) is often called the Orr–Sommerfeld equation. Its study for small ϵ, which physically corresponds to low viscosity or high velocity, is important for the answer to the practical question as to when small disturbances will cause the flow to become turbulent.

Again, the physical questions to be answered require that the equation (1.3-1) be analyzed in a domain which includes the point $x = 0$, which, in present-day terminology, would be called a turning point. That the point $x = 0$ plays an exceptional role is *a priori* plausible for several reasons. For one, near $x = 0$ the effect of the smallness of ϵ is compensated for by the smallness of x. Also, the "reduced" differential equation

$$x\psi(x,0)u_0'' + b(x,0)u_0 = 0, \tag{1.3-2}$$

obtained by setting $\epsilon = 0$ in (1.3-1), has multivalued solutions near $x = 0$, at least when $\psi(0,0) \neq 0$, while all solutions of the full equation (1.3-1) are there single-valued, provided ψ and b are.

The asymptotic theory of the Orr–Sommerfeld equation is much harder than that of equation (1.1-1), not only because it is of higher order, but also because the coefficients of (1.3-1) are not real, but holomorphic complex-valued functions, even for real x. Therefore, its solutions must be investigated in regions of the complex x-plane.

Among the many mathematicians who have struggled with this problem, only a few can be singled out in this brief account. Heisenberg [22], shortly before his fundamental contributions to Quantum Theory, wrote a long paper on equation (1.3-1) in which skillful formal calculations are supported by powerful physical intuition. His basic approach resembles in many respects that of Gans for the simpler equation (1.1-1), but I know no reason to conclude that he knew Gans' paper.

Heisenberg begins by constructing four functions which "almost" solve (1.3-1) for small ϵ. They are

$$\hat{u}_j(x,\epsilon) = x^{-5/4}\exp\left\{(-1)^{j-1}\frac{i}{\epsilon}\int_0^x (t\psi(t,0))^{1/2}\,dt\right\}, \qquad j = 1,2$$
$$\hat{u}_j(x,\epsilon) = \hat{u}_{j0}(x) + \hat{u}_{j1}(x)\epsilon, \qquad\qquad\qquad j = 3,4. \tag{1.3-3}$$

Here $\hat{u}_{j0}, \hat{u}_{j1}$ are certain functions. The $\hat{u}_{j0}$, $j = 3,4$, are two independent solutions of the reduced equation (1.3-2). The analogy of $\hat{u}_j$, $j = 1,2$, with the approximate solution of (1.1-1) in (1.1-2) is apparent.

Again, these four functions cannot possibly approximate a fundamental system of solutions in a full complex neighborhood of $x = 0$. Therefore, Heisenberg performs the stretching transformation

$$\tau = \epsilon^{-2/3}x \tag{1.3-4}$$

and gets the new differential equation

$$\frac{d^4u}{d\tau^4} + \tau\psi(\tau\epsilon^{2/3},\epsilon)\frac{d^2u}{d\tau^2} + \epsilon^{2/3}b(\tau\epsilon^{2/3},\epsilon)u = 0. \tag{1.3-5}$$

By virtue of a fundamental theorem on analytic linear differential equations, (1.3-5) possesses a fundamental system of solutions that are holomorphic in $\epsilon^{1/3}$ at $\epsilon = 0$, at least in finite regions of the τ-plane. The first approximations to these solutions are double integrals of Bessel functions of order $\frac{1}{3}$.

By a matching technique the particular "inner solutions" of (1.3-5) which are the continuations of the solutions $\hat{u}_j$, $j = 1,2,3,4$, of (1.3-1) can be explicitly approximated. The procedure is ingenious and essentially correct, but far from mathematically complete. For one, the proof that the functions $\hat{u}_j$ are approximate solutions in regions bounded away from $x = 0$ is lacking, and, second, any bounded neighborhood of $\tau = 0$ in the τ-plane corresponds to a neighborhood of $x = 0$ that shrinks to a point, as $\epsilon \to 0$, so that there might be a gap for small ϵ, between the domains of validity of the outer and the inner approximations.

Heisenberg's mathematics were further developed by Tollmien in a series of papers between 1929 and 1948 ([80], [81], etc.) and clarified in several respects by C. C. Lin in Ref. 45, and in several earlier papers and, to some extent, by two papers of mine [85], [86]. Only quite recently the contributions of W. H. Reid (see [9] for an exposition and for references) have introduced methods by which the gaps in the theory can be closed. To

Heisenberg, however, goes the credit for being the first to use outer and inner approximations systematically and to match them, in analogy to the techniques employed earlier for the simpler equation (1.1-1).

1.4. The So-Called WKB Method

Shortly after Schroedinger introduced into quantum mechanics the partial differential equation named after him, the efforts of applying it to the problem of a simple atomic collision led again to the ordinary differential equation (1.1-1) with a function φ that had one or more zeros in the interval of interest. Sometimes φ also had poles and often it also depended in a regular manner on ϵ. (In these applications ϵ is small because it involves Planck's constant.)

In 1926 three physicists, G. Wentzel [97], H. A. Kramers [34], and L. Brillouin [6], rediscovered Gans' results without being aware of them. Their methods differed in detail, and they added some new results regarding the relevant eigenvalue problem, which was the task of adjusting a parameter (usually the energy) occurring in φ so that there would exist a solution bounded in $-\infty < x < \infty$ (or else in intervals bounded by poles of φ) for small $\epsilon > 0$. This work was of prime importance in atomic physics, and it greatly enhanced the interest in such problems among mathematicians. The term "turning points" for the zeros of φ came into use at that time. It was probably introduced by Kramers in the German version "Umkehrpunkt" and refers to the fact that in the limit of classical mechanics a zero of φ corresponds to an obstacle in a collision, at which the moving particle has to turn back (just as total reflection of a ray in Gans' paper appeared as a limit phenomenon in the passage from physical to geometric optics). To the mathematicians the English terminology appealed for a different reason: If φ is analytic, paths that go around the zeros of φ in the complex x-plane are used in methods of connecting the right and left outer solutions without searching for complete information on the inner solutions. The expression "transition point" is older than "turning point" and is still in use, occasionally with a meaning somewhat more general than that of the latter.

The three papers of 1926 mentioned above have had such an impact on the physicists that the abbreviation "WKB method"—after the initials of their authors—is now in common use, not only to describe their treatments of that particular turning point problem but, more sweepingly, for any asymptotic solution of the differential equation (1.1-1) or more general equations. Even the approximation (1.1-2), known as early as 1837, is frequently called the WKB approximation. Soon after 1926 it was pointed out that H. Jeffreys had independently rediscovered Gans' method in 1924 [30]. Since then, it has frequently been referred to as the JWKB method.

Since 1926 many theoretical physicists and a number of mathematicians have developed refined and precise methods for solving eigenvalue problems for differential equations of the type (1.1-1). This body of work also often has the label WKB affixed to it. It is not a part of this account.

1.5. The Contribution of R. E. Langer

Only with the work of R. E. Langer ([36]–[39], etc.), did mathematically complete proofs enter turning point theory. His main motivation was his desire to put the asymptotic theory of equation (1.1-1) in regions containing zeros of φ on a precise mathematical basis. In subsequent papers he also studied some equations of higher order. Langer's first new idea was to extend Liouville's method of constructing a transformation of the given equation into one close to a simpler comparison transformation by using a comparison equation closer to the problem to be solved than equation (1.1-6).

To sketch Langer's procedure in his earlier contributions, consider again equation (1.1-1) under the hypothesis—more general than (1.2-1)—that

$$\varphi(x) = x^\alpha \psi(x), \qquad \psi(0) \neq 0, \tag{1.5-1}$$

where α is some positive (not necessarily integral) number and ψ is holomorphic at $x = 0$. Langer now chooses the transformation (1.1-4), (1.1-5) in such a way that the new differential equation becomes

$$\epsilon^2 \frac{d^2v}{d\xi^2} + \xi^\alpha v + \epsilon^2 \chi(\xi, \epsilon)v = 0. \tag{1.5-2}$$

For values of α such that χ is bounded, or at least integrable at $\epsilon = 0$, this equation differs so little from the comparison equation

$$\epsilon^2 \frac{d^2w}{d\xi^2} + \xi^\alpha w = 0 \tag{1.5-3}$$

that w can be hoped to be asymptotically equal to v, as $\epsilon \to 0$. Equation (1.5-3) can be completely solved by certain Bessel functions.

The appropriate functions a and ξ can be calculated by a simple modification of the procedure in section 1.1. One finds that one must choose

$$\xi(x) = \left[\frac{\alpha + 2}{2} \int_0^x t^{\alpha/2} \psi^{1/2}(t)\, dt \right]^{2/(\alpha+2)} \tag{1.5-4}$$

and

$$a(x) = \left(\frac{d\xi}{dx} \right)^{-1/2}; \tag{1.5-5}$$

then,

$$\chi = a'' a^3. \tag{1.5-6}$$

Observe that the multivalued function ξ in (1.5-4) has no branch point at $x = 0$, but that its values define a set of functions holomorphic in x at $x = 0$, and that

$$\xi(x) = kx + O(x^2), \qquad k \neq 0, \quad \text{as} \quad x \to 0,$$

with a constant k depending on the choice of branch in (1.5-4). Thus, χ is bounded at $x = 0$. This fact enables Langer to prove that equation (1.5-2) has solutions which are close, as $\epsilon \to 0$, to solutions of (1.5-3) not only in regions of the complex plane that are bounded away from the turning point, but even in certain sectors with vertex at $\xi = 0$, i.e., $x = 0$. Thus, the distinction of outer and inner solutions is unnecessary. The only required matching of solutions takes place at $x = 0$ and is comparatively easy.

Langer's method has been described here in a way that emphasizes the analogy with Liouville's approach. Actually, Langer proceeds slightly differently, but the difference is trivial: Rather than to transform the given equation into one close to equation (1.5-3), he transforms (1.5-3) into an equation close to the given one. In later papers Langer improves the approximations so that they are close to true solutions to any order of ϵ.

Langer's work was welcomed by the physicists because it made their connection problems so much easier to solve, and also because having a precise mathematical foundation to the theory, the common misunderstandings and the confusion about what was actually meant by statements on asymptotic approximation were more easily clarified.

Langer's comparison method is very satisfactory when it can be applied. It turns out, however, that this is a quite exceptional situation. Even the relatively simple Orr–Sommerfeld equation has never been transformed into an equation solvable by already well-known transcendental functions, at least not in regions abutting on, or containing, the turning point.

1.6. Remarks on Recent Trends

The recent development of the asymptotic theory with a small parameter is the main topic of this report. In this chapter only a few general remarks will be made. Since Langer's seminal papers of the 1930s, the literature has proliferated enormously. Langer's work on second-order equations has been successfully applied to equations reducible to Weber's equation for parabolic cylinder functions. This effort then developed into studies of problems involving several turning points. Many variants of such methods have been pursued by mathematical physicists.

When the more general techniques based on stretching and matching are applied, the connection of different fundamental systems of solutions given asymptotically in different but overlapping regions is an important, only very partially solved problem. One difficulty stems from the fact that asymptotic approximations with respect to a parameter never characterize a solution uniquely. One way to overcome this difficulty has been by derivation of "doubly asymptotic expansions" in unbounded domains, i.e., approximations that have asymptotic character not only for the passage to the limit as $\epsilon \to 0$, but which are also asymptotic with respect to x, as x tends to infinity in certain directions. Such information often does describe solutions in a unique fashion (see Chapter X).

For the extension of the theory to differential equations of higher order, most investigators have preferred to deal with the more general case of first-order systems of differential equations. In vector-matrix notation such systems can be written

$$\epsilon^h \frac{dy}{dx} = A(x,\epsilon)y \qquad (h \text{ a positive integer})$$

and this is the formulation used in most parts of this report. It turns out that even the *definition* of the concept of turning points is a matter of some difficulty for such general systems, as will be seen below.

Between 1837 and 1915 the development of the asymptotic theory had progressed in two directions. One was the extension to differential equations of any order. The other was the development of approximations of higher order in ϵ. This can be done for the simple equation (1.1-1) by replacing the Liouville transformations (1.1-4) and (1.1-5) by more complicated changes of variables that depend on ϵ, for instance by setting

$$u = a(x,\epsilon)v, \qquad \xi = \xi(x,\epsilon).$$

In the particular case that the differential equation is holomorphic in x and ϵ, it is often possible to choose

$$a(x,\epsilon) = \sum_{r=0}^{\infty} a_r(x)\epsilon^r, \qquad \xi(x,\epsilon) = \sum_{r=0}^{\infty} \xi_r(x)\epsilon^r$$

so that the new differential equation becomes (1.1-6) *without* a small correction term.

Here, however, an important new difficulty arises. The series are almost always divergent. Since those formalisms were developed during the historical period when one of the main concerns of mathematicians was to create a precise logical basis for the rich and fruitful manipulations of the 17th and 18th centuries, the tendency was to discard these divergent processes as useless. It is primarily the merit of Poincaré that this was not done, but that a vigorous theory of divergent but asymptotic series representations was developed. When applied to differential equations it leads to approximate solutions of great theoretical interest and numerical usefulness.

CHAPTER II
Formal Solutions

2.1. Introduction

The principal objects of study in this book are linear homogeneous ordinary differential equations that involve one scalar parameter. By introducing the higher derivatives of the dependent variable or variables as additional unknown functions, any such equation, or system of equations, can be rewritten as an equivalent first-order system:

$$\frac{dy_j}{dx} = \sum_{k=1}^{n} a_{jk}(x,\epsilon)y_k, \qquad j = 1, 2, \ldots, n. \tag{2.1-1}$$

The parameter has been designated by the letter ϵ, because it will be small in most situations to be encountered. If the $n \times n$ matrix $A(x,\epsilon)$ with entries $a_{jk}(x,\epsilon)$ and the column vector y with components y_j, $j = 1, 2, \ldots, n$, are introduced, equation (2.1-1) takes on the concise and convenient form

$$y' = A(x,\epsilon)y. \tag{2.1-2}$$

The prime indicates differentiation with respect to x.

In this book A will be an analytic function of x and ϵ. Most of the results that involve real variables only could easily be modified so as to be meaningful for functions possessing only a certain number of continuous derivatives.

The space of all matrices of n rows and m columns with entries that are holomorphic, i.e., regular analytic functions of one variable in a set D, will sometimes be denoted by $H_{nm}(D)$. The set D may be in $\mathbb{C}$ or on a Riemann surface over $\mathbb{C}$.

According to a basic existence theorem all solutions of the differential equation (2.1-2) are holomorphic functions of x at all values of x and ϵ where $A(x,\epsilon)$ is defined and holomorphic in x. The solutions are not necessarily holomorphic in ϵ at all points where $A(x,\epsilon)$ is holomorphic in ϵ, because the product of any solution $y(x,\epsilon)$ with any scalar function of ϵ alone, holomorphic or not, is again a solution. It is true, however, that a solution that is holomorphic in ϵ for one particular value of x depends holomorphically on ϵ at every point of the complex (x,ϵ)-space where A is holomorphic in both variables.

The most important concern in the study of analytic functions is probably the location and nature of their singularities. One is, therefore, led to the task of studying the structure of the solutions of equation (2.1-2) near points where A has singularities. If ϵ is held constant and A has there a singularity in x at some point $x = a$, the solutions have, in general, very complicated behavior at $x = a$. Even when the singularity of A with respect to x is an isolated one, the solutions are usually multivalued around $x = a$.

The behavior of the solutions of equation (2.1-2) near a singularity with respect to the parameter ϵ is, in most respects, more complicated than near a singularity in the independent variable. It is therefore a welcome fact that a solution y which is analytic in ϵ is a single-valued function of ϵ in the neighborhood of values of x and ϵ, where A has an isolated singularity with respect to ϵ.

More precisely, assume that A is holomorphic in x and ϵ for $|x - a| < x_1$, $0 < |\epsilon| < \epsilon_1$, but not necessarily at $\epsilon = 0$. Let y_0 be a constant vector independent of ϵ. Then the solution y determined by the initial condition $y(a,\epsilon) = y_0$ is holomorphic *in both variables* in $|x - a| < x_0$, $0 < |\epsilon| < \epsilon_0$.

That this is so follows immediately from the uniqueness theorem for differential equation. There is only one solution in $|x - a| < x_1$, for given ϵ, that satisfies $y(x,\epsilon) = y_0$. Hence, analytic continuation with respect to ϵ, for fixed x, around $\epsilon = 0$ cannot lead to a different solution upon return to the starting value of ϵ.

Even though most of the singularities in ϵ to be considered below will be of the isolated type, the simple fact stated above does not appear to play a helpful role in the existing literature. In fact, in most investigations ϵ is restricted to a ray or to a narrow sector in the ϵ-plane.

The simplest, and most important, situation occurs when A has for all x in a region a pole of fixed order with respect to ϵ at $\epsilon = 0$. By an obvious change of notation equation (2.1-2) can then be replaced by

$$\epsilon^h y' = A(x,\epsilon)y, \qquad (2.1\text{-}3)$$

where h is a positive integer. It will be assumed that A is holomorphic in x and ϵ for x in a set D to be specified later and for $|\epsilon| \leqslant \epsilon_0$, $\epsilon_0 > 0$. Then A admits a convergent expansion:

$$A(x,\epsilon) = \sum_{r=0}^{\infty} A_r(x)\epsilon^r, \qquad x \in D, \quad |\epsilon| < \epsilon_0, \qquad (2.1\text{-}4)$$

with coefficients A_r that are holomorphic in D.

If the factor ϵ^h in the left member of equation (2.1-3) were absent, one natural way to solve it near $\epsilon = 0$ would be to represent the unknown solutions as series in powers of ϵ. One can then insert that series into the differential equation, multiply, and rearrange according to powers of ϵ. The unknown coefficients of the series can then successively be calculated from the equations obtained by equating like powers of ϵ in the two members of the equation. The *a priori* knowledge that the solution sought does exist and is holomorphic in ϵ then guarantees that the formal process leads to a convergent series representation for a solution.

The presence of the singularity in ϵ at $\epsilon = 0$ makes such an approach inapplicable to equation (2.1-3). It is still true, however, that the equation can be solved, near $\epsilon = 0$, by certain skillful formal operations with power series. Two decisively new complications will now appear in this procedure: For one, the resulting series will, in general, be divergent. Second, the proof that the partial sums of the series are, in some sense and in certain regions, approximations to true solutions requires considerable effort. In this chapter only the formal solutions will be derived. Special cases of the technique to be developed have been known for almost a century. The presentation without restrictive hypotheses to be given here is essentially due to Turrittin [83]. By making use of ideas of Sibuya [72] and Wasow [87], the procedure can be simplified and presented in global rather than local form. The material collected in the Appendix, Chapter XII, will often be referred to.

The structure of the leading matrix in the series expansion (2.1-4) is of great importance in the construction to be described. It may be assumed that

$$A_0(x) \not\equiv 0, \tag{2.1-5}$$

since, otherwise, equation (2.1-3) could be divided by ϵ.

The first step in the construction to be described consists in a transformation of the differential equation which reduces $A_0(x)$ to its Jordan canonical form. This operation involves some not quite obvious considerations on the ring of holomorphic functions. The next section, while not directly concerning differential equations, is necessary for an understanding of the limitations of the formal theory. More on such matters can be found in Chapter XII.

2.2. The Jordan Form of Holomorphic Functions

Let M be an $n \times n$ matrix-valued function of the complex variable x with entries that are holomorphic in a region $D \subset \mathbb{C}$. Such a matrix will be said to be holomorphic in D. For every $x \in D$ the matrix $M(x)$ has a corresponding Jordan matrix $J(x)$ which is unique except for the order of its blocks. There exist then matrix-valued functions T on D such that $T(x)$ is invertible and

$$T^{-1}(x)M(x)T(x) = J(x). \tag{2.2-1}$$

There are many such matrix functions T, and usually none of them is holomorphic in all of D.

For every $x \in D$ there exist n, not necessarily distinct, eigenvalues of $M(x)$, say, $\lambda_j(x)$, $j = 1, 2, \ldots, n$. They are the zeros of the characteristic polynomial

$$\phi(\lambda, x) := \det(\lambda I - M(x)). \tag{2.2-2}$$

It follows from the Weierstrass Preparation Theorem and the theory of algebraic functions that the numbers $\lambda_j(x)$ are the values of analytic functions which, in general, are not single-valued. The set S_1 of those points in D where the multiplicity of some eigenvalues changes is discrete. In the remaining set, $D - S_1$, the multiplicities of the eigenvalues are constant, say, $m_1, m_2, \ldots, m_p$. In the neighborhood of every point of $D - S_1$ the eigenvalues are the values of p distinct holomorphic functions. If these functions are analytically continued along a closed curve in $D - S_1$, they will return to the starting point with some permutation of their order.

In the neighborhood of a point $x_1 \in S_1$ the values of the functions λ_j admit convergent expansions of the form

$$\sum_{r=0}^{\infty} c_r(x - x_1)^{r/q}, \tag{2.2-3}$$

depending on j, where q is some positive integer. Of course, q may be replaced by any multiple kq, if, at the same time, the sequence $\{c_r\}$ is replaced by

$$\{\tilde{c}_r\} = c_0 \underbrace{00 \ldots 0}_{k-1} c_1 \underbrace{00 \ldots 0}_{k-1} c_2 00 \ldots .$$

(See, e.g., [23] for the basic properties of algebraic functions.)

The points of S_1 are not always branch points in the usual sense of that term. The multiplicity of the eigenvalues of

$$\begin{pmatrix} x & 1 \\ 0 & 0 \end{pmatrix},$$

for instance, changes at $x = 0$, but the eigenvalue functions, x and 0 have no branch points there.

The Jordan matrix $J(x)$ is the direct sum of Jordan blocks. Let $\kappa(x)$ be the number of such blocks, and denote by $d_1(x), d_2(x), \ldots, d(x)_{\kappa(x)}$ the degrees of the elementary divisors of $J(z)$. Then each block has the form

$$\lambda_j(x)I + H_{d_j(x)},$$

where I is the identity matrix with the dimension $d_j(x)$ of that block, and

$$H_{d_j(x)} = \begin{bmatrix} 0 & 1 & 0 & 0 & \cdot & \cdot & \cdot & 0 & 0 \\ 0 & 0 & 1 & 0 & \cdot & \cdot & \cdot & 0 & 0 \\ \hline 0 & 0 & 0 & 0 & \cdot & \cdot & \cdot & 0 & 1 \\ 0 & 0 & 0 & 0 & \cdot & \cdot & \cdot & 0 & 0 \end{bmatrix}.$$

H_k will be called a "shifting" matrix of dimension k. At any point x where some of the $d_j(x)$ are not constant in a full neighborhood, the matrix J must have a discontinuity. The points of S_1 may be—but need not be—such discontinuities. There may be additional discontinuities, as in the following trivial example: The matrix

$$M(x) = \begin{pmatrix} 0 & x \\ 0 & 0 \end{pmatrix}$$

has the Jordan form

$$J(x) = \begin{cases} \begin{pmatrix} 0 & 1 \\ 0 & 0 \end{pmatrix}, & x \neq 0 \\ \begin{pmatrix} 0 & 0 \\ 0 & 0 \end{pmatrix}, & x = 0, \end{cases}$$

which is discontinuous at $x = 0$, although the multiplicity of the eigenvalues is 2 throughout.

Let S_2 be the points of $D - S_1$ over which J is discontinuous.

Lemma 2.2-1. *The set S_2 is discrete.*

PROOF. The points of S_2 are the points of D where the multiplicities of the eigenvalues of $M(x)$ are constant, while the degrees of some elementary divisors of $M(x)$ change. These degrees change exactly at the points where the degrees of some invariant factors of the matrix $\lambda I - M(x)$ change, since these two sets of numbers determine each other uniquely. The invariant factors can be calculated as follows: Consider the not identically zero minors of order j, $j = 1, 2, \ldots, n$, of the matrix $\lambda I - M(x)$. They are polynomials in λ with coefficients holomorphic for $x \in D$. Remove from $D - S_1$ all points where the leading coefficient of any of these minors has a zero. (If the minor is of degree zero in λ, the "leading coefficient" is the minor itself.) These zeros form a discrete set $\hat{S}$. Define $D_0 :\equiv 1$, and let D_j, $j = 1, \ldots, n$, denote the greatest common divisor, in the ring of polynomials in λ, of all minors of $\lambda I - M(x)$ of order j normalized to be *monic* polynomials. D_j has holomorphic coefficients in $D - S_1 - \hat{S}$. By the standard theory of the Jordan form of matrices, the invariant factors of $\lambda I - M(x)$ are the values of the functions

$$i_j = D_{n-j}/D_{n-j-1}, \qquad j = 0, 1, \ldots, n - 1,$$

which are monic polynomials of λ with coefficients holomorphic in $D - S_1 - \hat{S}$. The degrees of these polynomials in λ are constant in $D - S_1 - \hat{S}$ and so are, therefore, the sizes of the blocks in the Jordan form of $M(x)$. Hence, $S_2 \subset \hat{S}$ so that S_2 is a discrete set, as was to be proved. (See, e.g., Ref. 23, Ch. VI, for the algebraic facts used here.)

For the example preceding this lemma, the minors of $\lambda I - M(x)$ of order 1 are λ and $-x$; λ^2 is the only minor of order 2. Therefore, the set $\hat{S}$ consists of the one point $x = 0$. Also, S_1 is the empty set. For $x \neq 0$ one finds $D_1 = 1$, $D_2 = \lambda^2$ and $i_0(x) = 1$, $i_1(x) = \lambda^2$. At $x = 0$, on the other hand, $D_1(0) = \lambda$, $D_2(0) = \lambda^2$, $i_1(0) = \lambda$, $i_0(0) = \lambda$, so that the degrees of the i_j are discontinuous at $x = 0$.

Theorem 12.2-2 of the Appendix gives sufficient conditions under which the Jordan matrix of a matrix holomorphic in D is holomorphically similar to it in D. An example in Section 12.2 also indicates that those conditions are almost necessary. From now on it will be assumed that D is such that A_0 satisfies those conditions there. This is stated in the hypothesis below.

Hypothesis II-1. *The Jordan matrix for A_0 is holomorphic in D, and the multiplicity of the eigenvalues of $A_0(x)$ is constant in D.*

This condition is, in particular, satisfied if D is simply connected and contains no points of the sets S_1 and S_2 for $M = A_0$.

2.3. A Formal Block Diagonalization

The formal simplification of the given differential equation (2.1-3) begins with the reduction of the leading coefficient matrix A_0 in (2.1-4) to its Jordan form by a transformation

$$y = Tv. \tag{2.3-1}$$

This produces a new differential equation

$$\epsilon^h v' = Bv \tag{2.3-2}$$

with

$$B = T^{-1}AT - \epsilon^h T^{-1}T'. \tag{2.3-3}$$

Thanks to Hypothesis II-1 and Theorem 12.2-2, the matrix T can be chosen holomorphic in D. The matrix B admits a series expansion

$$B(x,\epsilon) = \sum_{r=1}^{\infty} B_r(x)\epsilon^r, \tag{2.3-4}$$

with coefficients B_r holomorphic in D. Of course, T and the B_r may fail to be holomorphic on the boundary of D. Here is a very simple example:

$$A_0(x) = \begin{bmatrix} 0 & x & 0 & x \\ 1 & 0 & 0 & 0 \\ 0 & 0 & 0 & 1 \\ 0 & 0 & x & 0 \end{bmatrix}.$$

Formula (2.2-2) becomes

$$\phi(\lambda, x) = (\lambda^2 - x)^2$$

and

$$\lambda_1(x) \equiv \lambda_2(x) \equiv \sqrt{x}, \qquad \lambda_3(x) \equiv \lambda_4(x) = -\sqrt{x}.$$

The algebraic function $\sqrt{x}$ and $-\sqrt{x}$ must be uniquely defined by assigning them fixed determinations at some point (other than zero). With

$$T(x) = \begin{bmatrix} (\sqrt{x})^3/2 & x/2 & -(\sqrt{x})^3/2 & x/2 \\ x/2 & 0 & x/2 & 0 \\ 0 & 1 & 0 & 1 \\ -x & \sqrt{x} & -x & -\sqrt{x} \end{bmatrix}$$

one has

$$T^{-1}(x)A_0(x)T(x) = \begin{bmatrix} \sqrt{x} & 1 & 0 & 0 \\ 0 & \sqrt{x} & 0 & 0 \\ 0 & 0 & -\sqrt{x} & 1 \\ 0 & 0 & 0 & -\sqrt{x} \end{bmatrix} = J(x).$$

The set S_1 consists here of one point 0, and S_2 is empty. The two distinct eigenvalues $\sqrt{x}$ and $-\sqrt{x}$, each of multiplicity 2 for $x \neq 0$, coalesce into one eigenvalue of order 4 at $x = 0$.

If $A_0(x)$ has at all points of D only one distinct eigenvalue, the "shearing" technique to be described in Section 2.4 is the next simplifying transformation. If $A_0(x)$ has more than one eigenvalue, recall that the region D was defined so that two eigenvalues can coincide at a point in D only if they are identically equal. Let the n eigenvalues—as functions in D—be labeled so that

$$\lambda_1 = \lambda_2 = \cdots = \lambda_m \neq \lambda_{m+k}, \qquad k = 1, 2, \ldots, n - m, \qquad (2.3\text{-}5)$$

and let the blocks of J be ordered so that the blocks pertaining to $\lambda_1, \lambda_2, \ldots, \lambda_m$ come first. Then the Jordan matrix $B_0 = J$ is the direct sum of two matrices of order m and $n - m$, respectively, which have no eigenvalue in common in any point of D.

The next step in the successive simplification of the coefficient matrix of the differential equation is a formal transformation which block-diagonalizes all matrices $B_1, B_2, \ldots$ into two diagonal blocks of orders m and $n - m$. The technique for accomplishing this was first described by Sibuya [72]. It has wider applicability than the present context and will, therefore, be proved in a more general form, in which it is not required that D satisfy Hypothesis II-1 nor that B_0 be in Jordan form. In Theorem 2.3-1 the notation has been changed back from B_r to A_r.

Theorem 2.3-1. *Let $\sum_{r=0}^{\infty} A_r \epsilon^r$ be a formal series in which the A_r are $n \times n$-dimensional matrix-valued functions of the complex variable x holomorphic in a region $D \subset \mathbb{C}$. Assume that A_0 is block diagonal:*

$$A_0 = A_0^{11} \oplus A_0^{22}$$

and that $A_0^{11}(x)$ and $A_0^{22}(x)$ have no eigenvalue in common for any $x \in D$.

Then there exists a series $\sum_{r=0}^{\infty} P_r \epsilon^r$ such that the formal transformation

$$y = \left(\sum_{r=0}^{\infty} P_r \epsilon^r \right) v \tag{2.3-6}$$

takes the formal differential equation

$$\epsilon^h \frac{dy}{dx} = \left(\sum_{r=0}^{\infty} A_r \epsilon^r \right) y \tag{2.3-7}$$

into an equation

$$\epsilon^h \frac{dv}{dx} = \left(\sum_{r=0}^{\infty} B_r \epsilon^r \right) v \tag{2.3-8}$$

with the following properties.

(i) *All P_r and B_r are holomorphic in D;*
(ii) *$P_0 = I$;*
(iii) *All B_r are block-diagonal:*

$$B_r = B_r^{11} \oplus B_r^{22},$$

with B_r^{11} having the same dimension as A_0^{11}.

PROOF. Insertion of (2.3-6) into (2.3-7) and identification with (2.3-8) produces the formal relation

$$\epsilon^h \sum_{r=0}^{\infty} P_r' \epsilon^r = \left(\sum_{r=0}^{\infty} A_r \epsilon^r \right) \left(\sum_{r=0}^{\infty} P_r \epsilon^r \right) - \left(\sum_{r=0}^{\infty} P_r \epsilon^r \right) \left(\sum_{r=0}^{\infty} B_r \epsilon^r \right).$$

Expansion and formal identification of like powers of ϵ in the two members of (2.3-8) leads to the recursion formulas

$$B_0 = A_0, \tag{2.3-9}$$

$$A_0 P_r - P_r A_0 = B_r - H_r, \qquad r > 0, \tag{2.3-10}$$

with

$$H_r := \sum_{\substack{\alpha+\beta=r \\ 0<\alpha<r}} (P_\beta B_\alpha - B_\alpha P_\beta) + A_r - P_{r-h}'. \tag{2.3-11}$$

Here it is understood that $P_k = 0$ for $k < 0$.

Next, write A_r, B_r, P_r, and H_r in partitioned form

$$A_r = \begin{bmatrix} A_r^{11} & A_r^{12} \\ A_r^{21} & A_r^{22} \end{bmatrix}, \qquad B_r = \begin{bmatrix} B_r^{11} & B_r^{12} \\ B_r^{21} & B_r^{22} \end{bmatrix}, \qquad \text{etc.,} \tag{2.3-12}$$

where A_r^{11}, B_r^{11}, etc., have dimension $m \times m$. The aim is to determine the P_r so that all B_r become block-diagonal:

$$B_r = \begin{pmatrix} B_r^{11} & 0 \\ 0 & B_r^{22} \end{pmatrix}, \qquad r \geq 0. \tag{2.3-13}$$

As Sibuya has shown, this can even be done by matrices, P_r, of the special form

$$P_r = \begin{pmatrix} 0 & P_r^{12} \\ P_r^{21} & 0 \end{pmatrix}, \qquad r > 0. \tag{2.3-14}$$

In fact, with B_r, P_r so chosen, conditions (2.3-10) become

$$\begin{bmatrix} 0 & A_0^{11}P_r^{12} - P_r^{12}A_0^{22} \\ A_0^{22}P_r^{21} - P_r^{21}A_0^{11} & 0 \end{bmatrix} = \begin{bmatrix} B_r^{11} - H_r^{11} & -H_r^{12} \\ -H_r^{21} & B^{22} - H_r^{22} \end{bmatrix}. \tag{2.3-15}$$

For $r = 1$, formula (2.3-11) reduces to $H_1 = A_1$ and, therefore, one must choose $B_1^{11} = A_1^{11}$, $B_1^{22} = A_1^{22}$ to satisfy (2.3-15) for $r = 1$. Also, P_1^{12} must be a solution of the equation

$$A_0^{11}P_1^{12} - P_1^{12}A_0^{22} = -H_1^{12}, \tag{2.3-16}$$

which is a system of $m(n - m)$ linear nonhomogeneous algebraic equations for the entries of P_1^{12}. Now, it is a well-known fact of elementary matrix theory that the eigenvalues of the linear operator L on the matrices X of dimension $m(n - m)$ defined by $LX := A_0^{11}X - XA_0^{22}$ are the differences of the eigenvalues of A_0^{11}, A_0^{22}. By assumption, those differences are nowhere zero in D. Hence, P_1^{12} in (2.3-16) can be calculated by Cramer's rule and is holomorphic in D. The matrix P_1^{21} can be found analogously from (2.3-15). Once P_1 and B_1 are known, H_2 can be calculated from (2.3-11) and it is clear how to continue the calculation of the P_r and B_r by induction. Thus, Theorem 2.3-1 is proved.

Now, return to the notation and the hypotheses before Theorem 2.3-1 was stated. That theorem then implies that the original system (2.1-3) can be formally split into two uncoupled systems of order m and $n - m$, respectively, with leading matrices J^{11} and J^{22}, say. If J^{22} still has more than one distinct eigenvalue, the same decomposition process can be applied. Note that the hypothesis of the original power series (2.1-4) for A being convergent has nowhere been used in these formal calculations.

The at-most n successive transformations of the original differential equation can be assembled into one transformation. The result obtained is restated below in a slightly changed notation, more convenient for reference in later sections.

Theorem 2.3-2. *Let $\sum_{r=0}^{\infty} A_r \epsilon^r$ be a formal series in which the A_r are $n \times n$-dimensional matrix-valued functions of the complex variable x holomorphic in a region $D \subset \mathbb{C}$. Denote by $J(x)$ the Jordan form of $A_0(x)$. If Hypothesis II-1 is satisfied, then there exists a series $\sum_{r=0}^{\infty} P_r \epsilon^r$ such that the* *formal transformation*

$$y = \left(\sum_{r=0}^{\infty} P_r \epsilon^r \right) v \tag{2.3-17}$$

takes the differential equation

$$\epsilon^h \frac{dy}{dx} = \left(\sum_{r=0}^{\infty} A_r \epsilon^r \right) y \tag{2.3-18}$$

into an equation

$$\epsilon^h \frac{dv}{dx} = \left(\sum_{r=0}^{\infty} B_r \epsilon^r \right) v \tag{2.3-19}$$

with the following properties.

(i) *All P_r and B_r are holomorphic in D;*
(ii) *$\det P_0(x) \neq 0$ for all $x \in D$;*
(iii) *All B_r are block diagonal. The orders of these blocks are the multiplicities of the eigenvalues of A_0 in D, and each block has only one distinct eigenvalue.*
(iv) *$B_0 = J$.*

The simplest case is the one in which $p = n$. Then the problem has been formally reduced to a diagonal system, which can be formally solved by quadratures and straightforward manipulations with series.

As a rather trivial illustration of the foregoing theory, the Liouville–Green approximation to the solutions of the scalar differential equation $\epsilon^2 u'' - a(x)u = 0$ will now be derived in this manner. This is not the simplest way of doing it, because the method described here was chosen for its generality, not for its simplicity in special cases.

Set

$$y = \begin{pmatrix} y_1 \\ y_2 \end{pmatrix} := \begin{pmatrix} u \\ \epsilon u' \end{pmatrix}.$$

The scalar differential equation for u then becomes the system

$$\epsilon y' = \begin{bmatrix} 0 & 1 \\ a & 0 \end{bmatrix} y. \tag{2.3-20}$$

If $a \neq 0$, the transformation

$$y = \begin{bmatrix} 1 & 1 \\ \sqrt{a} & -\sqrt{a} \end{bmatrix} v$$

changes it into

$$\epsilon v' = \left[\begin{pmatrix} \sqrt{a} & 0 \\ 0 & -\sqrt{a} \end{pmatrix} - \frac{a'}{4a} \begin{pmatrix} 1 & -1 \\ -1 & 1 \end{pmatrix} \epsilon \right] v,$$

which is equation (2.3-2) for this problem. Continuing the diagonalization process one sees that

$$H_1 = -\frac{a'}{4a}\begin{pmatrix} 1 & -1 \\ -1 & 1 \end{pmatrix}.$$

Because $H_1 = B_1$, formula (2.3-15) becomes, for $r = 1$ (and with the notation changed from B to C),

$$\begin{bmatrix} 0 & 2\sqrt{a}\,P_1^{12} \\ -2\sqrt{a}\,P_1^{21} & 0 \end{bmatrix} = \begin{bmatrix} C_1^{11} + \dfrac{a'}{4a} & -\dfrac{a'}{4a} \\ \dfrac{a'}{4a} & C_1^{22} + \dfrac{a'}{4a} \end{bmatrix}.$$

Therefore,

$$P_1 = \begin{bmatrix} 0 & -\dfrac{a'}{8(\sqrt{a})^3} \\ -\dfrac{a'}{8(\sqrt{a})^3} & 0 \end{bmatrix}, \qquad C_1 = -\frac{a'}{4a}I.$$

Combining the two transformations performed so far, one sees that a transformation of the form

$$y = \left[\begin{bmatrix} \dfrac{1}{\sqrt{a}} & \dfrac{1}{-\sqrt{a}} \\ -\sqrt{a} & \sqrt{a} \end{bmatrix} - \begin{bmatrix} \dfrac{1}{-\sqrt{a}} & \dfrac{1}{\sqrt{a}} \\ -\sqrt{a} & \sqrt{a} \end{bmatrix}\frac{a'}{8(\sqrt{a})^3}\epsilon + \cdots\right]w \qquad (2.3\text{-}21)$$

takes system (2.3-20) into

$$\epsilon w' = \left[\begin{bmatrix} \sqrt{a} & 0 \\ 0 & -\sqrt{a} \end{bmatrix} - \frac{a'}{4a}I\epsilon + \cdots\right]w. \qquad (2.3\text{-}22)$$

The dots indicate terms of higher order in ϵ, which could be calculated explicitly in succession. The symbol $\sqrt{a}$ stands for a determination of the square root whose definition can be fixed uniquely by its value at one point and by analytic continuation on the Riemann surface of the square root function.

Integration of (2.3-22) produces the matrix solution

$$W(x,\epsilon) = \left(a(x)^{-1/4}I + \cdots\right)\begin{bmatrix} e^{(1/\epsilon)\int\sqrt{a(x)}\,dx} & 0 \\ 0 & e^{-(1/\epsilon)\int\sqrt{a(x)}\,dx} \end{bmatrix}. \qquad (2.3\text{-}23)$$

The dots indicate a series of positive powers of ϵ. A return to the variable y by means of the transformation (2.3-21) shows that there exists a formal matrix series solution Y of system (2.3-20) of the form

$$Y(x, \epsilon) = \left[\begin{bmatrix} (a(x))^{-1/4} & (a(x))^{-1/4} \\ (a(x))^{1/4} & -(a(x))^{1/4} \end{bmatrix} + \cdots \right]$$

$$\times \begin{bmatrix} e^{(1/\epsilon) \int \sqrt{a(x)} \, dx} & 0 \\ 0 & e^{-(1/\epsilon) \int \sqrt{a(x)}} \end{bmatrix}. \qquad (2.3\text{-}24)$$

Without the dots this is, indeed, the Liouville–Green approximation.

The eigenvalues $\sqrt{a(x)}$, $-\sqrt{a(x)}$ of $A_0(x)$ for that differential equation coalesce at the zeros of $a(x)$. These points do not belong to D. At these points formula (2.3-24) becomes meaningless.

2.4. Parameter Shearing: Its Nature and Purpose

The decomposition method described in the preceding section has reduced the problem of satisfying systems of the form (2.1-3) by formal series to the case that the leading term of the coefficient matrix has only one distinct eigenvalue and is in Jordan form. This simplification has been purchased by limiting the arguments to certain subregions of the region in which the coefficients of the differential equation are holomorphic. Also, the series that constitute the coefficients of the new differential equations are, in general, divergent. For the further analysis of the problem it is convenient to return to the initial notation and to assume that the given problem already has the form of one of the blocks in equation (2.3-19).

Assume, therefore, that the coefficients of the given differential equation

$$\epsilon^h y' = \left(\sum_{r=0}^{\infty} A_r \epsilon^r \right) y \qquad (2.4\text{-}1)$$

are holomorphic on a certain region $D \in \mathbb{C}$ and that $A_0(x)$ has only the one eigenvalue $\lambda(x)$. Moreover, $A_0(x)$ is in Jordan form.

An elementary but important further simplification consists of the transformation

$$y = \exp\left\{ \epsilon^{-h} \int^x \lambda(t) \, dt \right\} y^*. \qquad (2.4\text{-}2)$$

The only change in the differential equation (2.4-1) caused by this substitution is the replacement of $A_0(x)$ by the matrix $A_0(x) - \lambda(x)I$. This matrix has zero as its only eigenvalue. Such matrices are called nilpotent. Accordingly, no generality is lost by adding the condition below.

Hypothesis II-2 (Nonrestrictive). *The leading matrix $A_0(x)$ in (2.4-1) is nilpotent and in Jordan form for all $x \in D$.*

Because of Hypotheses II-1 and II-2, the matrix A_0 is now independent of x.

In the very exceptional case that $A(x, \epsilon) \equiv A_0$, the solution of the differential equation can be immediately accomplished by elementary means. It has the form described in the Main Theorem 2.8-1, and this case can be left aside from now on.

Transformations of the form (2.3-17) with an invertible leading matrix P_0 induce a similarity transformation of A_0, and they do not destroy the property of Hypothesis II-2 introduced above. To carry the reduction of the differential equation further, transformations that may have singularities at $\epsilon = 0$ must be admitted. It will be shown that one can change the differential equation so as to introduce more than one eigenvalue into the leading coefficient matrix by means of quite simple transformations $y = Tv$ in which T is a function of ϵ alone. This matrix T is, however, not necessarily holomorphic or holomorphically invertible at $\epsilon = 0$. Once this has been achieved, the block-diagonalization method of Section 2.3 can be applied again to the new differential equation, which results in a further reduction of the order.

Very simple examples show that one must expect the appearance of fractional powers of ϵ in the formal solutions of some differential equations. For instance, the system

$$\epsilon y' = \begin{bmatrix} 0 & 1 \\ \epsilon & 0 \end{bmatrix} y$$

has the explicit elementary solution matrix

$$Y(x) = \begin{pmatrix} 1 & 1 \\ \epsilon^{1/2} & -\epsilon^{1/2} \end{pmatrix} \exp\left\{ \begin{pmatrix} 1 & 0 \\ 0 & -1 \end{pmatrix} \epsilon^{-1/2} x \right\}.$$

It turns out that the aim of changing the leading coefficient matrix into one with more than one eigenvalue can be achieved by transformations with matrices of the simple form

$$T(\epsilon) = \mathrm{diag}(1, \epsilon^{\alpha}, \epsilon^{2\alpha}, \ldots, \epsilon^{(n-1)\alpha}), \tag{2.4-3}$$

provided fractional values of α are permitted. This was recognized by Turrittin [83].

By Hypothesis II-2 the matrix A_0 is independent of x and has the block-diagonal structure

$$A_0 = \mathrm{diag}(H_1, H_2, \ldots, H_s), \tag{2.4-4}$$

where the H_j are "shifting" matrices

$$H_j = \begin{bmatrix} 0 & 1 & 0 & 0 & \cdot & \cdot & \cdot & 0 & 0 \\ 0 & 0 & 1 & 0 & \cdot & \cdot & \cdot & 0 & 0 \\ \cdot & \cdot & \cdot & \cdot & \cdot & \cdot & \cdot & \cdot & \cdot \\ 0 & 0 & 0 & 0 & \cdot & \cdot & \cdot & 0 & 1 \\ 0 & 0 & 0 & 0 & \cdot & \cdot & \cdot & 0 & 0 \end{bmatrix},$$

not necessarily all of the same order. These orders are the degrees of the elementary divisors of A_0.

If $T(\epsilon)$ has the special form (2.4-3), then

$$T^{-1}(\epsilon)A_0 T(\epsilon) = \epsilon^\alpha A_0,\qquad (2.4\text{-}5)$$

so that the structure of A_0 is then not decisively changed by the transformation

$$y = T(\epsilon)v\qquad (2.4\text{-}6)$$

of the differential equation (2.4-1). Turrittin [83] recognized that as the transformation (2.4-3), (2.4-6) moves the entries of the matrices A_r in (2.4-1) in different ways, depending on their position in the matrix, the leading matrix of the transformed differential equation will differ from A_0, at least for appropriate values of α. In fact, as will be shown below, one can always find a rational value of α for which the new differential equation is "simpler" in a sense still to be defined.

Turrittin coined the term "shearing transformation" for (2.4-6), presumably because of its effect on a matrix series $V(\epsilon) = \sum_{r=0}^{\infty} V_r \epsilon^r$. If α is a positive integer, the series of $T(\epsilon)V(\epsilon)$ is obtained from the one for $V(\epsilon)$ by moving each row by α units to the right with respect to the one above, just as shearing of a material moves its strata sideways with respect to each other.

2.5. Simplification by a Theorem of Arnold

The result of a transformation of the form (2.4-3), (2.4-6) is still too complicated for a detailed study. A preliminary transformation which makes the coefficient matrix "sparser" (i.e., introduces more zeros) was applied by Turrittin at this point of his investigation. Here, a related, more substantial simplification will be described, which is of interest in itself.

In Ref. 3, V. E. Arnold constructed a holomorphic similarity transformation for holomorphic matrices which takes them into an especially simple form, almost—but not quite—canonical. This is described in Section 12.4 of the Appendix. In Ref. 95 I used Arnold's result for a formal simplification of differential equations of the form (2.4-1). In the present context these local results must be extended so as to apply globally.

The scalar d and the matrices Γ_j used below are defined in the Appendix, Section 12.4.

Theorem 2.5-1. *If the matrices A_r are holomorphic in a region D and if A_0*

satisfies condition II-2, *then the formal differential equation*

$$\epsilon^h y' = \left(\sum_{r=0}^{\infty} A_r(x)\epsilon^r \right) y \tag{2.5-1}$$

can be changed into

$$\epsilon^h v' = \left[A_0 + \sum_{j=1}^{d} \left(\sum_{r=1}^{\infty} \rho_{jr}(x)\epsilon^r \right) \Gamma_j \right] v \tag{2.5-2}$$

by means of a formal transformation

$$y = \left(\sum_{r=0}^{\infty} P_r(x)\epsilon^r \right) v \tag{2.5-3}$$

in which $P_0 = I$ and the P_r, as well as the ρ_{jr} are holomorphic in D.

PROOF.   For abbreviation, the symbols A, B, P will be used for the series appearing in (2.5-1), (2.5-2), and (2.5-3), respectively. In analogy with the proof of Theorem 2.3-1 one sees that what has to be proved is that the formal identity

$$\epsilon^h P' = AP - PB \tag{2.5-4}$$

can be satisfied by appropriate choice of the P_r and the ρ_{jr}. Identifying like powers of ϵ in the two members of (2.5-4) after having multiplied out and rearranged the series, one first finds for $r = 0$ the condition

$$A_0 P_0 - P_0 A_0 = 0,$$

which will be satisfied by choosing $P_0 = I$. The subsequent conditions then take the form

$$A_0 P_r - P_r A_0 - \sum_{j=1}^{d} \rho_{jr}\Gamma_j = F_r, \qquad r = 1, 2, \ldots \tag{2.5-5}$$

with

$$F_r = P'_{r-h} - A_r + \sum_{\substack{\alpha+\beta=r \\ \alpha,\beta>0}} \left(A_\alpha P_\beta - P_\beta \sum_{j=1}^{d} \rho_{j\alpha}\Gamma_j \right), \qquad r \geq 1 \tag{2.5-6}$$

$(P_k := 0,$ for $k < 0)$. In particular,

$$F_1 = -A_1. \tag{2.5-7}$$

If F_r is known, equation (2.5-5) can be solved (not uniquely) for P_r and the $\rho_{jr}, j = 1, 2, \ldots, d$, thanks to Lemma 12.4-4. Moreover, if F_r is holomorphic in D, there exist solutions that are holomorphic in D, because the left member of the system has constant coefficients. Equation (2.5-7) implies that F_1 is holomorphic in D. Since F_r depends only on the P_k, ρ_{ik} with $k < r$, all F_r, and, hence, all P_r and ρ_{jr} can be taken as holomorphic functions in D. Thus, the theorem is proved.

2.6. Parameter Shearing: Its Application

In light of Theorem 2.5-1 it may be assumed that the transformation described there has already been performed. That is, to Hypothesis II-2 one may add—without losing generality—

Hypothesis II-3. *All coefficients A_r, $r > 0$, in the equation (2.5-1) have the form*

$$A_r(x) = \sum_{j=1}^{d} \rho_{jr}(x)\Gamma_j, \qquad r > 0. \tag{2.6-1}$$

A differential equation (2.5-1) whose coefficients satisfy Hypotheses II-2 and II-3 will be said to be in Arnold's form. The fewer blocks there are in the partition of A_0, or, in other words, the fewer elementary divisors A_0 possesses, the sparser the matrices A_r will be.

Figure 2.1 may help the reader visualize the situation. It illustrates the case that A_0 has three elementary divisors. The heavy lines indicate the last rows or first columns of the blocks of the A_r. Only *they* may contain nonzero entries.

Following the program outlined in Section 2.4 the differential equation (2.5-1) which is supposed to already be in Arnold's form (2.5-2), will now be subjected to the transformation (2.4-6) with $T(\epsilon)$ as in (2.4-3). The exponent α is, as yet, undetermined. After division by ϵ^{α}, the transformed differential equation becomes

$$\epsilon^{h-\alpha}v' = \left[A_0 + \sum_{r=1}^{\infty} \sum_{j=1}^{d} \rho_{jr}(x)\Gamma_j \epsilon^{r-(1+\mu_j-\nu_j)\alpha} \right]v, \tag{2.6-2}$$

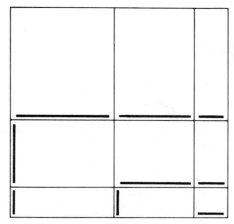

Figure 2.1. Arnold's form for a matrix.

where μ_j and ν_j are the row and column number, respectively, of the nonzero entry of Γ_j. Here, formula (2.4-5) has been used.

Now consider the set Σ of those positive rational numbers of the form $r/(1 + \mu_j - \nu_j)$, $r = 1, 2, \ldots$, $j = 1, 2, \ldots, d$ for which the factor ρ_{jr} in (2.6-2) is not identically zero. Whenever $\alpha \in \Sigma$ in (2.6-2), the coefficient matrix will have nonzero entries independent of ϵ in addition to those present in A_0. Two quite different cases have to be distinguished.

Case I.   *No element of Σ lies in the interval $(0, h)$.*

In this case set

$$\alpha = h. \tag{2.6-3}$$

Then the right-hand member of (2.6-2) contains no negative powers of ϵ. In fact, if one had $r - (1 + \mu_j - \nu_j)h < 0$ for a term actually present in (2.6-2), then it would follow that $0 < r/(1 + \mu_j - \nu_j) < h$, contrary to the assumption. The transformed differential equation (2.6-2) with $\alpha = h$ has no longer a positive power of ϵ multiplying the derivative in the left-hand member, and the powers of ϵ occurring on the right-hand side are non-negative. The problem of solving such an equation formally by series in powers of ϵ is not difficult. It will be taken up in Section 2.8.

Case II.   *Σ contains numbers in $(0, h)$.*

All positive values of $r/(1 + \mu_j - \nu_j)$ exceed the number $1/(n + 1)$, hence Σ has a minimum. Choose this minimum as the value of α in (2.6-2). Then no negative powers of ϵ are present in the right member, but at least one term contains ϵ with the exponent 0. Of course, α may be a fraction, say,

$$\alpha = p/q \quad (p, q \text{ relatively prime and positive}). \tag{2.6-4}$$

The substitution

$$\tilde{\epsilon} = \epsilon^{1/q} \tag{2.6-5}$$

then produces a problem with integral powers of $\tilde{\epsilon}$. If the terms are reordered according to powers of $\tilde{\epsilon}$, the leading matrix will now have nonzero entries in addition to those of A_0. These additional entries all lie on or below the diagonal, since above the diagonal $r - (1 + \mu_j - \nu_j)\alpha > 0$.

In general, the new leading matrix will have at least two distinct eigenvalues. The method of Sections 2.2 and 2.3 can then be applied again, and the problem will have been split into two or more similar problems of lower order. There is, however, one important difference between the situation in Sections 2.2, 2.3, and the present one: The leading matrix of the new differential equation may fail to be holomorphically similar to its Jordan form in all of D. It may be necessary to replace D by a subregion $D^{(1)} \subset D$ which does not contain any points of the discrete set of points in D

analogous to the set $S_1 \cup S_2$ introduced in Section 2.2 in the discussion of Hypothesis II-1 for A_0.

There remains to be investigated the exceptional situation when the leading coefficient of the "sheared" differential equation again has only one distinct eigenvalue. First, however, here is a simple illustration of the shearing technique. Let

$$\epsilon y' = \begin{pmatrix} 0 & x \\ \epsilon & 0 \end{pmatrix} y. \tag{2.6-6}$$

The transformation

$$y = \begin{pmatrix} 1 & 0 \\ 0 & x^{-1} \end{pmatrix} z$$

takes this differential equation into

$$\epsilon z' = \begin{pmatrix} 0 & 1 \\ \epsilon x & \epsilon x^{-1} \end{pmatrix} z$$

which is in Arnold's form. The point $x = 0$ is an exceptional point. This was to be expected, since the leading coefficient matrix

$$A_0(x) = \begin{pmatrix} 0 & x \\ 0 & 0 \end{pmatrix}$$

has a discontinuity in its Jordan form at $x = 0$. The shearing transformation

$$z = \begin{pmatrix} 1 & 0 \\ 0 & \epsilon^\alpha \end{pmatrix} v$$

leads to the new differential equation

$$\epsilon^{1-\alpha} v' = \begin{pmatrix} 0 & 1 \\ \epsilon^{1-2\alpha} x & \epsilon^{1-\alpha} x^{-1} \end{pmatrix} v.$$

The correct value for α is $\alpha = \frac{1}{2}$, and then,

$$\epsilon^{1/2} v' = \begin{pmatrix} 0 & 1 \\ x & \epsilon^{1/2} x^{-1} \end{pmatrix} v.$$

The next transformation is

$$v = \begin{pmatrix} 1 & 1 \\ x^{1/2} & -x^{1/2} \end{pmatrix} w,$$

which diagonalizes the new leading coefficient matrix. One finds

$$\epsilon^{1/2} w' = \left[\begin{pmatrix} x^{1/2} & 0 \\ 0 & x^{-1/2} \end{pmatrix} - \frac{1}{4} x^{-1} \begin{pmatrix} 1 & -1 \\ -1 & 1 \end{pmatrix} \epsilon^{1/2} \right] w.$$

Now follows the successive diagonalization of the higher-order terms in the coefficient matrix, which will not be carried out here because it is a routine calculation.

Again, I emphasize that the purpose of the foregoing calculation is to give an illustration of a method. For the actual solution of the differential equation the transformation $x = \epsilon^{1/3}t$, $y_1 = u_1$, $y_2 = \epsilon^{1/3}u_2$ is preferable, since it leads to

$$\frac{du}{dt} = \begin{pmatrix} 0 & t \\ 1 & 0 \end{pmatrix} u, \qquad u = \begin{pmatrix} u_1 \\ u_2 \end{pmatrix}$$

and hence to Airy's equation

$$\frac{d^2 u_2}{dt^2} = tu_2 .$$

2.7. Parameter Shearing: The Exceptional Case

In this section the differential equation will be assumed to have the form of (2.5-2). Only Case II, when

$$0 < \alpha < h, \tag{2.7-1}$$

is of concern here. Let the matrices Γ_j be labeled so that those with their nonzero entry on or below the diagonal, i.e., the ones with $\nu_j \leqslant \mu_j$, come first. Let there be d' of such matrices Γ_j. Then the sheared differential equation (2.6-2) has, after rearrangement according to powers of ϵ, a coefficient matrix with a leading term of the form

$$\tilde{A}_0(x) = A_0 + \sum_{1 \leqslant j \leqslant d'} \sigma_j(x)\Gamma_j , \tag{2.7-2}$$

with $d' \leqslant d$ and $\nu_j \leqslant \mu_j$ for $j = 1, 2, \ldots, d'$. Not all σ_j are the zero function.

The only situation not yet analyzed is the one in which $\tilde{A}_0(x)$ shares with $A_0(x)$ the property of possessing only one distinct eigenvalue. A first observation is that whenever A_0 has a nonzero entry *on the main diagonal*, the number α must have been a positive integer, since for that entry $\mu_j = \nu_j$ and therefore $\alpha = r$ for some value of r. Then, $\tilde{\epsilon} = \epsilon$ and the exponent $\tilde{h}$ of $\tilde{\epsilon}$ in the transformed differential equation is less than h. By hypothesis, the leading matrix $\tilde{A}_0(x)$ has only one distinct eigenvalue. Therefore, the next step would be a transformation of type (2.4-2) which reduces that eigenvalue to zero. After that, the differential equation will have to be transformed to its Arnold form, which will be followed by the next shearing. If the result is again a problem with a leading matrix that has only one eigenvalue and a nonzero entry on the main diagonal, repetition of the argument produces a problem with a still lower value of h. This can happen only a finite number of times before h has been reduced to zero.

There remains the possibility that $\tilde{A}_0(x)$ has *no* nonzero entries in its diagonal, and that it has only one distinct eigenvalue. Since the trace of $\tilde{A}_0$ is the sum of its eigenvalues, one concludes then that $\tilde{A}_0(x)$ shares with A_0

the property of being nilpotent. However, if α is a fraction, ϵ has to be replaced by $\tilde{\epsilon}^q$ [use (2.6-5)] in (2.6-2), and $(h - \alpha)q$ may well be larger than h, so that one might fear to have complicated rather than simplified the differential equation by the shearing transformation.

It turns out that, nevertheless, the differential equation simplifies in some sense under the shearing in the situation under scrutiny. In fact, if a repetition of shearings followed by transformations of the leading coefficient to its Jordan form always produces nilpotent leading matrices, the leading matrix will eventually have one Jordan block only. This was proved by Turrittin in Ref. 83. An adaptation of the argument to the parameterless asymptotic theory can be found in Section 19 of Ref. 90. The details of the proof are somewhat long and intricate. Therefore, only a sketch will be given here. The full argument can be found in the references mentioned.

Observe first, that by the definition of the region $D^{(1)}$ in Section 2.6, the Jordan matrix $\tilde{A}_0$ is holomorphic in $D^{(1)}$ and that the degrees of its elementary divisors are constant there. Since $\tilde{A}_0(x)$ is nilpotent, all matrices $\sigma_j(x)\Gamma_j$ in (2.7-2) for which the nonzero entry of Γ_j lies in a diagonal block of the partition must be identically zero. Choose a value of x that corresponds to a point of $D^{(1)}$ where all the σ_j that are not identically zero are different from zero. This condition excludes only a discrete set of points. If $\tilde{A}_0(x)$ has nonzero entries in some rows where A_0 has only zeros, one sees easily that $\tilde{A}_0(x)$ has more linearly independent columns than A_0. This means that $\tilde{A}_0(x)$ has higher rank than A_0, and hence, that $\tilde{J}(x)$, the Jordan matrix for $\tilde{A}_0(x)$, has fewer blocks than A_0.

If the rows that are zero in A_0 are also zero in $A_0(x)$, Turrittin's method calls for an analysis of the matrix $\lambda I - \tilde{A}_0(x)$, and of the greatest common divisors $D_j(\lambda, x)$ of its minors of order j. These polynomials in λ and their relations to the invariant factors of $\tilde{A}_0(x)$ have already been mentioned in the proof of Lemma 2.2-1. Their degrees, d_j, form a nondecreasing sequence so that $1 = d_0 \leqslant d_1 \leqslant \cdots \leqslant d_n = n$. The differences $d_{n-j} - d_{n-j-1}$ are the degrees of the invariant factors.

Now, a careful inspection of the minors of $\lambda I - \tilde{A}_0(x)$ shows that the presence of the additional nonzero entries which distinguish $\tilde{A}_0(x)$ from A_0 never increases the degrees of the greatest common divisors of the minors of each order, and that one such degree at least actually decreases. One concludes that such a sequence of shearings and reductions to Jordan form cannot indefinitely continue. After a finite number of steps the Jordan matrix for the lead matrix would have only one nonconstant invariant factor. Since, by assumption, the matrix is also nilpotent, it must be a simple $n \times n$ shifting matrix. The differential equation can then be taken into Arnold's form.

One more shearing then introduces nonzero entries into the last row of that shifting matrix. If one of those is in the main diagonal, α is an integer and the exponent h has been reduced. If not, $\tilde{A}_0(x)$ must have more than one eigenvalue and the *order of the problem* can be reduced.

2.8. Formal Solution of the Differential Equation

Here is a summary of the types of transformations encountered so far. They are listed here as changes of variables from y to v. Actually, each transformation leads to a new dependent variable. They must be performed in a sequence, as described above. After the first block-diagonalization, each affects only a subset of the n components of the n-dimensional vectorial-dependent variable, but each such transformation can be regarded as a change of variables in n dimensions: The other components remain unchanged.

(i) Reduction of the leading coefficient matrix to its Jordan form by a transformation of the form

$$y = T(x)v. \tag{2.8-1}$$

(ii) Separation of distinct eigenvalues of the leading matrix by a block-diagonalization of all terms in the series for the coefficient matrix:

$$y = \left(I + \sum_{r=1}^{\infty} P_r(x)\epsilon^r\right)v. \tag{2.8-2}$$

(iii) Splitting off exponential factors by setting

$$y = \exp\left\{\epsilon^{-h}\int^x \lambda(t)\,dt\right\}v.$$

(iv) Reduction to Arnold's form by transformations of the same type as in (ii).

(v) Shearing transformations:

$$y = \operatorname{diag}(1, \epsilon^{\alpha}, \epsilon^{2\alpha}, \ldots, \epsilon^{(n-1)\alpha})v.$$

(vi) Changing the parameter by setting

$$\epsilon = \tilde{\epsilon}^q.$$

The complete desired reduction to a set of uncoupled differential equations, each of them either of dimension one or without the factor ϵ^h in the left-hand number, can be achieved by a *finite* sequence of such transformations. Transformations of type ii lower the order of the systems that remain to be simplified. Hence, they can occur at most n times. For the same reason, only a finite number of shearings that result in introducing new eigenvalues in the lead matrix can be present. Shearings that do not introduce more than one eigenvalue may lower the exponent h of ϵ in the left number. This can obviously happen only a finite number of times in succession. After a transformation of type iii, to make the new lead matrix nilpotent, that lead matrix never has more Jordan blocks than before. Therefore, a finite number of shearings that do not introduce new eigenvalues will always lead to a problem with one Jordan block in the lead matrix, as was explained above. Transformations of type i, iii, v, or vi occur always before or after a transformation of type ii or iv, and never twice in a row. Hence the total number of transformations is finite.

The matrix of the combined transformation is the product of those described above in order from left to right. Because of the successively refined block structure of these matrices as you go from left to right, the matrices corresponding to transformations of type iii, which are scalar in each block, commute with all those to their right. Therefore, if the n-dimensional dependent variable after the complete reduction of the problem is denoted by the letter w, the whole formal transformation from y in (2.3-18) to the differential equation for w has the form—in a notation independent of the previous ones—

$$y = \left(\sum_{r=0}^{\infty} P_r(x) \epsilon^{r/m} \right) e^{\tilde{Q}(x,\epsilon)} w, \qquad (2.8\text{-}3)$$

where m is some positive integer, and $\tilde{Q}$ is a diagonal matrix-valued function of the form

$$\tilde{Q}(\cdot, \epsilon) = \sum_{j=1}^{mj-1} \tilde{Q}_j \epsilon^{(j-mh)/m}.$$

The matrices P_r and $\tilde{Q}_j$ are holomorphic in x in a certain region D. The matrix $P_0(x)$ is, in general, not invertible, because of the effect of shearing transformations. But the formal series in powers of $\epsilon^{1/m}$ for $\det(\sum_{r=0}^{\infty} P_r(x) \times \epsilon^{r/m})$ does not vanish identically.

The system of differential equations for w consists of uncoupled scalar equations and of systems of differential equations without a positive power of ϵ in the left-hand member.

The scalar equations can easily be solved by quadratures. Assume, temporarily, that the given differential equation is scalar, and set $\sigma := \epsilon^{1/m}$, $H := hm$. Then one deals with the formal scalar differential equation

$$\sigma^H w' = \left(\sum_{r=0}^{\infty} A_r \sigma^r \right) w. \qquad (2.8\text{-}4)$$

To solve it, perform the transformation

$$w = \exp\left\{ \int \sum_{j=0}^{H-1} A_r(t) \sigma^{j-H} \, dt \right\} \tilde{w} \qquad (2.8\text{-}5)$$

which leads, after cancellation of σ^H from both members, to the differential equation

$$\tilde{w}' = \left(\sum_{r=H}^{\infty} A_r \sigma^{r-H} \right) \tilde{w}. \qquad (2.8\text{-}6)$$

Formally, (2.8-6) can be satisfied by the series in powers of σ obtained by setting

$$\tilde{w}(x,\sigma) = \exp\left\{ \sum_{r=H}^{\infty} \left(\int^{x} A_r(t)\, dt \right) \sigma^{r-H} \right\}$$

$$= \exp\left\{ \int^{x} A_H(t)\, dt \right\} \exp\left\{ \sum_{r=H+1}^{\infty} \left(\int^{x} A_r(t)\, dt \right) \sigma^{r-H} \right\}, \quad (2.8\text{-}7)$$

expanding the last factor by the series for the exponential function, and then collecting like powers of σ. This can be done because each power of σ occurs only a finite number of times in the expansion of the last exponential function. The lower endpoints of the integrals can be chosen at any fixed point in D. The result of this calculation is therefore far from unique.

Insertion of the series for $\tilde{w}$ into w in (2.8-5) yields an expansion of the form

$$w(x,\sigma) = \left(\sum_{r=0}^{\infty} B_r(x)\sigma^r \right) \exp\left\{ \sum_{j=0}^{H-1} \int^{x} A_r(t)\, dt\, \sigma^{j-H} \right\}. \quad (2.8\text{-}8)$$

The functions B_r are holomorphic in D.

The other type of differential equation now to be solved explicitly can also be written in the form (2.8-4) if one sets $H = 0$ and interprets the A_r as matrices. It has formal solutions that differ from (2.8-8) only by the absence of the exponential factor.

The solutions of the several scalar equations and systems can be assembled into one matrix solution W of dimension $n \times n$ for the differential equation resulting from (2.3-18) by the formal transformation (2.8-3). This matrix has the form

$$W(x,\epsilon) = \left(\sum_{r=0}^{\infty} W_r(x)\epsilon^{r/m} \right) e^{\tilde{Q}(x,\epsilon)}, \quad (2.8\text{-}9)$$

$$\tilde{Q}(x,\epsilon) = \sum_{j=0}^{k-1} \tilde{Q}_j(x)\epsilon^{(j-k)/m}. \quad (2.8\text{-}10)$$

The $\tilde{Q}$ are diagonal matrices, and k is a non-negative integer. (If $k = 0$, $\tilde{Q} :\equiv 0$.) From its origin in the expansion of exponential functions in (2.8-7) it follows that $\det W_0(x) \neq 0$ for all $x \in D$.

Finally, one has to return from W to the corresponding matrix solution Y of (2.3-18) by the transformation (2.8-3). At long last the formal solution of the differential equation (2.1-3) is at hand and is described in the theorem below.

Theorem 2.8-1. *Let $\sum_{r=0}^{\infty} A_r \epsilon^r$ be a formal series in which the A_r are $n \times n$-dimensional matrix-valued functions of the complex variable x, holomorphic in a region $D \subset \mathbb{C}$, and ϵ is a complex parameter. Let h be a positive* *integer. Then the formal differential equation*

$$\epsilon^h \frac{dy}{dx} = \left(\sum_{r=0}^{\infty} A_r(x)\epsilon^r \right) y \tag{2.8-11}$$

can be satisfied by replacing y with a formal expression of the form

$$\left(\sum_{r=0}^{\infty} Y_r(x)\epsilon^{r/m} \right) e^{Q(x,\epsilon)}. \tag{2.8-12}$$

The symbols in (2.8-12) are defined as follows:

(i) *The Y_r are $n \times n$ matrix-valued functions locally holomorphic in D, except for a set S of isolated points, the same for all r, where they have branch points or poles. The Q_j have the same properties as the Y_r, and k is a non-negative integer ($Q := 0$, if $k = 0$).*
(ii) *The formal series for $\det[\sum_{r=0}^{\infty} Y_r(x)\epsilon^{r/m}]$ in powers of $\epsilon^{1/m}$ is not identically zero.*

Corollary. *If $Q \equiv 0$, and if $\sum_{r=0}^{\infty} A_r(x)\epsilon^r$ converges, then $\sum_{r=0}^{\infty} Y_r(x)\epsilon^{r/m}$ converges except, possibly, at the points where the Y_r are not holomorphic.*

PROOF. If $Q \equiv 0$, no eigenvalues other than zero are met in the reduction process that led to the formal solution (2.8-12). The successive steps are, therefore, all of the form i, iv, v or vi. Only steps of the form iv involve infinite series in powers of ϵ. However, the use of the whole infinite series in (2.6-2) was only a convenience. In the transformation (2.5-5) the series could be replaced by a partial sum of enough terms to calculate the proper exponent α in (2.4-3). Hence, the full reduction process leads, after finitely many steps, to a differential equation of the same order as the original one, but with $h = 0$. This means that the solutions are holomorphic, except at the singularities of Y_r, if any. This proves the corollary.

2.9. Some Comments and Warnings

The algorithm for the formal solution of the differential equations of the type considered here has the advantage of applying in all cases and to reveal the general structure of the solution. More often than not, however, it is not the best way of calculating the actual approximations. This is particularly true when the given differential equation is a scalar equation of order greater than one. Then it may be preferable to avoid the passage to an equivalent first-order system.

The simple equation

$$\epsilon^2 u'' = xu \tag{2.9-1}$$

is equivalent to the system

$$\epsilon y' = \begin{pmatrix} 0 & 1 \\ x & 0 \end{pmatrix} y \tag{2.9-2}$$

with $y := \left(\begin{smallmatrix} u \\ \epsilon u' \end{smallmatrix} \right)$, but the transformation

$$x = \epsilon^{2/3} t \tag{2.9-3}$$

of the independent variable opens a better approach. It reduces the problem to the parameterless equation

$$\frac{d^2 u}{dt^2} = tu,$$

which is the well-known Airy Equation whose properties, including its divergent power series expansions have been thoroughly studied by means of special methods.

Also, there are many first-order systems that are equivalent to one and the same scalar equation. Some may lend themselves better to a formal analysis than others. For instance, the "Orr–Sommerfeld" equation

$$\epsilon^2 u^{(4)} - x\psi(x, \epsilon)u'' - b(x, \epsilon)u = 0 \tag{2.9-4}$$

is best left in scalar form if only formal results are needed. One sees readily that (2.9-4) can be satisfied by power series in ϵ whose leading terms are solutions of

$$x\psi(x, 0)u_0'' + b(x, 0)u_0 = 0. \tag{2.9-5}$$

In some equivalent systems the analogous fact may be much harder to recognize. The transformation

$$y := \begin{bmatrix} y_1 \\ y_2 \\ y_3 \\ y_4 \end{bmatrix} = \begin{bmatrix} u \\ u' \\ u'' \\ \epsilon u''' \end{bmatrix}$$

takes (2.9-5) into

$$\epsilon y' = A(x, \epsilon)y$$

with

$$A(x, \epsilon) = \begin{bmatrix} 0 & \epsilon & 0 & 0 \\ 0 & 0 & \epsilon & 0 \\ 0 & 0 & 0 & 1 \\ b(x, \epsilon) & 0 & x\psi(x, \epsilon) & 0 \end{bmatrix},$$

in which the relevance of equation (2.9-5) for small ϵ is not immediately apparent.

By the principle of analytic continuation, expression (2.8-12) satisfies the differential equation (2.8-11) formally in any extension of the region D where the Y_r and Q remain holomorphic. The continuation may lead back to the same points of D with different values for the Y_r and Q. Then these functions must be defined on an appropriate Riemann domain rather than on a region in $\mathbb{C}$. In D, the singularities of Y_r and Q are points where A_0 or the analogous leading coefficient matrices in the later stages of the process of formal solution fail to satisfy Hypothesis II-1. In the proof of the fact that no other singularities occur in D, Theorem 12.2-2 was used essentially. It was proved for regions in $\mathbb{C}$ only, but the whole theory of the Appendix applies to Riemann regions, as well as to regions in $\mathbb{C}$. Therefore, in a region in which the A_r of (2.8-11) are holomorphic the only singularities of the possibly multivalued functions Y_r and Q in (2.8-12) are the points where the leading coefficient matrices of the several transformed differential equations occurring in the proof of Theorem 2.8-1 fail to be holomorphic.

CHAPTER III
Solutions Away From Turning Points

3.1. Asymptotic Power Series: Definition of Turning Points

The formal, usually divergent series in powers of ϵ, or of some root of ϵ, which appear in the theorems of Chapter II, are related to true solutions of the differential equation in the sense that they are asymptotic representations of analytic functions which are such true solutions when substituted for those series. This fact will be established in the present section.

The concept of asymptotic representations by power series was first clearly defined by Poincaré, who recognized their fundamental role in the study of singularities of differential equations. As the relevant properties of asymptotic power series have been described in the literature in many places, for instance in Refs. 10, 66, and 90, only a very brief summary, adapted to the context of this book, will be given here.

Definition 3.1-1. Let a_r, $r = 0, 1, \ldots$, be functions of x holomorphic in a region $D \subset \mathbb{C}$. Let E be a region of the complex plane having the origin as a boundary point. A complex-valued function f defined in $D \times E$ will be said to be asymptotically represented in D by the series $\sum_{r=0}^{\infty} a_r(x)\epsilon^r$, as ϵ tends to zero in E, if for all non-negative integers N

$$\lim_{\substack{\epsilon \to 0 \\ \epsilon \in E}} \left[f(x,\epsilon) - \sum_{r=0}^{N} a_r(x)\epsilon^r \right] \epsilon^{-N} = 0 \qquad \text{for} \quad x \in D.$$

If, for each N the limit is approached uniformly for $x \in D^* \subset D$, the asymptotic representation is called uniform in D^*.

The usual notation for the asymptotic relation defined above is

$$f(x,\epsilon) \sim \sum_{r=0}^{\infty} a_r(x)\epsilon^r \quad \text{for} \quad x \in D, \text{ as } \epsilon \to 0 \text{ in } E. \quad (3.1\text{-}1)$$

In most respects asymptotic series may be manipulated by the same rules as convergent series, but there are the following two important exceptions.

(i) The representation (3.1-1) is unique in one logical direction only: No function f has more than one expansion in a region D, but infinitely many functions have the same expansion as f.

(ii) Termwise differentiation is not always correct, although it is so in most applications.

A very useful fact is a result that is often called the "Borel–Ritt" theorem after two of its at least three independent discoverers. Roughly speaking it says that *any* series $\sum_{r=0}^{\infty} a_r(x)\epsilon^r$ is the asymptotic expansion of some function (and therefore of infinitely many functions). The version of that theorem stated below is tailored to its applications in the present context. Several variants can be proved by similar techniques (see, e.g., Refs. 4 and 90). In Chapter VI a more general form of this theorem will be proved, therefore, no proof will be given at this point.

Theorem 3.1-1 (Borel–Ritt). *Let a_r, $r = 0, 1, \ldots$ be functions holomorphic and bounded in a region $D \subset \mathbb{C}$. Let E be a sector*

$$E := \{\epsilon \in \mathbb{C} / \alpha < \arg \epsilon < \beta, 0 < |\epsilon| \leqslant \epsilon_0, 0 < \alpha < \beta < 2\pi\}. \quad (3.1\text{-}2)$$

Then there exists a function f of x and ϵ holomorphic for $x \in D$, $\epsilon \in E$, having for all $x \in D$ the asymptotic expansion

$$f(x,\epsilon) \sim \sum_{r=0}^{\infty} a_r(x)\epsilon^r \quad \text{as } \epsilon \to 0 \text{ in } E, \text{ for } x \in D. \quad (3.1\text{-}3)$$

The expansion is uniform in every compact subdomain of D. The relation (3.1-3) may be indefinitely differentiated termwise with respect to both variables.

The preceding theorem is remarkably general: The functions a_r may grow arbitrarily fast with r, and the constants α, β, ϵ_0 are arbitrary as long as $0 < \beta - \alpha < 2\pi$. Even that condition can be eliminated by defining E as a sector of a Riemann surface with a branch point at $\epsilon = 0$. It is obvious that $\epsilon = 0$ cannot be an interior point of E, except when the series converges. For, when the relation $\lim_{\epsilon \to 0} f(x,\epsilon) = a_0(x)$ is true in a full neighborhood of $\epsilon = 0$, then f can be defined so as to be a holomorphic function of ϵ at $\epsilon = 0$, by setting $f(x,0) := a_0(x)$.

The stage has been reached, finally, at which a precise definition of the term "turning point" can be given.

Definition 3.1-2. A point $x_0 \in D$ is called a turning point for the differential equation

$$\epsilon^h y' = A(x, \epsilon) y$$

with A holomorphic in $D \times E$, if none of the formal matrix solutions described in Theorem 2.8-1 is an asymptotic representation of a fundamental solution in a full neighborhood of x_0.

The expression "asymptotic representation" is used here in a somewhat more general sense than in Definition 3.1-1. What is meant, of course, is that if $y = Y(x, \epsilon)$ is any matrix solution (2.8-11), then one *never* has

$$Y(x, \epsilon) e^{-Q(x, \epsilon)} \sim \sum_{r=0}^{\infty} Y_r(x) \epsilon^{r/m} \qquad \text{as} \quad \epsilon \to 0 \quad \text{in } E,$$

uniformly for x in a *full* neighborhood of $x = x_0$, with Y_r as in (2.8-12).

This definition describes in mathematical language the essence of the customary loose use of the term "turning point." The definition is not quite satisfactory because of its negative character. This question will be discussed again later, when more facts on the asymptotic validity of formal solutions have been proved.

Points of D in whose neighborhood the formal solutions are not termwise holomorphic are, of course, turning points, since the true solutions are there holomorphic. In particular, the branch points of the eigenvalues which appeared in the constructions described in Chapter II are always turning points.

If two eigenvalues of $A_0(x)$ coalesce at a point where they have no branch point, there may or may not be a turning point there. Take, for instance, the system

$$\epsilon y' = \begin{pmatrix} x & \epsilon \psi(x, \epsilon) \\ 0 & 0 \end{pmatrix} y \tag{3.1-4}$$

with

$$\psi(x, \epsilon) = \sum_{r=0}^{\infty} \psi_r(x) \epsilon^r.$$

The formal procedures of Chapter II here involve successive divisions by x and differentiations which, in general, introduce poles at $x = 0$ into the coefficients of the formal solution. Then, $x = 0$ is surely a turning point. Exceptionally, these poles may be absent, such as when $\psi(x, \epsilon) \equiv x$. Then one fundamental matrix solution is

$$Y(x, \epsilon) = \begin{pmatrix} 1 & -\epsilon \\ 0 & 1 \end{pmatrix} \exp \left\{ \frac{1}{\epsilon} \begin{pmatrix} x^2/2 & 0 \\ 0 & 0 \end{pmatrix} \right\}.$$

Accordingly, $x = 0$ is not a turning point for equation (3.1-4) when $\psi(x, \epsilon) \equiv x$. A more detailed discussion of equation (3.1-4) can be found in Section 3.4.

3.2. A Method for Proving the Analytic Validity of Formal Solutions: Preliminaries

The procedure followed in this account of first constructing formal solutions of the differential equation and then showing that the formal solutions are asymptotic representations of actual solutions is very common, and almost everywhere in the literature the analytic validity is established by basically the same technique: The formal calculations yield a function which "almost solves" the differential equation, meaning that substitution of the function into the equation produces a "small" error term. It must then be shown that this error term changes the solution by a small amount, or, in a different way of expressing oneself, that the solution operator is continuous. This is usually done by rewriting the differential equation with the error term as an equivalent Volterra integral equation with the help of the variation of parameters formula. The existence of a solution of the integral equation can be established by some fixed point theorem. The solution is then appraised by inequalities which prove that it is close to a solution of the unperturbed differential equation.

If, as in Chapter II, the formal procedure has produced an expression involving an infinite power series, the Borel–Ritt theorem presents a natural approach for finding an approximate solution. The extension of that theorem from scalar to vector or matrix-valued functions is straightforward. Let there be given the formal matrix solution (2.8-12) and assume that the Y_r are holomorphic functions in D. There exists then (not uniquely) a matrix $\hat{U}(x, \epsilon)$, holomorphic over $D \times E$, with the asymptotic expansion

$$\hat{U}(x,\epsilon) \sim \sum_{r=0}^{\infty} Y_r(x)\epsilon^{r/m} \qquad \text{as} \quad \epsilon \to 0 \quad \text{in } E. \qquad (3.2\text{-}1)$$

The set E was defined in formula (3.1-2). To the formal solution (2.8-12) there corresponds the function $\hat{U}e^{Q}$ which, one hopes, will turn out to be close to a true solution of the given differential equation

$$\epsilon^h Y' = A(x,\epsilon)Y. \qquad (3.2\text{-}2)$$

The fact that $\sum_{r=0}^{\infty} Y_r(x)\epsilon^{r/m}e^{Q}$ is a formal solution of (3.2-2) is equivalent to the statement that $\sum_{r=0}^{\infty} Y_r(x)\epsilon^{r/m}$ is a formal solution of the differential equation

$$\epsilon^h V' = AV - V\epsilon^h Q', \qquad (3.2\text{-}3)$$

which results from (3.2-2) by the transformation

$$Y = Ve^{Q}. \qquad (3.2\text{-}4)$$

This means that

$$\epsilon^h \hat{U}' = A\hat{U} - \hat{U}\epsilon^h Q' + \Omega \qquad (3.2\text{-}5)$$

with

$$\Omega(x,\epsilon) \sim 0 \qquad \text{as} \quad \epsilon \to 0 \quad \text{in } E, \quad \text{for} \quad x \in D. \qquad (3.2\text{-}6)$$

Another way of expressing the basic property of $\hat{U}$ and of $\hat{U}e^Q$ is to introduce the matrix

$$A^* := \epsilon^h(\hat{U}e^Q)'(\hat{U}e^Q)^{-1}$$
$$= \epsilon^h\hat{U}'\hat{U}^{-1} + \epsilon^h\hat{U}Q'\hat{U}^{-1}. \qquad (3.2\text{-}7)$$

(The asterisk does not indicate the adjoint matrix.) Then $Y^* := \hat{U}e^Q$ is a true solution of the differential equation

$$\epsilon^h Y^{*\prime} = A^* Y^* \qquad (3.2\text{-}8)$$

and comparison of (3.2-5) with (3.2-7) shows that

$$A^* - A = \Omega \hat{U}^{-1} =: \Omega^* \sim 0 \qquad \text{in } D \quad \text{as} \quad \epsilon \to 0 \quad \text{in } E, \qquad (3.2\text{-}9)$$

provided that all Y_r are holomorphic in D and that the leading term of the formal series for $\det(\sum_{r=0}^{\infty}Y_r\epsilon^{r/m})$ has no zeros in D.

For further analysis, vectorial notation is preferable to dealing with full matrix solutions, since it yields sharper results. Set

$$Q = \text{diag}(q_1, q_2, \ldots q_n) \qquad (3.2\text{-}10)$$

and denote the column vectors of $\hat{U}$ in (3.2-1) by $\hat{u}_1, \hat{u}_2, \ldots, \hat{u}_n$. Then the columns of the matrix $\hat{U}e^Q$ are the vectors $\hat{u}_ke^{q_k}$, $k = 1, 2, \ldots, n$, and the asymptotic relation (3.2-1) becomes

$$u_k(x,\epsilon) \sim \sum_{r=0}^{\infty} y_{rk}(x)\epsilon^{r/m}, \qquad k = 1, 2, \ldots, n, \qquad (3.2\text{-}11)$$

where y_{rk} denotes the kth column of Y_r. By (3.2-5) the differential equation for $\hat{u}_k$ is

$$\epsilon^h\hat{u}_k' = (A - \epsilon^h q_k'I)\hat{u}_k + \omega_k, \qquad (3.2\text{-}12)$$

where ω_k is the kth column of Ω. Analogously, the columns v_k of a solution V of (3.2-3) satisfy the differential equation

$$\epsilon^h v_k' = (A - \epsilon^h q_k'I)v_k \qquad (3.2\text{-}13)$$

and

$$y_k := v_ke^{q_k} \qquad (3.2\text{-}14)$$

is then a vector solution of (3.2-2) when substituted for Y.

The aim is to show that among the solutions of (3.2-13) there is one that is asymptotically represented by the series in (3.2-11). This is tantamount to proving that (3.2-12) has a solution which is asymptotic to zero, because $\hat{u}_k - \hat{v}_k$ is a solution of (3.2-12) if $\hat{u}_k$ and $\hat{v}_k$ satisfy (3.2-12) and (3.2-13), respectively.

3.3. A General Theorem on the Analytic Validity of Formal Solutions

To carry out the program described above, i.e., to show that the vectorial differential equation

$$\epsilon^h z' = \left(A - \epsilon^h q_k' I \right) z + \omega_k , \tag{3.3-1}$$

which is satisfied by $z = \hat{u}_k$, also has solutions that are asymptotic to zero, rewrite (3.3-1) in the form

$$\epsilon^h z' = \left(A^* - \epsilon^h q_k' I \right) z - \Omega^* z + \omega_k , \tag{3.3-2}$$

in which (3.2-9) has been used, and apply the formula of variation of parameters. By formula (3.2-8) the homogeneous differential equation

$$\epsilon^h Z^{*\prime} = \left(A^* - \epsilon^h q_k' I \right) Z^* \tag{3.3-3}$$

has the fundamental matrix solution

$$Z^*(x,\epsilon) = \hat{U}(x,\epsilon) \exp\{ Q(x,\epsilon) - q_k(x,\epsilon) I \}, \tag{3.3-4}$$

while the solution of the corresponding differential equation with A in the place of A^* is not yet known. Any solution of the vectorial integral equation

$$z(x,\epsilon) = Z^*(x,\epsilon)c + \int^x Z^*(x,\epsilon)\left[Z^*(t,\epsilon) \right]^{-1}$$
$$\times \left[-\Omega^*(t,\epsilon)z(t,\epsilon) + \omega_k(t,\epsilon) \right] \epsilon^{-h} dt \tag{3.3-5}$$

solves (3.3-2), i.e., (3.3-1) by the formula for variation of parameters. Here, c is an arbitrary constant vector and the lower end point of the integral need not be the same for each of the n components of the vector in the integrand. Changing these end points amounts to changing c.

The kernel $Z^*(x,\epsilon)(Z^*(t,\epsilon))^{-1}$ in (3.3-5) has the form

$$\hat{U}(x,\epsilon) \exp\{ Q(x,\epsilon) - Q(t,\epsilon) - (q_k(x,\epsilon) - q_k(t,\epsilon))I \}\left[\hat{U}(t,\epsilon) \right]^{-1}$$

by (3.3-4). This can be slightly abbreviated into

$$Z^*(x,t)\left[Z^*(t,\epsilon) \right]^{-1} = \hat{U}(x,\epsilon) e^{\tilde{Q}_k(x,\epsilon) - \tilde{Q}_k(t,\epsilon)}\left[\hat{U}(t,\epsilon) \right]^{-1} \tag{3.3.-6}$$

by defining

$$q_{jk} := q_j - q_k , \qquad j,k = 1,2,\ldots,n,$$
$$\tilde{Q}_k := \mathrm{diag}\{ q_{1k}, q_{2k}, \ldots, q_{nk} \}. \tag{3.3-7}$$

The n paths of the integrals for the n components of the vector in (3.3-5) may be denoted by γ_{jk}, $j = 1,2,\ldots,n$, and assembled into a symbolic vector

$$\Gamma_x = (\gamma_{1x}, \gamma_{2x}, \ldots, \gamma_{nx}).$$

The constant c in (3.3-5) will then be chosen as zero. The integral equation can now be written

$$z(x,\epsilon) = \hat{U}(x,\epsilon) \int_{\Gamma_x} e^{\tilde{Q}_k(x,\epsilon) - \tilde{Q}_k(t,\epsilon)} \big[K(t,\epsilon) z(t,\epsilon) + k(t,\epsilon) \big] dt \quad (3.3\text{-}8)$$

with

$$K := -\hat{U}^{-1} \Omega \epsilon^{-h}, \qquad k := \hat{U}^{-1} \omega_k \epsilon^{-h}. \quad (3.3\text{-}9)$$

The jth component of the vectorial integrand in (3.3-8) contains the scalar exponential factor $\exp\{q_{jk}(x,\epsilon) - q_{jk}(t,\epsilon)\}$. It is clear that the sign of the function $\operatorname{Re}\{q_{jk}(x,\epsilon) - q_{jk}(t,\epsilon)\}$ along the paths of integration is decisive for the asymptotic nature of z. This explains the relevance of the definition below.

Definition 3.3-1. Let $x \in D$ and $\epsilon \in E$. A directed arc in D in the t-plane ending at $t = x$ is called progressive with respect to the function q_{jk}, the point x, and the domain E, if $\operatorname{Re} q_{jk}(t,\epsilon)$ is a nonincreasing function of t on that arc.

The next definition characterizes a class of domains for which the proof that formal solutions are asymptotic to true solutions is particularly easy.

Definition 3.3-2. The set $D^* \subset D$ is called accessible with respect to q_k for $\epsilon \in E$, if there are n points t_j in D^*, $j = 1, 2, \ldots, n$, such that every point $x \in D^*$ can be connected in D^* with the points t_j by paths that are progressive with respect to q_{jk}, $j = 1, 2, \ldots, n$, and are uniformly bounded in length.

Theorem 3.3-1. *If the functions Y_r, Q_j in the formal solution (2.8-12) are holomorphic in D^* and if D^* is compact and accessible with respect to q_k for all $\epsilon \in E$, then the differential equation*

$$\epsilon^h y' = A(x,\epsilon) y \quad (3.3\text{-}10)$$

possesses a vector solution of the form

$$y = v_k e^{q_k},$$

where v_k admits, uniformly in D^*, the asymptotic expansion

$$v_k(x,\epsilon) \sim \sum_{r=0}^{\infty} y_{rk}(x) \epsilon^{r/m} \qquad \text{as} \quad \epsilon \to 0 \quad \text{in } E. \quad (3.3\text{-}11)$$

Here y_{rk} is the kth column vector of the matrix Y_r in Theorem 2.8-1, and q_k is the kth diagonal entry of the matrix Q in that theorem.

PROOF. For fixed $\epsilon \in E$ let $\mathscr{B}$ be the space consisting of all vector-valued functions of x of dimension n whose components are holomorphic in D^*. If

$v \in \mathscr{B}$ has the components v_μ, $\mu = 1, 2, \ldots, n$, define the norm $\|v\|$ by

$$\|v\| := \max_{x \in D^*} \left(\sum_{\mu=1}^{n} |v_\mu(x)| \right).$$

With this norm $\mathscr{B}$ is a Banach space. If M is a matrix whose columns m_μ are elements of $\mathscr{B}$, then the associated matrix norm is

$$\|M\| = \max_{\mu = 1, \ldots, n} \|m_\mu\|.$$

The right-hand member of (3.3-8) defines an operator S on z which maps $\mathscr{B}$ into itself. If $v, w \in \mathscr{B}$, then

$$\|Sw - Sv\| \leqslant \|\hat{U}\| \|K\| \sup_{x \in D^*} \sup_{j=1,2,\ldots,n} \left(\int_{\gamma_{jx}} |dt| \right) \|w - v\|, \quad (3.3\text{-}12)$$

because the paths γ_{jk} are progressive. As $\epsilon \to 0$ in E, the norm $\|\hat{U}\|$, which is a function of ϵ, remains bounded. The factor $\|K\|$ is asymptotic to zero. Using also the assumption that D^* is bounded, one concludes from (3.3-12) that

$$\|Sw - Sv\| \leqslant c_1 |\epsilon| \|w - v\| \qquad \text{for} \quad \epsilon \in E \qquad (3.3\text{-}13)$$

with some constant c_1 independent of ϵ.

(The obvious stronger inequality with an arbitrary power of ϵ in the right member is not needed.) Therefore, S is a contraction mapping for $|\epsilon| < c_1^{-1}$, $\epsilon \in E$. For such values of ϵ the integral equation (3.3-8) possesses a unique solution z in $\mathscr{B}$ by the contraction mapping theorem.

That this solution z is asymptotic to zero is now an immediate consequence of the fact that k in (3.3-9) is asymptotic to zero. For (3.3-13) implies that $\|Sz\| \leqslant c_1 |\epsilon| \|z\|$ and, returning to (3.3-8), one sees that in D^*

$$\|z\| \leqslant c_1 |\epsilon| \|z\| + C_N |\epsilon|^N$$

for all sufficiently small $\epsilon \in E$ and all $N \geqslant 0$ with some constants C_N. For $|\epsilon| \leqslant c_2 < c_1^{-1}$ it follows that $\|z\| = O(\epsilon^N)$, as was to be proved.

The fact that the integral equation (3.3-8) has only one solution does not imply that this is the only solution of the differential equation which has the given formal solution as its asymptotic expansion in D^*. If, for instance $v_k e^{q_k}$ is multiplied by $1 + \exp\{\epsilon^{-h_1}\}$ with $h_1 > h$, the duly modified solution has the same asymptotic expansion as the original one.

3.4. A Local Asymptotic Validity Theorem

The general theorem 3.3-1 reduces the problem of finding asymptotic solutions of the differential equations considered here to the task of finding accessible domains. The result to be proved now is strictly local: D^* may have to be taken very small, and ϵ may have to be limited to a narrow sector of the ϵ-plane.

For the sake of simplicity, ϵ will be even restricted to a small segment of the positive real axis

$$0 < \epsilon \leqslant \epsilon_0. \tag{3.4-1}$$

The extension of the results to a narrow sector of the ϵ-plane is a minor technical matter. The domain D^* will be the closure of a neighborhood of a point x_1 in D. This point will be subjected to a condition which, while still relatively mild, excludes other points besides those where the Y_r or Q_j fail to be holomorphic. The condition is that the point x_1 be what I shall call "asymptotically simple" in the sense of the following definition.

Definition 3.4-1. A point $x_1 \in D$ will be called asymptotically simple with respect to q_k if those of the functions dq_{jk}/dx, $j = 1, 2, \ldots, n$, which are not identically zero are holomorphic and different from zero at $x = x_1$ for all sufficiently small $\epsilon > 0$.

Remember that the coefficients of the polynomials dq_j/dx of $\epsilon^{-1/m}$ are the successive eigenvalues of lead matrices that appear in the formal reduction process described in Chapter II. The coefficients of $dq_{jk}/dx = dq_j/dx - dq_k/dx$ [cf. formula (3.3-7)] are therefore differences of such eigenvalues and the condition in Definition 3.4-1 means that at $x = x_1$ the eigenvalues corresponding to the highest power of $\epsilon^{-1/m}$ are all distinct unless they are identically equal. The simplest case is the one when all eigenvalues of $A_0(x)$ are distinct at x_1. Then x_1 is clearly asymptotically simple.

Theorem 3.4-1. Let k $(1 \leqslant k \leqslant n)$ be given and assume that $x_1 \in D$ is an asymptotically simple point with respect to q_k. Then there is a disk

$$D^* := \{ x/0 < |x - x_1| \leqslant d \}$$

which is accessible with respect to q_k.

PROOF. The function q_{jk} is a polynomial in $\epsilon^{-1/m}$ with analytic coefficients. If q_{jk} is not identically zero, let κ_{jk} be the degree of this polynomial and define

$$\rho_{jk}(x) := \lim_{\epsilon \to 0+} \epsilon^{\kappa_{jk}/m} q_{jk}(x, \epsilon). \tag{3.4-2}$$

The curve defined by the equation

$$\mathrm{Re}\, \rho_{jk}(x) = \mathrm{Re}\, \rho_{jk}(x_1) \tag{3.4-3}$$

has the tangent vector $(\mathrm{Im}(d\rho_{jk}/dx), \mathrm{Re}(d\rho_{jk}/dx))$, which is not zero at $x = x_1$, if this point is simple with respect to q_k. Therefore, if S is a sufficiently short straight segment containing x_1 and having a direction that is not tangent to any of those curves $\operatorname{Re}\rho_{jk}(x) = \operatorname{Re}\rho_{jk}(x_1)$ for the given k, then the q_{jk} strictly increases or decreases along S or else it is identically zero, provided ϵ is sufficiently small. Now, one can construct a rhombus Σ with center at x_1, two opposite vertices a_1 and a_2, and angles so small that the nonzero functions $\operatorname{Re}q_{jk}$ change monotonically along all segments from a_1 and a_2 into the rhombus. Then every point in Σ can be connected with a_1 or a_2 by a path progressive with respect to q_k. The disk D^* can then be taken inside Σ (Fig. 3.1).

Taken together with Theorem 3.3-1, Theorem 3.4-1 yields immediately the main local results on asymptotic validity.

Theorem 3.4-2.   *If the point $x_1 \in D$ is asymptotically simple with respect to q_k, then the corresponding column of the formal matrix solution (2.8-12) is an asymptotic representation of a true vector solution in a neighborhood of x_1.*

Theorem 3.4-3.   *If the point $x_1 \in D$ is asymptotically simple for all q_k, then the formal expression (2.8-12) represents asymptotically a fundamental matrix solution of the differential equation (3.2-2) in a full neighborhood of x_2. This is always the case when all eigenvalues of $A(x_1,0)$ are distinct.*

At points that are not asymptotically simple the coefficients of the formal solutions are likely to have branch points or poles, and those points are then turning points. However, as example (3.1-4) with $\psi(x,\epsilon) \equiv x$ shows, this is not always so. There, $x = 0$ is not simple, for $q_{12}(x,\epsilon) = x^2/2\epsilon$, and equation (3.4-3) becomes $\operatorname{Re}(x^2) = 0$. It defines two straight lines intersecting at $x = 0$, not a simple curve, and the construction in the proof of Theorem 3.4-1 breaks down if $x_1 = 0$.

It is a tempting conjecture that *only* the points where some of the coefficients of the formal solutions described in Theorem 2.8-1 have singularities are turning points. If this is true it will be a very welcome simplification of the theory. I know of no proof of this conjecture, nor have I been

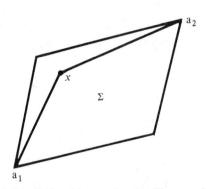

Figure 3.1. Paths of integration for Theorem 3.4-1.

able to construct a counterexample. The matter deserves further investigation. Equation (3.1-4) can be completely solved by quadratures and integrations by part. One fundamental matrix solution of (3.1-4) is explicitly given by

$$
\begin{bmatrix} 1 & \int^{x} e^{(x^2 - t^2)/2\epsilon} \epsilon^{-1} \psi(t, \epsilon)\, dt \\ 0 & 1 \end{bmatrix} \exp\left\{ \frac{1}{\epsilon} \begin{bmatrix} x^2/2 & 0 \\ 0 & 0 \end{bmatrix} \right\}.
$$

In the special case that $\psi(x, \epsilon) = x^s$, s a non-negative integer, one finds by repeated integrations by parts that

$$
I_s(x, \epsilon) := \int_0^{x} e^{(x^2 - t^2)/2\epsilon} \epsilon^{-1} t^s\, dt = p_s(x, \epsilon) + q_s(\epsilon) I_\kappa(x, \epsilon), \qquad (3.4\text{-}4)
$$

where p_s and q_s are polynomials and $\kappa = 1$ or 0 depending on whether s is odd or even. Now,

$$
I_1(x, \epsilon) = e^{x^2/2\epsilon} - 1.
$$

Since

$$
\begin{pmatrix} e^{x^2/2\epsilon} \\ 0 \end{pmatrix}
$$

is a vector solution of (3.1-4), the matrix

$$
\begin{bmatrix} 1 & p_s(x, \epsilon) - q_s(\epsilon) \\ 0 & 1 \end{bmatrix} \exp\left\{ \frac{1}{\epsilon} \begin{bmatrix} \frac{x^2}{2} & 0 \\ 0 & 0 \end{bmatrix} \right\}
$$

is also a fundamental matrix solution of (3.1-4) for $\psi(x, \epsilon) = x^s$, s odd. It has the form of the formal solutions in Theorem (2.8-1), and this formal solution has no singularities at $x = 0$. As p_s and q_s are polynomials in ϵ, $x = 0$ is not a turning point. This result can be generalized to show that if ψ in (3.1-4) is an odd function of x, formal fundamental solutions without singularities at $x = 0$ exist. An inspection of the remainder term then shows that $x = 0$ is then not a turning point.

On the other hand, if s is even, then $\kappa = 0$ in (3.4-4), and

$$
I_0(x, \epsilon) = \int_0^{x} e^{(x^2 - t^2)/2\epsilon}\, dt \qquad (3.4\text{-}5)
$$

does not have an asymptotic series in powers of ϵ with holomorphic coefficients in a neighborhood of $x = 0$.

While the foregoing example does not settle the question as to whether the formal solutions have always singularities at turning points, it adds to the probability of the conjecture.

The natural occurrence of the error integral in (3.4-5) also suggests that the insistence on asymptotic representations in terms of powers and exponential functions alone is an artificial restriction. In Chapters V and VI it will be shown that an analysis of the behavior at turning points is often best performed by introducing well-known transcendental functions into the representation.

3.5. Remarks on Points That Are Not Asymptotically Simple

Points that are not asymptotically simple but at which the coalescing eigenvalues are holomorphic occur in many differential equations. Usually they are turning points. The differential equation for the parabolic cylinder function is the best known example. More generally, whenever $A_0(x)$ is an Hermitian matrix, a theorem of Rellich states that all its eigenvalues are holomorphic, even at points where their multiplicities change. Such points are, thus, not asymptotically simple for the differential equation.

In the absence of a satisfactory theory of the asymptotic nature of solutions near points that are not asymptotically simple, it may be of some value to analyze further a simple but instructive (and not trivial) example, namely, again the differential equation (3.1-4). The results reported below were proved by Professor R. Schäfke, to whom I am grateful for communicating them to me.

The question as to whether there are cases where the formal solutions of Theorem 2.8-1 have holomorphic coefficients at some point but are, nevertheless, not asymptotic expansions of solutions in a full neighborhood of that point is of particular interest. If this could never happen, then the not quite satisfactory Definition 3.1-2 of turning points could be simplified by saying that a turning point is a point where the formal solutions have singularities.

For example (3.1-4), Schäfke has shown that for

$$\psi(x,\epsilon) = e^{-1/\sqrt{\epsilon}} \tag{3.5-1}$$

a formal fundamental solution is

$$Y(x,\epsilon) := (I + 0\epsilon + 0\epsilon + \cdots)\exp\left\{\frac{1}{\epsilon}\begin{pmatrix} x^2/2 & 0 \\ 0 & 0 \end{pmatrix}\right\}$$

(this is pretty obvious) while there exists no analytic solution

$$Y(x,\epsilon) = \hat{Y}(x,\epsilon)\exp\left\{\frac{1}{\epsilon}\begin{pmatrix} x^2/2 & 0 \\ 0 & 0 \end{pmatrix}\right\}$$

with

$$\hat{Y}(x,\epsilon) \sim I \qquad \text{as} \quad \epsilon \to 0+ \quad \text{in} \quad |x| \leq x_1.$$

However, the function in (3.5-1) is not holomorphic in ϵ at $\epsilon = 0$, as was assumed to be the case in the formulation of Definition 3.1-2. No counterexample with completely holomorphic data has yet been given.

It was shown in Section III.4 that $x = 0$ is no turning point of equation (3.1-4) for

$$\psi(x,\epsilon) = x^{2m+1},$$

while it *is* a turning point if

$$\psi(x,\epsilon) = x^{2m}.$$

Generalizing this observation one can write

$$\psi(x,\epsilon) = \sum_{r=0}^{\infty} \psi_r(\epsilon) x^r \tag{3.5-2}$$

and calculate formal solutions by superposition. If the ψ_r are formal series in powers of ϵ, Schäfke has shown that the formal solutions of the type in Theorem 2.8-1 have no singularity at $x = 0$ if and only if the relation

$$\sum_{r=0}^{\infty} \psi_{2r}(\epsilon)(2r-1)(2r-3)\ldots 1 \cdot \epsilon^r = 0 \tag{3.5-3}$$

holds identically in ϵ. The left member of (3.5-3) must be interpreted as the formal series in powers of ϵ obtained by replacing $\psi_{2r}(\epsilon)$ by its series and rearranging the double summation according to powers of ϵ.

Condition (3.5-3) is obviously satisfied when ψ is, for all ϵ, an odd function of x, but not only then. An example is the even function

$$\psi(x,\epsilon) = \epsilon - x^2$$

for which a solution—formal and analytic—of the differential equation (3.1-4) is

$$Y(x,\epsilon) = \begin{pmatrix} 1 & \epsilon x \\ 0 & 1 \end{pmatrix} \exp\left\{ \frac{1}{\epsilon}\begin{pmatrix} x^2/2 & 0 \\ 0 & 0 \end{pmatrix}\right\}.$$

Even when the coefficients of the formal fundamental matrix solution have no singularities at $x = 0$, the series $\sum_{r=0}^{\infty} Y_r(x)\epsilon^{r/m}$ of (2.8-12) may be divergent. An example is

$$\psi(x,\epsilon) = x/(1-x^2).$$

Whether the series converges or not, for the case that its coefficients are holomorphic at $x = 0$, Schäfke has proved that when ψ is holomorphic in both variables for $|x| \leq x_0$ $|\epsilon| \leq \epsilon_0$, there does exist then a matrix $\hat{Y}(x,\epsilon)$ with

$$\hat{Y} \sim \sum_{r=0}^{\infty} Y_r(x)\epsilon^{r/m}, \quad \text{as} \quad \epsilon \to 0+,$$

uniformly in a neighborhood of $x = 0$ and so that

$$Y(x,\epsilon) = \hat{Y}(x,\epsilon)\exp\left\{ \frac{1}{\epsilon}\begin{pmatrix} x^2/2 & 0 \\ 0 & 0 \end{pmatrix}\right\}$$

is a fundamental matrix solution of the differential equation. Therefore, if there exist differential equations of the type considered here with a *holomorphic* coefficient matrix at $x = 0$ and with a termwise holomorphic formal solution there, which, nevertheless, does not asymptotically represent a true solution at $x = 0$, equation (3.1-4) is not one of them.

CHAPTER IV

Asymptotic Transformations of Differential Equations

4.1. Asymptotic Equivalence

A decisive part of the method of the first chapter was the transformation of a formal differential equation

$$\epsilon^h y' = \left(\sum_{r=0}^{\infty} A_r(x)\epsilon^r \right) y \tag{4.1-1}$$

into another one of the same type,

$$\epsilon^h z' = \left(\sum_{r=0}^{\infty} B_r(x)\epsilon^r \right) z, \tag{4.1-2}$$

by a formal transformation

$$y = \left(\sum_{r=0}^{\infty} P_r(x)\epsilon^r \right) z, \qquad P_0(x) \text{ invertible.} \tag{4.1-3}$$

The A_r, B_r, P_r were holomorphic in some region D of $\mathbb{C}$ or of a certain Riemann surface. It was then shown that, if

$$A(x,\epsilon) \sim \sum_{r=0}^{\infty} A_r(x)\epsilon^r \qquad \text{as} \quad \epsilon \to 0, \quad \epsilon \in E, \tag{4.1-4}$$

and if $B(x,\epsilon)$ is a matrix for which

$$B(x,\epsilon) \sim \sum_{r=0}^{\infty} B_r(x)\epsilon^r, \tag{4.1-5}$$

then there often exists a matrix $P(x,\epsilon)$ with

$$P(x,\epsilon) \sim \sum_{r=0}^{\infty} P_r(x)\epsilon^r, \tag{4.1-6}$$

such that the analytic differential equation

$$\epsilon^h y' = A(x, \epsilon) y \tag{4.1-7}$$

is transformed into

$$\epsilon^h z' = B(x, \epsilon) z \tag{4.1-8}$$

by the change of variables

$$y = P(x, \epsilon) z. \tag{4.1-9}$$

The conditions under which the foregoing was proved need not be repeated here in detail, except for the essential fact that the regions in question were not allowed to contain turning points. Now, it is the central part of the so-called *uniform* theory of turning points that such transformations can sometimes also be constructed at turning points. In this chapter this subject will be treated from a fairly general point of view.

Formal differential equations of the form (4.1-1) and (4.1-2) that are related by a transformation such as (4.1-3) will be called *formally equivalent* in the region D. The differential equations (4.1-7) and (4.1-8) are then appropriately called *asymptotically equivalent* for $x \in D$, as $\epsilon \to 0$ in a sector S of the ϵ-plane. Finally, if the series (4.1-4) is a *convergent* representation of $A(x, \epsilon)$ it may happen—though rarely—that B and P are also holomorphic at $\epsilon = 0$. Then, one may speak of *holomorphic equivalence*.

It is easy to show that these concepts of equivalence become trivial when $h = 0$. In fact, the relation of the three matrices involved is, for any h,

$$\epsilon^h P' = AP - PB. \tag{4.1-10}$$

If A and B have convergent series in powers of ϵ, *and* if $h = 0$, this is a system of n^2 scalar differential equations for the n^2 entries of the matrix P with coefficients holomorphic for $x \in D$, $\epsilon = 0$. Let U and V be the fundamental matrix solutions of $U' = AU$, $V' = -VB$, for which at some point $x_1 \in D$ the initial conditions $U(x_1, 0) = I$, $V(x_1, 0) = I$ are satisfied. Then U and V are holomorphic and invertible in x and ϵ for $(x, \epsilon) \in D \times E$. The matrix $P = UV$ satisfies (4.1-10) with $h = 0$ and is invertible in $D \times E$.

The algorithm just mentioned exists whether the series for A and B converge or not. Hence, *any* two formal differential equations, (4.1-1) and (4.2-2), are formally equivalent *when* $h = 0$. If A and B have the asymptotic series expansion (4.1-4) and (4.1-5) and if (4.1-3) is a *formal* transformation linking the two differential equations, then if $h = 0$, the P_r can be calculated successively. By the Borel–Ritt theorem there then exists a matrix $\hat{P}$ with the asymptotic expansion

$$\hat{P}(x, \epsilon) \sim \sum_{r=0}^{\infty} P_r(x) \epsilon^r,$$

and one has

$$\hat{P}' = A\hat{P} - \hat{P}B + \Omega \tag{4.1-11}$$

with

$$\Omega \sim 0. \qquad (4.1\text{-}12)$$

Because of the continuous dependence of the solutions of holomorphic differential equations on the data, the equation

$$Z' = AZ - ZB + \Omega$$

also possesses solutions that are asymptotic to zero. If $Z \sim 0$ is such a solution, then $P = \hat{P} - Z$ is a solution of (4.1-10) for $h = 0$. Moreover, $P \sim \sum_{r=0}^{\infty} P_r \epsilon^r$, so that the conclusion is, again, that *any two* differential equations whose coefficients have asymptotic expansions are *asymptotically equivalent*, provided $h = 0$.

When $h > 0$, the foregoing arguments break down, of course, and the equivalence relations actually divide each set of differential equations into more than one equivalence class. In the neighborhood of points that are not turning points, a few relatively simple observations can be made. Here is an example.

Theorem 4.1-1. *Assume the A and B in (4.1-7) and (4.1-8) possess for $x \in D$ the asymptotic series (4.1-4) and (4.1-5) and that the corresponding formal differential equations are formally equivalent in D. If D contains no turning points, the differential equations are asymptotically equivalent in every sufficiently small region $D^* \subset D$.*

PROOF. Let (4.1-3) be the formal transformation of (4.1-1) into (4.1-2). By the Borel–Ritt theorem there exists a matrix $\tilde{P}$ with the expansion

$$\tilde{P}(x,\epsilon) \sim \sum_{r=0}^{\infty} P_r(x)\epsilon^r \qquad \text{as} \quad \epsilon \to 0, \quad \text{for} \quad |x| \leqslant x_0, \qquad (4.1\text{-}13)$$

if $x_0 > 0$ is not too large. The transformation

$$Y = \tilde{P}\tilde{z} \qquad (4.1\text{-}14)$$

takes (4.1-7) into a differential equation

$$\epsilon^h \tilde{z}' = \tilde{B}(x,\epsilon)\tilde{z} \qquad (4.1\text{-}15)$$

with

$$\tilde{B} \sim B. \qquad (4.1\text{-}16)$$

Thus, (4.1-15) is asymptotically equivalent to (4.1-7) and it only remains to be proved that (4.1-8) and (4.1-15) are equivalent. In the original notation this amounts to showing that the theorem is true when $A \sim B$. In that case the *formal* equivalence is established by (4.1-3) with $P_0 \equiv I$, $P_r \equiv 0$, $r > 0$. Now, as $x = 0$ is not a turning point, the formal solutions of Theorem 2.8-1 which are the same for the two differential equations, correspond to fundamental matrix solutions

$$Y = \hat{Y}e^Q, \qquad Z = \hat{Z}e^Q$$

for the differential equations (4.1-7) and (4.1-8), respectively, where $\hat{Y}$ and $\hat{Z}$ are functions with the same asymptotic expansion in powers of $\epsilon^{1/m}$, provided D^* is sufficiently small. Define P by

$$P(x,\epsilon) := YZ^{-1} = \hat{Y}\hat{Z}^{-1}.$$

Then

$$P(x,\epsilon) \sim I,$$

and the transformation $y = Pz$ takes Z into Y. Therefore, it takes (4.1-7) into (4.1-8), because a differential equation is uniquely determined by one of its fundamental matrix solutions. This proves the theorem.

Roughly speaking it says that, locally, formal equivalence implies asymptotic equivalence, except possibly at turning points.

4.2. Formal Invariants

Every equivalence relation in a set leads to the search for invariants, i.e., for quantities associated with the elements of the set that have the same value for all elements in the same equivalence class. Accordingly, one may ask for the formal, the asymptotic, and the holomorphic invariants of the differential equations of the form (4.1-1) or (4.1-7). In this section some results on *formal* invariants and related formal properties are collected.

Consider again the formal differential equation (4.1-1) and its formal solution

$$\left(\sum_{r=0}^{\infty} Y_r(x)\epsilon^{r/m} \right) e^{Q(x,\epsilon)} \tag{4.2-1}$$

as established in Theorem 2.8-1. *From now on the notation will be shortened by using letters for formal series and by indicating that meaning by employing bold characters.* For instance,

$$\hat{\mathbf{Y}}(x,\epsilon) := \sum_{r=0}^{\infty} Y_r(x)\epsilon^{r/m}, \qquad \det \hat{\mathbf{Y}}(x,\epsilon) \neq \mathbf{0}, \tag{4.2-2}$$

$$\mathbf{Y}(x,\epsilon) := \hat{\mathbf{Y}}(x,\epsilon)e^{Q(x,\epsilon)}. \tag{4.2-3}$$

Remember that Q was a diagonal matrix whose diagonal entries were polynomials in $\epsilon^{-1/m}$ without constant term and with analytic coefficients. The formal matrix solution (4.2-3) will be called a *basic formal matrix solution* ("BFM-solution").

For what follows it is important to define the expression "basic formal matrix solution" with more precision.

Definition 4.2-1. Let the coefficients A_r of the differential equation (4.1-1) be holomorphic in a region $D \subset \mathbb{C}$. Let $Y_r, r = 0, 1, \ldots, Y_0 \not\equiv 0$ be analytic functions of x that are single-valued on a Riemann domain $\mathcal{M}$ over D and have at most a discrete set of singularities there. Let m be a positive integer. Denote by Q a diagonal $n \times n$ matrix-valued polynomial of $\epsilon^{-1/m}$ without constant terms whose coefficients satisfy the same conditions as the Y_r. Then, an expression $Y(x, \epsilon)$ of the form (4.2-2), (4.2-3) is called a basic formal matrix solution of the formal differential equation (4.1-1) if the formal power series $\hat{Y}$ satisfies the formal differential equation

$$\epsilon^h \hat{Y}' = A\hat{Y} - \epsilon^h \hat{Y} Q'.$$

The differential equation for $\hat{Y}$ above is, of course, equivalent to the differential equation $\epsilon^h Y' = AY$. It is preferable for a definition because it deals only with power series in ϵ.

There exist many BFM-solutions for the same equation (4.2-1). For instance, the order of the diagonal entries of Q is arbitrary. There is, however, the theorem below.

Theorem 4.2-1. *The diagonal entries of the matrix dQ/dx, where Q is the matrix in formula (4.2-3) for a basic formal matrix solution of equation (4.2-1), are formal invariants of the differential equation.*

PROOF. Let $\hat{Y}e^Q, \hat{Y}^* e^{Q^*}$ be two BFM-solutions of (4.1-1). Let $A(x, \epsilon)$ be a matrix, holomorphic for $x \in D$, $\epsilon \in S$, having (4.1-4) as its asymptotic expansion. Such a matrix always exists by the Borel–Ritt theorem. In the neighborhood of a point that is not a turning point there exist then, by Theorem 2.8-1, matrices $\hat{Y}(x, \epsilon), \hat{Y}^*(x, \epsilon)$ with the asymptotic series expansions $\hat{Y}, \hat{Y}^*$, there, such that

$$Y := \hat{Y}e^Q, \qquad Y^* := \hat{Y}^* e^{Q^*} \tag{4.2-4}$$

are fundamental matrix solutions of (4.1-7). Hence, the matrix

$$C := e^{-Q} \hat{Y}^{-1} \hat{Y}^* e^{Q^*} \tag{4.2-5}$$

depends on ϵ only, not on x. The matrix $\hat{Y}^{-1} \hat{Y}^*$ has an asymptotic series in powers of $\epsilon^{1/p}$, where p is some positive integer, and $\det(\hat{Y}^{-1} \hat{Y}^*)$ is not identically asymptotic to zero and nowhere equal to zero.

Each entry of (4.2-5) has the form

$$\psi(x, \epsilon) e^{\phi(x, \epsilon)}, \tag{4.2-6}$$

with ψ and ϕ depending on which entry is taken. The function ψ has an asymptotic series in ascending powers of $\epsilon^{1/p}$, and ϕ is a polynomial in $\epsilon^{-1/p}$ without constant term. Since C does not depend on x, the derivative of (4.2-6) vanishes identically, i.e.,

$$\psi' + \psi\phi' \equiv 0. \tag{4.2-7}$$

Hence, either

$$\psi \sim 0 \tag{4.2-8}$$

or ψ'/ψ has an asymptotic series in non-negative powers of $\epsilon^{1/p}$. In the latter case the identity

$$\phi' = -\psi'/\psi,$$

which follows from (4.2-7), is possible only if both members vanish identically, since a polynomial in $\epsilon^{-1/p}$ without constant term cannot, otherwise, have an asymptotic expansion in ascending non-negative powers of $\epsilon^{1/p}$.

Now, consider the entries in one column of (4.2-5). They are of the form (4.2-6), and each ϕ has the form $q_j^* - q_k$ with the same k, but with j assuming all values $j = 1, 2, \ldots, n$. Here q_j^*, q_k denote the diagonal entries of Q^* and Q, respectively. At least one of these differences must have a zero derivative, since, otherwise, all corresponding functions ψ in (4.2-6) would be asymptotic to zero, which is impossible because $\det(\hat{Y}^{-1}\hat{Y}^*)$ is not asymptotic to zero. This proves that every $q_j^{*\prime}$ is equal to some q_k'. By the symmetry of the argument, each q_j' is equal to some $q_k^{*\prime}$. Thus, the diagonal entries of Q are uniquely determined by the differential equation. If the coefficient matrix in (4.1-3) is denoted by P, the expression $P\hat{Y}e^Q$ is a formal BFM-solution of (4.1-2) after the product $P\hat{Y}$ has been expanded. This completes the proof of Theorem 4.2-1.

The matrix Q' determines Q only to within the addition of an arbitrary diagonal matrix that is independent of x and is a polynomial in $\epsilon^{-1/m}$. All matrices Q obtained from one of them by the addition of such a diagonal matrix occur among the BFM-solutions of a given differential equation. One can normalize Q by requiring that at some fixed point $x_1 \in D$ one have $Q(x_1, \epsilon) = 0$ for all ϵ. This will be assumed to be done for the rest of this section.

To make matters more precise, the notation for the diagonal entries of Q will be changed by using the same subscript for all q_j whose derivatives are identically equal. Since Q has been normalized as explained above, these q_j are themselves equal. Then, for the appropriate ordering of the columns of Y,

$$Q(x, \epsilon) = \text{diag}(q_1(x, \epsilon)I_1, q_2(x, \epsilon)I_2, \ldots, q_p(x, \epsilon)I_p). \qquad (4.2\text{-}9)$$

The order of each identity matrix I_j is the multiplicity of q_j.

Let (4.2-1) be a BFM-solution of (4.1-2) with Q normalized in (4.2-9). The expression

$$\hat{Y}e^Q C \qquad (4.2\text{-}10)$$

with an arbitrary formal matrix series

$$C(\epsilon) := \sum_{r=0}^{\infty} C_r \epsilon^{r/m} \qquad (4.2\text{-}11)$$

also satisfies the differential equation (4.1-1) formally. However, (4.2-10) is not, in general, a BFM-solution. If C commutes with Q this is, of course, true, but it is not immediately clear that there are no additional BFM-solutions. That this *is* the case is the statement of the next theorem.

Theorem 4.2-2. *If*

$$\mathbf{Y} = \hat{\mathbf{Y}} e^{Q} \qquad (4.2\text{-}12)$$

is a BFM-solution of the differential equation (4.1-1) with Q as in (4.2-9), then

$$\mathbf{Z} = \hat{\mathbf{Z}} e^{Q} \qquad (4.2\text{-}13)$$

is another such BFM-solution of (4.1-1), if and only if

$$\hat{\mathbf{Z}}(x, \epsilon) = \hat{\mathbf{Y}}(x, \epsilon)\mathbf{C}(\epsilon), \qquad (4.2\text{-}14)$$

where the formal series $\mathbf{C}(\epsilon)$ is independent of x and has the following properties.

(i) $\mathbf{C}(\epsilon) = \sum_{j=1}^{P \oplus} \mathbf{C}_j(\epsilon)$.
(ii) $\mathbf{C}_j(\epsilon) = \sum_{r=\kappa}^{\infty} C_{jr}\epsilon^{r/m}$, *(not all $C_{j\kappa}$ are zero) with constant square matrices C_{jr} of the dimension of I_j. The integer κ may be positive, zero, or negative.*
(iii) *The series $\hat{\mathbf{Y}}\mathbf{C}$ begins with a not identically zero term independent of ϵ.*
(iv) $\det \mathbf{C}(\epsilon) \not\equiv 0$.

PROOF. That $\mathbf{Z}$ in (4.2-13) is a BFM-solution, if conditions (i) to (iv) are fulfilled, is almost obvious: $\mathbf{YC}$ is a solution, $\mathbf{C}$ commutes with e^{Q}, and $\hat{\mathbf{Z}} = \hat{\mathbf{Y}}\mathbf{C}$ has the form required for $\hat{\mathbf{Z}}e^{Q}$ to be a BFM-solution.

Conversely, if (4.2-12) as well as (4.2-13) are BFM-solutions and $\mathbf{C}$ is defined by

$$\mathbf{C}(x, \epsilon) := \hat{\mathbf{Y}}^{-1}(x, \epsilon)\hat{\mathbf{Z}}(x, \epsilon), \qquad (4.2\text{-}15)$$

then $\mathbf{C}$ has the form

$$\mathbf{C}(x, \epsilon) = \epsilon^{r_0/m} \sum_{r=0}^{\infty} C^r(x)\epsilon^{r/m}, \qquad C^0(x) \not\equiv 0. \qquad (4.2\text{-}16)$$

(Here, the superscript r is an index, not an exponent.) The integer r_0 is not necessarily positive, but it is finite because $\det \hat{\mathbf{Y}}(x, \epsilon) \not\equiv 0$ by Theorem 2.8-1. From the fact that $\mathbf{Y}$ and $\mathbf{Z}$ satisfy (4.1-1) one deduces, by verification, that $\mathbf{C}$, as defined in (4.2-15), satisfies the formal differential equation

$$\mathbf{C}' = Q'\mathbf{C} - \mathbf{C}Q'. \qquad (4.2\text{-}17)$$

Now, recall that Q is a polynomial in $\epsilon^{-1/m}$ without constant term. Leaving aside the vacuous case $Q \equiv 0$, one can write

$$Q(x, \epsilon) = \sum_{j=0}^{H-1} Q_j(x)\epsilon^{(j-H)/m}, \qquad 1 \leq H \leq mh, \qquad (4.2\text{-}18)$$

with $Q_0(x) \not\equiv 0$.

Insertion of (4.2-16) and (4.2-18) into (4.2-17) produces the recursive sequence of equations

$$Q_0'C^0 - C^0Q_0' = 0,$$

(4.2-19)

$$Q_0'C^r - C^rQ_0' = \sum_{\substack{\alpha+\beta=r \\ \alpha<r}} (C^\alpha Q_\beta' - Q_\beta'C^\alpha) + \frac{dC^{r-Hm}}{dx}, \qquad r > 0,$$

with $C^k :\equiv 0$ for $k < 0$. The first of these formulas implies that C^0 is block diagonal with the size of the blocks as in (4.2-9). The summation in the formula for C^1 reduces to one term, which is zero, because C^0 is block-diagonal and therefore commutes with Q_1'. If $Hm = 1$, then $dC^{1-Hm}/dx = dC^0/dx$, which was already proved to be block diagonal. Hence, the left member of (4.2-19) with $r = 1$ must also be block diagonal if $mH = 1$. However, the diagonal blocks are also zero, because of (4.2-9). Thus, the left member of (4.2-19) is zero in this case, and it follows that C^0 is independent of x. The equation for C^1 has now been reduced to

$$Q_0'C^1 - C^1Q_0' = 0$$

so that C^1 must be block diagonal. By induction one proves that all C^r are block-diagonal and independent of x, at least when $Hm = 1$.

If $Hm > 1$, the matrices dC^{r-mH}/dx are zero, by definition, for $0 \leqslant r < mH$, and one concludes from (4.2-19) that $C^0, C^1, \ldots, C^{Hm-1}$ are block diagonal. Equation (4.2-19) for $r = Hm$ becomes

$$Q_0'C^{Hm} - C^{Hm}Q_0' = \frac{dC^0}{dx}.$$

It follows that the left member is block diagonal and, hence, even zero. Thus, C^0 is independent of x and C^{Hm} is block diagonal. The induction now proceeds as in the case $Hm = 1$.

Parts (i) and (ii) of the theorem have now been proved. Parts (iii) and (iv) are immediate consequences of the definition of $\mathbf{C}$ in (4.2-15).

4.3. Formal Circuit Relations with Respect to the Parameter

When fractional powers of ϵ occur in the formal solutions, the fact that the data involve only integral powers of ϵ implies certain useful symmetry relations for the formal solutions, which will be studied in this section. The results of this section are trivial or vacuous when m in (4.2-1), etc., is equal to one.

The first observation is that as the determination of the function $\epsilon^{1/m}$, which is multivalued if $m > 1$, was left arbitrary, a BFM-solution is changed into another one if $\epsilon^{1/m}$ is replaced by $\epsilon^{1/m}\epsilon^{2\pi i/m}$. Here it will be assumed that m is the smallest integer in terms of which the BFM-solution can be written. Thus, if $\mathbf{Y}$ is one given such solution, replacing $\epsilon^{1/m}$ by $\epsilon^{1/m} e^{2\pi i \mu / m}$, $\mu = 0, 1, 2, \ldots, m-1$ in that series, generates m, generally distinct, BFM-solutions, which may be called $\mathbf{Y}^{(\mu)}$, $\mu = 0, 1, \ldots, m-1$, with $\mathbf{Y}^0 := \mathbf{Y}$.

Now consider the blocks $q_j(x, \epsilon) I_j$, $j = 1, 2, \ldots, p$, of $Q(x, \epsilon)$, as in (4.2-9). Since the diagonal entries of Q' are formal invariants, $Q(x, \epsilon)$ must also contain all those blocks $q_j(x, \epsilon e^{2\pi i \mu / m}) I_j$ that are distinct from $q_j(x, \epsilon) I_j$. From now on it will be assumed that all the distinct blocks so generated from one of them are placed consecutively in Q in such a way that moving ϵ once around $\epsilon = 0$ in positive direction changes each block cyclically into the next, except for the last block, which becomes the first. The direct sum of these blocks will be called a "superblock," and a BFM-solution with such an ordering of the diagonal entries in Q will be called a "standard formal matrix solution" (SFM-solution).

Let Π be the $n \times n$ permutation matrix describing that cyclic permutation within each superblock. Denote by $Q^{(0)}(:= Q), Q^{(1)}, \ldots, Q^{(m-1)}$ the m diagonal matrices obtained from Q by $0, 1, \ldots, m-1$ positive rotations around the origin of the ϵ-plane. Then

$$Q^{(\mu+1)} = \Pi^{-1} Q^{(\mu)} \Pi, \qquad \mu = 0, 1, \ldots, m-1. \tag{4.3-1}$$

Let $\hat{\mathbf{Y}}^{(\mu)}$ (with $\hat{\mathbf{Y}}^{(0)} := \hat{\mathbf{Y}}$), $\mu = 0, 1, \ldots, m-1$, be the series obtained from $\hat{\mathbf{Y}}$ by μ circuits around $\epsilon = 0$, then

$$\mathbf{Y}^{(\mu)} = \hat{\mathbf{Y}}^{(\mu)} e^{Q^{(\mu)}} = \hat{\mathbf{Y}}^{(\mu)} \Pi^{-\mu} e^{Q} \Pi^{\mu}.$$

Therefore,

$$\mathbf{Y}^{(\mu)} \Pi^{-\mu} = \hat{\mathbf{Y}}^{(\mu)} \Pi^{-\mu} e^{Q} \tag{4.3-2}$$

is an SFM-solution with the same Q-matrix as $\mathbf{Y}$. By applying Theorem 4.2-2 to the two solutions $\mathbf{Y}$ and the one in (4.3-2), one concludes that

$$\hat{\mathbf{Y}}^{(\mu)} \Pi^{-\mu} = \hat{\mathbf{Y}} \Gamma^{(\mu)}, \tag{4.3-3}$$

where $\Gamma^{(\mu)}$ is a certain block diagonal matrix-valued function of ϵ alone with the properties of $\mathbf{C}$ in Theorem 4.2-2. Also, $\Gamma^{(0)} = I$.

The simple observations that led to formula (4.3-3) are sufficient to prove a theorem of some interest.

Theorem 4.3-1. *Every formal equivalence class contains differential equations for whose BFM-solutions, $\hat{\mathbf{Y}} e^{Q}$, the series $\hat{\mathbf{Y}}$ is independent of x.*

PROOF. Let $\epsilon^h \mathbf{Z}' = \mathbf{A} \mathbf{Z}$ be a given differential equation of the type considered and let $\mathbf{Z} = \hat{\mathbf{Z}} e^{Q}$ be one of its formal SFM-solutions. Let $x = x_1$ be a point at which the series $\hat{\mathbf{Z}}(x, \epsilon)$ has holomorphic coefficients. Then

$$\mathbf{P}(x, \epsilon) := \hat{\mathbf{Z}}(x_1, \epsilon)(\hat{\mathbf{Z}}(x, \epsilon))^{-1}$$

has holomorphic coefficients near $x = x_1$. The matrix $\mathbf{P}(x, \epsilon)$ is a power series in $\epsilon^{1/m}$. Since $\mathbf{P}(x_1, \epsilon) \equiv I$, this series $\mathbf{P}(x, \epsilon)$ contains no negative powers of $\epsilon^{1/m}$. If $\epsilon^{1/m}$ is replaced by $\epsilon^{1/m}e^{2\pi i/m}$, $\hat{\mathbf{Z}}(x,\epsilon)$ becomes $\hat{\mathbf{Z}}^{(1)}(x,\epsilon)$ $= \hat{\mathbf{Z}}(x,\epsilon)\mathbf{\Gamma}^{(1)}(\epsilon)\Pi$ by (4.3-3), therefore $\mathbf{P}(x,\epsilon)$ remains unchanged. This means that $\mathbf{P}(x,\epsilon)$ contains no fractional powers of ϵ. Therefore, the formal transformation $y = \mathbf{P}z$ takes the given differential equation into one in the same formal equivalence class. One of the fundamental matrices of the new differential equation is

$$\mathbf{Y}(x,\epsilon) = \mathbf{P}(x,\epsilon)\mathbf{Z}(x,\epsilon) = \hat{\mathbf{Z}}(x_1,\epsilon)e^{Q(x,\epsilon)}.$$

By (4.2-14) all other BFM-solutions of the same differential equation share with $\mathbf{Y}(x,\epsilon)$ the property that $\hat{\mathbf{Z}}$ is independent of x. This proves Theorem 4.3-1.

The structure of the matrices $\mathbf{\Gamma}^{(\mu)}$ in (4.3-3) may be quite complicated. Observe, in particular, that $\mathbf{\Gamma}^{(\mu+1)}$ is, in general, different from the series obtained from $\mathbf{\Gamma}^{(\mu)}$ by circling the origin once. Nor is it true that $\mathbf{\Gamma}^{(2)}$ $= [\mathbf{\Gamma}^{(1)}]^2$, etc.

To construct SFM-solutions with relatively simple matrices $\mathbf{\Gamma}^{(\mu)}$, let $\mathbf{Y} = \hat{\mathbf{Y}}e^{Q}$ be one SFM-solution. Then all other SFM-solutions with the same Q have the form

$$\mathbf{Z} = \hat{\mathbf{Z}}e^{Q}, \tag{4.3-4}$$

with

$$\hat{\mathbf{Z}} = \hat{\mathbf{Y}}\mathbf{C}, \tag{4.3-5}$$

$\mathbf{C}$ being as in Theorem 4.2-2. More generally,

$$\hat{\mathbf{Z}}^{(\mu)} = \hat{\mathbf{Y}}^{(\mu)}\mathbf{C}^{(\mu)}, \qquad \mu = 0, 1, \dots, m - 1, \tag{4.3-6}$$

where $\mathbf{C}^{(\mu)}(\epsilon)$ is obtained from $\mathbf{C}(\epsilon)$ by μ circuits about $\epsilon = 0$. Since the matrix $\mathbf{\Gamma}^{(\mu)}$ in (4.3-3) depends on the choice of the solution $\mathbf{Y}$, formula (4.3-3) and the corresponding formula for $\mathbf{Z}$ will now be written

$$\hat{\mathbf{Y}}^{(\mu)} = \hat{\mathbf{Y}}\mathbf{\Gamma}_y^{(\mu)}\Pi^{\mu}, \qquad \hat{\mathbf{Z}}^{(\mu)} = \hat{\mathbf{Z}}\mathbf{\Gamma}_z^{(\mu)}\Pi^{\mu}. \tag{4.3-7}$$

Insertion of (4.3-7) into (4.3-6) and use of (4.3-5) leads to the relation

$$\mathbf{\Gamma}_z^{(\mu)} = \mathbf{C}^{-1}\mathbf{\Gamma}_y^{(\mu)}(\Pi^{\mu}\mathbf{C}^{(\mu)}\Pi^{-\mu}), \qquad \mu = 0, 1, \dots, m - 1, \tag{4.3-8}$$

between $\mathbf{\Gamma}_z^{(\mu)}$ and $\mathbf{\Gamma}_y^{(\mu)}$.

If $\mathbf{Y}$, and therefore $\mathbf{\Gamma}_y^{(\mu)}$, are given, the matrices $\mathbf{\Gamma}_z^{(\mu)}$ depend on the choice of $\mathbf{C}$, which, so far, may be any matrix of the form described in Theorem 4.2-2.

The aim of the calculations that follow is to determine $\mathbf{C}$ in such a way that the $\mathbf{\Gamma}_z^{(\mu)}$ become as simple as possible, or, in other words, that the circuit relations for the $\hat{\mathbf{Z}}^{(\mu)}$ are as simple as possible.

To avoid unnecessarily involved formulas the arguments will first be carried out for $\mu = 1$ only. In that case, (4.3-8) can also be written

$$\Gamma_z = C^{-1}\Gamma_y(\Pi C \Pi^{-1}) \tag{4.3-9}$$

with $\Gamma_y := \Gamma_y^{(1)}$, $\Gamma_z := \Gamma_z^{(1)}$. All matrices in (4.3-9) are block diagonal in the partition corresponding to the superblocks defined above. If there are g such superblocks they can be identified by the subscript ν, $\nu = 1, 2, \ldots, g$, and one can write

$$Q = \sum_{\nu=1}^{g} {}^{\oplus} Q_\nu$$

$$\Gamma_y = \sum_{\nu=1}^{g} {}^{\oplus} \Gamma_{y\nu}, \qquad \Gamma_z = \sum_{\nu=1}^{g} {}^{\oplus} \Gamma_{z\nu},$$

$$C = \sum_{\nu=1}^{g} {}^{\oplus} C_\nu \qquad \Pi = \sum_{\nu=1}^{g} {}^{\oplus} \Pi_\nu$$

Each superblock Q_ν, $\Gamma_{y\nu}$, $\Gamma_{z\nu}$, C_ν is itself block diagonal in such a way that the νth of the g superblocks consists of m_ν smaller diagonal blocks of the same size $d_\nu \times d_\nu$. In more precise notation:

$$Q_\nu(x, \epsilon) = \sum_{k=0}^{m_\nu - 1} {}^{\oplus} q_\nu(x, \epsilon e^{2k\pi i}) I_\nu, \qquad \nu = 1, 2, \ldots, g$$

$$\Gamma_{y\nu} = \sum_{k=0}^{m_\nu - 1} {}^{\oplus} \Gamma_{y\nu k}, \qquad \Gamma_{z\nu} = \sum_{k=0}^{m_\nu - 1} {}^{\oplus} \Gamma_{z\nu k} \tag{4.3-10}$$

$$C_\nu = \sum_{k=0}^{m_\nu - 1} {}^{\oplus} C_{\nu k}.$$

The integer m_ν is a factor of m and the νth superblock has the dimension $d_\nu m_\nu$. Also, $\sum_{j=1}^{g} m_\nu d_\nu = n$. By writing $\epsilon e^{2k\pi i}$ in the first of these formulas it is meant that in the polynomial in $\epsilon^{1/m}$ which is q_ν, one has to replace $\epsilon^{1/m}$ by $\epsilon^{1/m} e^{2\pi i k/m}$. The identity matrix I_ν has dimension $d_\nu \times d_\nu$.

The permutation matrix Π is block diagonal in the coarse partition into superblocks, but these superblocks Π_ν are not block diagonal. However, each matrix $\Pi_\nu C_\nu \Pi_\nu^{-1}$ *is* block diagonal. It is obtained from C_ν by moving the first of the m_ν diagonal blocks into the last position and moving the other $m_\nu - 1$ such blocks one block up. Both members of formula (4.3-8) are block diagonal in the finer partition. That formula can, therefore, be written in the more explicit form

$$\Gamma_{z\nu}^{(\mu)} = C_\nu^{-1}\Gamma_{y\nu}^{(\mu)}(\Pi_\nu^\mu C_\nu^{(\mu)}\Pi_\nu^{-\mu}), \qquad \nu = 1, 2, \ldots, g, \quad \mu = 1, 2, \ldots, m_\nu - 1,$$
$$\tag{4.3-11}$$

where $\Gamma_{z\nu}^{(\mu)}$ and $\Gamma_{y\nu}^{(\mu)}$ are the νth superblock of $\Gamma_z^{(\mu)}$ and $\Gamma_y^{(\mu)}$, respectively.

Now assume that the solution Y, and therefore the matrices $\Gamma_{y\nu}^{(\mu)}$ are given. Then (4.3-11) represents $\Gamma_{z\nu}^{(\mu)}$ in terms of the matrices C_ν and $C_\nu^{(\mu)}$.

The aim is to choose C_ν so simple that the $\Gamma_{z\nu}^{(\mu)}$ become as simple as possible. Consider first the case $\mu = 1$. I claim that for a suitable choice of C_ν the matrices $\Gamma_{z\nu k}^{(1)}$ for fixed ν will all be equal [see formula (4.3-10)]. For $\mu = 1$ the m_ν equations (4.3-11) are

$$\Gamma_{z\nu k}^{(1)} = C_{\nu k}^{-1}\Gamma_{y\nu k}^{(1)}C_{\nu,k-1} \quad \text{for} \quad k = 1, 2, \ldots, m_\nu - 1;$$

$$\Gamma_{z\nu 0}^{(1)} = C_{\nu 0}^{-1}\Gamma_{y\nu 0}^{(1)}C_{\nu,m_\nu - 1}. \tag{4.3-12}$$

If

$$\Gamma_{z\nu k}^{(1)} = K_\nu^{(1)}, \qquad k = 0, 1, \ldots, m_\nu - 1, \tag{4.3-13}$$

then the $C_{\nu k}$, $k = 1, 2, \ldots, m_\nu - 1$, can be successively calculated in terms of $C_{\nu 0}$ by means of the equations

$$C_{\nu k} = \Gamma_{y\nu k}^{(1)}C_{\nu,k-1}\left(K_\nu^{(1)}\right)^{-1}, \qquad k = 1, 2, \ldots, m_\nu - 1. \tag{4.3-14}$$

Substitution of these equations into the last equation of (4.3-12) produces the condition

$$K_\nu^{(1)} = C_{\nu 0}^{-1}\left(\Gamma_{y\nu 0}^{(1)}\Gamma_{y\nu,m_\nu-1}^{(1)} \cdots \Gamma_{y\nu 1}^{(1)}\right)C_{\nu 0}\left(K_\nu^{(1)}\right)^{-m_\nu+1}$$

that is,

$$\left(K_\nu^{(1)}\right)^{m_\nu} = C_{\nu 0}^{-1}\left(\Gamma_{y\nu 0}^{(1)}\Gamma_{y\nu,m_\nu-1}^{(1)} \cdots \Gamma_{y\nu 1}^{(1)}\right)C_{\nu 0}, \tag{4.3-15}$$

from which $K_\nu^{(1)}$ can be calculated. The $d_\nu \times d_\nu$ matrix $C_{\nu 0}$ may be any formally invertible series in powers of $\epsilon^{1/m}$.

It must still be shown that this $K_\nu^{(1)}$ is then a series in ascending positive powers of $\epsilon^{1/m}$. To this end, return to formula (4.3-3) with $\mu = 1$ and take determinants on both sides. As $\det \hat{Y}$ and $\det \hat{Y}^{(1)}$ are scalar series that begin with nonzero terms of the same non-negative degree in $\epsilon^{1/m}$, the series $\Gamma^{(1)}$ (i.e., $\Gamma_y^{(1)}$) begins with a nonzero constant term. The $\Gamma_{y\nu k}^{(1)}$, $\nu = 1, 2, \ldots, g$; $k = 0, 1, \ldots, m_\nu - 1$ are the diagonal blocks of $\Gamma^{(1)}$ and it follows that $K_\nu^{(1)}$ is, indeed, a series in non-negative powers of $\epsilon^{1/m}$.

The lemma below can now be stated—and it will be proved—for any μ, not only $\mu = 1$.

Lemma 4.3-1. *If*

$$Y(x,\epsilon) = \hat{Y}(x,\epsilon)e^{Q(x,\epsilon)} \tag{4.3-16}$$

is an SFM-solution of (4.1-1) then

$$\hat{Y}(x, \epsilon e^{2\pi i\mu}) = \hat{Y}(x,\epsilon)\Gamma^{(\mu)}(\epsilon)\Pi^\mu \tag{4.3-17}$$

where the matrices $\Gamma^{(\mu)}(\epsilon)$ have the block structure of $C(\epsilon)$ in Theorem 4.2-2 and the permutation matrix Π was defined before (4.3-1). The $\Gamma^{(\mu)}$ depend on the choice of the solution Y in (4.3-16). There exist certain special SFM-solutions for which all m_ν diagonal blocks $\Gamma_{\nu k}^{(\mu)}$, $k = 0, 1, \ldots, m_\nu - 1$ of the νth superblock $\Gamma_\nu^{(\mu)}$ are equal to the same matrix $K_\nu^{(\mu)}$.

PROOF. The only part of the lemma still to be proved is its last sentence for $\mu > 1$. To do this, replace ϵ in the identity (4.3-17) with $\mu = 1$ by $\epsilon e^{2\pi i}$ and find

$$\hat{Y}(x, \epsilon e^{4\pi i}) = \hat{Y}(x, \epsilon e^{2\pi i})\Gamma^{(1)}(\epsilon e^{2\pi i})\Pi$$

or, by the definition of $\hat{Y}^{(\mu)}$ and by formula (4.3-7),

$$\hat{Y}^{(2)}(x, \epsilon) = \hat{Y}^{(1)}(x, \epsilon)\Gamma^{(1)}(\epsilon e^{2\pi i})\Pi.$$

On the other hand, for $\mu = 2$ formula (4.3-17) can also be written

$$\hat{Y}^{(2)}(x, \epsilon) = \hat{Y}\Gamma^{(2)}(\epsilon)\Pi^2,$$

so that

$$\Gamma^{(2)}(\epsilon) = \Gamma^{(1)}(\epsilon)\Pi\Gamma^{(1)}(\epsilon e^{2\pi i})\Pi^{-1}. \qquad (4.3\text{-}18)$$

By what has been proved already, each superblock of $\Gamma^{(1)}$ for the special SFM-solution is the direct sum of *identical* smaller blocks, the ones called $K_\nu^{(1)}$ in (4.3-13). Because of this special form one has $\Pi^{-1}\Gamma^{(1)}\Pi = \Gamma^{(1)}$. Therefore, (4.3-18) reduces to

$$\Gamma^{(2)}(\epsilon) = \Gamma^{(1)}(\epsilon)\Gamma^{(1)}(\epsilon e^{2\pi i}).$$

More generally, one finds

$$\Gamma^{(\mu)}(\epsilon) = \Gamma^{(1)}(\epsilon)\Gamma^{(1)}(\epsilon e^{2\pi i}) \ldots \Gamma^{(1)}(\epsilon e^{2(\mu-1)\pi i}), \qquad (4.3\text{-}19)$$

which completes the proof of Lemma 4.3-1 for any μ.

For $\mu = m$ formula (4.3-19) becomes

$$\sum_{\mu=0}^{m-1} \Gamma^{(1)}(\epsilon e^{2\pi i\mu}) = I$$

Let the special SFM-solutions for which Lemma 4.3-1 is true be called SSFM-solutions, for short. If Y and Z are both SSFM solutions of the same differential equation, (4.3-15) simplifies to

$$K_{z\nu}^{m_\nu} = C_{\nu 0}^{-1}K_{y\nu}^{m_\nu}C_{\nu 0}. \qquad (4.3\text{-}20)$$

Here $K_{z\nu}$ and $K_{y\nu}$ are the matrices called K_ν, above, one for the SSFM-solution Y, the other for the solution Z. This proves that $K_{y\nu}^{m_\nu}$ and $K_{z\nu}^{m_\nu}$ are similar if Y and Z are SSFM-solutions of the same differential equation and the order in which the superblocks are arranged is the same.

Now, if the differential equation for Y is changed into a formally equivalent one by setting

$$y = P(x, \epsilon)v$$

where

$$P(x, \epsilon) = \sum_{r=0}^{\infty} P_r(x)\epsilon^r, \qquad \det P_0(x) \neq 0,$$

and if (4.3-10) is an SSFM-solution, then

$$(\mathbf{P}^{-1}\hat{\mathbf{Y}})^{(\mu)} = \mathbf{P}^{-1}\hat{\mathbf{Y}}^{(\mu)} = \mathbf{P}^{-1}\hat{\mathbf{Y}}\Gamma_y^{(\mu)}\Pi^{\mu},$$

so that $\Gamma_v^{(\mu)} = \Gamma_y^{(\mu)}$, because of (4.3-3).

Combining this last result with (4.3-20) results in the lemma below:

Lemma 4.3-2. *If* $\mathbf{Y}$ *is an SSFM-solution of* (4.1-1) *and the matrices* $\Gamma^{(\mu)}$, $\mu = 0, 1, \ldots, m-1$ *are defined as in* (4.3-3) *then the Jordan form of* $[\Gamma^{(\mu)}]^m$ *is a formal invariant.*

CHAPTER V

Uniform Transformations at Turning Points: Formal Theory

5.1. Preparatory Simplifications

The formal theory of Chapter II essentially dealt with global simplifying transformations of the given differential equation in regions from which all potential turning points were removed. In this chapter domains containing turning points will be considered, and the first question is: How far can the differential equation be simplified by *formal* transformations with well understood asymptotic properties in such regions. The essence of Langer's method belongs here. His reduction of certain second-order equations with turning points led him to differential equations so simple as to be solvable by classical special functions ([36], [31], [38], [39] and other papers).

One important general reduction result valid in regions from which turning points were not excluded was introduced in Chapter II already; namely, Theorem 2.3-1. If this theorem is combined with Theorem 12.3-1 of the Appendix, one has the following result.

Theorem 5.1-1. *Consider the formal differential equation*

$$\epsilon^h y' = \mathbf{A} y \tag{5.1-1}$$

with

$$\mathbf{A}(x, \epsilon) = \sum_{r=0}^{\infty} A_r(x) \epsilon^r \tag{5.1-2}$$

and $A_r \in H_{nn}(D)$. *Let the characteristic polynomial* ϕ *of* A_0 *be factored into* $p \geqslant 1$ *monic polynomials* ϕ_j *of* λ *of positive degrees with coefficients in* $H_{11}(D)$:

$$\phi = \phi_1 \phi_2 \ldots \phi_p, \tag{5.1-3}$$

such that no two of these polynomial factors have a common zero for any $x \in D$, *and that no* ϕ_j, $j = 1, 2, \ldots, p$, *is so factorable. Then the formal differential equation is formally equivalent in D to a differential equation of the same type with the additional property that all terms of the coefficient matrix are the direct sum of p matrices. No two of the p diagonal blocks of the coefficient matrix have for* $\epsilon = 0$ *any eigenvalues in common at any point* $x \in D$.

In Chapter VI this formal theorem will be supplemented by an analytic result on true block diagonalization. At present, the formal investigation will be directed at further simplification of the p formal, formally uncoupled, differential equations obtained by applying Theorem 5.1-1. Therefore, no generality will be lost in the remainder of this section if the assumption below is introduced.

Hypothesis V-1.   *The characteristic polynomial*

$$\phi(\lambda, x) := \det(\lambda I - A_0(x))$$

of A_0 is not holomorphically factorable in D.

Whether or not this hypothesis is satisfied depends, of course, on the region D. The characteristic polynomial $\phi(\lambda, x) = \lambda^2 - x(x - 1)$ of

$$A_0(x) = \begin{pmatrix} 0 & 1 \\ x(x - 1) & 0 \end{pmatrix}$$

has the factorization $\phi(\lambda, x) = (\lambda - x^{1/2}(x - 1)^{1/2}) \cdot (\lambda + x^{1/2}(x - 1)^{1/2})$, which is holomorphic in simply connected regions that do not contain the points $x = 0$, $x = 1$, but not, e.g., in neighborhoods of the origin.

This example also shows that a region satisfying Hypothesis V-1 may contain more than one turning point. This occurrence will not be studied in this section. To eliminate it, the assumption below will be imposed which is a restriction on the size of D.

Hypothesis V-2.   *There is a point $x_0 \in D$ such that in $D - \{x_0\}$ all eigenvalues of $A_0(x)$ have constant multiplicity.*

Without loss of generality it may be assumed that $x_0 = 0$. One consequence of Hypothesis V-1 is that $A_0(0)$ has only one distinct eigenvalue, because the eigenvalues that tend to the same limit at $x = 0$ are the zeros of a polynomial in λ, which is a factor of the characteristic polynomial and has holomorphic coefficients.

The change of variable

$$y = \exp\left\{ \frac{1}{\epsilon^h n} \int_0^x \operatorname{tr} A(t, \epsilon)\, dt \right\} v \tag{5.1-4}$$

then produces a differential equation for v whose coefficient matrix, $A(x, \epsilon) - (1/n)(\operatorname{tr} A(x, \epsilon))I$, has trace zero for all x and is nilpotent at $\epsilon = 0$, $x = 0$. A further change of the dependent variable, this time with a constant matrix, transforms $A_0(0)$ into its Jordan form. Returning to the previous notation it will be assumed from now on that the differential equation has the property below.

Hypothesis V-3 (Nonrestrictive). $\operatorname{tr} A(x, \epsilon) \equiv 0$, and $A_0(0)$ is nilpotent and in Jordan form.

It must not be forgotten that (5.1-4) is not an equivalence transformation in the sense of Section 4.1. The occurrence of a factor that is an exponential function of ϵ in (5.1-4) has changed the asymptotic character of the solution decisively.

As in Chapter II, there is a formal equivalence transformation based on Arnold's theory which, in general, introduces more zero entries into the coefficient matrix. Arnold's method is based on the Implicit Function Theorem and is, therefore, intrinsically local. The arguments so much resemble those in the proof of Theorem 2.5-1 that the result will be stated but not proved.

Theorem 5.1-2. *If the differential equation* (5.1-1), (5.1-2) *satisfies the Hypotheses V-1 and V-2, then it is, in a sufficiently small neighborhood D_0 of $x = 0$, formally equivalent to*

$$\epsilon^h \mathbf{v}' = \left[A_0(0) + \sum_{j=1}^{d} \left(\sum_{r=1}^{\infty} \rho_{jr} \epsilon^r \right) \Gamma_j \right] \mathbf{v}, \qquad (5.1\text{-}5)$$

with $\rho_{jr} \in H_{11}(D_0)$. The integer d and the matrices Γ_j are defined in Section 12.4.

Even after all these substantial simplifications, there remains a bewildering variety of types of turning point problems, very few of which have been studied. A systematic list of all possible occurrences will not be attempted here. Only for equations of order two do there exist theories of some—but not complete—generality. The enormous difficulties encountered in the analysis of the Orr–Sommerfeld equation mentioned in Section 2.9 and in Chapter I are an indication of how far removed our present knowledge is from a general turning point theory for $n > 2$.

Accordingly, only some parts of the general formal analysis will be presented here for systems of arbitrary order n. Later, the discussion will be limited to the case $n = 2$.

Taking advantage of Theorem 5.1-2 the problem can be taken in the form (5.1-5). The restrictive *assumption that $A_0(0)$ has only one elementary divisor will be added* to Hypotheses V-1 and V-2. The given differential equations (5.1-1) and (5.1-2) then become

$$\epsilon^h v' = \begin{bmatrix} 0 & 1 & \cdots & 0 & 0 \\ 0 & 0 & & 0 & 1 \\ \mathbf{a}_n(x,\epsilon) & \mathbf{a}_{n-1}(x,\epsilon) & \cdots & \mathbf{a}_2(x,\epsilon) & \mathbf{a}_1(x,\epsilon) \end{bmatrix} v \quad (5.1\text{-}6)$$

with

$$\mathbf{a}_j(x,\epsilon) = \sum_{r=0}^{\infty} a_{jr}(x)\epsilon^r, \quad j = 1, 2, \ldots, n,$$

$$a_{j0}(0) = 0, \quad j = 1, 2, \ldots, n. \quad (5.1\text{-}7)$$

Also, because $\operatorname{tr} A(x,\epsilon) \equiv 0$ by Hypothesis V-3,

$$a_1(x,\epsilon) \equiv 0. \quad (5.1\text{-}8)$$

Observe that (5.1-6) is equivalent to the formal nth order scalar differential equation $\epsilon^h u^{(n)} = \mathbf{a}_n(x,\epsilon)u + \cdots + \mathbf{a}_1(x,\epsilon)u^{(n-1)}$.

One last preparatory simplification:

Lemma 5.1-2. *If*

$$a_{n0}(x) = x^\beta a_{n0}^*(x), \quad a_{n0}^*(0) \neq 0,$$

β a positive integer, there exists a change of variables from (x, v) to (ξ, w) independent of ϵ which transforms (5.1-6) into the differential equation

$$\epsilon^h \frac{dw}{d\xi} = \begin{bmatrix} 0 & 1 & \cdot & \cdot & \cdot & 0 & 0 \\ & & & \cdot & \cdot & \cdot & \\ 0 & 0 & \cdot & \cdot & \cdot & 0 & 1 \\ \mathbf{b}_n(\xi,\epsilon) & \mathbf{b}_{n-1}(\xi,\epsilon) & \cdot & \cdot & & & \mathbf{b}_1(\xi,\epsilon) \end{bmatrix} w \quad (5.1\text{-}9)$$

of the same type as (5.1-6) and for which, in addition,

$$b_n(\xi,0) = \xi^\beta. \quad (5.1\text{-}10)$$

However, $\mathbf{b}_1(\xi,\epsilon) \not\equiv 0$. Only $b_1(\xi,0) = 0$ can be claimed.

PROOF. Set

$$\xi = \phi(x), \quad v = \Omega(x)w := \left[\operatorname{diag}\left(1, \omega(x), \ldots, \omega^{n-1}(x)\right) \right]w$$

with analytic functions ϕ, ω as yet undetermined. Then (5.1-6) becomes

$$\epsilon^h \frac{dw}{d\xi} = \mathbf{B}(\xi,\epsilon)w$$

with

$$\mathbf{B}(\xi,\epsilon) = \frac{dx}{d\xi} \begin{bmatrix} 0 & \omega(x) & \cdots & 0 & 0 \\ 0 & & \cdots & \cdots & 0 & \omega(x) \\ & & \cdots & \cdots & & \\ \omega^{1-n}(x)\mathbf{a}_n(x,\epsilon) & \cdots & \cdots & & \omega(x)\mathbf{a}_2(x,\epsilon) & \mathbf{a}_1(x,\epsilon) \end{bmatrix}$$

$$- \epsilon^h \frac{dx}{d\xi} \Omega^{-1}(x)\Omega'(x).$$

To prove the lemma, ω and ϕ must be determined so that

$$\frac{dw}{d\xi}\,\omega(x) = 1 \quad \text{and} \quad \omega^{1-n}(x)a_n(x,0)\frac{dx}{d\xi} = \xi^\beta.$$

Elimination of ω leads to

$$\left(\frac{d\xi}{dx}\right)^n \xi^\beta = x^\beta a_{n0}^*(x),$$

a differential equation for ϕ, which can be solved by elementary means. One finds that

$$\xi = \left[\frac{n+\beta}{n}\int_0^x \left[t^\beta a_n^*(t)\right]^{1/n} dt\right]^{n/(\beta+n)} =: \phi(x)$$

is one solution, and then that

$$\omega(x) = \left[a_n(x,0)\phi^{-\beta}(x)\right]^{1/n}. \tag{5.1-11}$$

To show that ϕ is holomorphic and one-to-one in $|x| \leqslant x_0$, observe that since $a_n^*(0) \neq 0$, one has

$$\int_0^x \left[t^\beta a_n^*(t)\right]^{1/n} dt = a_n^*(0)^{1/n}\frac{\beta+n}{n}x^{(\beta+n)/n}\left[1 + O(x)\right],$$

and therefore

$$\xi = \phi(x) = a_n^*(0)^{1/(\beta+n)}x\left[1 + O(x)\right],$$

with $O(x)$ standing for a function holomorphic at $x = 0$. This function depends on the choice in the determination of the fractional power of $a_n^*(0)$. Thus, ϕ has no branch point at $x = 0$, and $\phi'(0) \neq 0$. It then follows from (5.1-11) that ω is holomorphic at $x = 0$ as well.

5.2. A Method for Formal Simplification in Neighborhoods of a Turning Point

Taking into account the several simplifications performed in the preceding section, the remaining *formal* problem can be formulated, in the original notation, as follows.

Let there be given the formal differential equation

$$\epsilon^h y' = \mathbf{A}(x,\epsilon)y, \tag{5.2-1}$$

with

$$\mathbf{A}(x,\epsilon) = \sum_{r=0}^\infty A_r(x)\epsilon^r \tag{5.2-2}$$

of the special structure

$$A(x,\epsilon) = \begin{bmatrix} 0 & 1 & \cdot & & 0 \\ 0 & 0 & & \cdot & 0 \\ 0 & 0 & \ldots & & \cdot & 1 \\ a_n(x,\epsilon) & a_{n-1}(x,\epsilon) & \ldots & a_2(x,\epsilon) & a_1(x,\epsilon) \end{bmatrix}, \quad (5.2\text{-}3)$$

where

$$a_j(x,\epsilon) = \sum_{r=0}^{\infty} a_{jr}(x)\epsilon^r, \qquad j = 1, 2, \ldots, n, \qquad (5.2\text{-}4)$$

$$a_{n0}(x) = x^{\beta}, \qquad\qquad \beta \text{ a positive integer}, \qquad (5.2\text{-}5)$$

$$a_{j0}(0) = 0, \qquad\qquad j = 1, 2, \ldots, n, \qquad (5.2\text{-}6)$$

$$a_{10}(x) \equiv 0. \qquad\qquad (5.2\text{-}7)$$

The aim is to simplify the differential equation formally as much as possible by means of transformations

$$y = P(x,\epsilon)z \quad \text{with} \quad P(x,\epsilon) = \sum_{r=0}^{\infty} P_r(x)\epsilon^r, \qquad \det P_0(x) \neq 0, \quad (5.2\text{-}8)$$

in which the P_r are to be holomorphic in a fixed neighborhood of $x = 0$. Such transformations were called formal equivalence transformation in Chapter IV.

One result in this direction was proved in Ref. 89. It is stated here in its general form, but it will not be proved here for $n > 2$.

Theorem 5.2-1. *If* $A(x,\epsilon)$ *is as in* (5.2-2)–(5.2-7) *and if* $h = 1$, $\beta = 1$, *the differential equation* (5.2-1) *is formally equivalent to*

$$\epsilon z' = A_0(x)z \qquad (5.2\text{-}9)$$

in a region containing the turning point $x = 0$.

For $n > 2$ the usefulness of that theorem is limited by the sad fact that even the simplified equation (5.2-9) is usually rather intractable. Instead of describing the proof for all n, I shall therefore limit myself to the case $n = 2$. The method of the proof of Theorem 5.2-1 is the same, except that the algebra for general n is more involved (see [89].) On the other hand, the comparative simplicity of the case of order two facilitates generalizations to values of β other than $\beta = 1$.

Insertion of (5.2-8) into (5.2-1) and comparison of coefficients produces a differential equation

$$\epsilon z' = B(x,\epsilon)z \qquad (5.2\text{-}10)$$

with

$$B(x,\epsilon) = \sum_{r=0}^{\infty} B_r(x)\epsilon^r, \qquad (5.2\text{-}11)$$

where A_r, P_r, B_r are related by the formulas

$$A_0 P_0 - P_0 B_0 = 0$$

$$A_0 P_r - P_r B_0 = P'_{r-1} - \sum_{\substack{\mu+\nu=r \\ \mu>0}} (A_\mu P_\nu - P_\nu B_\mu), \qquad r \geqslant 1.$$

(5.2-12)

It is natural to choose

$$B_0(x) = A_0(x).$$

(5.2-13)

Then, when $n = 2$, one has

$$A_0(x) = \begin{pmatrix} 0 & 1 \\ x^\beta & 0 \end{pmatrix},$$

(5.2-14)

by (5.2-3), (5.2-5), and (5.2-7). The right members of (5.2-12) depend on the choice of $P_0, P_1, \ldots, P_{r-1}$ and $B_1, B_2, \ldots, B_r$ so that, as on previous occasions, one hopes to determine the P_r and B_r as holomorphic functions successively for $r = 1, 2, \ldots$, as was done in Section 2.3. Here, as there, the main difficulty lies in the fact that the linear operator L defined by

$$LX := A_0(x)X - XA_0(x)$$

is singular: its nullspace has positive dimension. In Section 2.3 the difficulty was overcome by making use of the fact that $A_0(0)$ had more than one distinct eigenvalue. Here, $A_0(0)$ is nilpotent, and the recursion formula must be tackled more directly.

The dimension d of the nullspace of L is given in Section 12.4. For the matrix (5.2-14) one finds $d = 2$ for all x. This can also be shown directly by verifying that the elements I and $A_0(x)$ of the nullspace are independent for all x and that, as a direct calculation will show, $LX = 0$ implies

$$X(x) = p(x)I + q(x)A_0(x),$$

(5.2-15)

with some scalar factors $p(x), q(x)$.

For $r > 0$, equations (5.2-12) have solutions only if the right members satisfy certain consistency conditions. To formulate those conditions it is convenient to introduce a scalar product in the four-dimensional vector space of all 2×2 matrices by means of the definition

$$(X, Y) := \operatorname{tr}(XY^*),$$

(5.2-16)

where $Y^* := \bar{Y}^T$. One verifies directly that (5.2-16) is, indeed, a scalar product. Moreover, the calculation below proves that the operator L^*, adjoint to L in this inner product space, is given by

$$L^* X = A_0^* X - X A_0^*.$$

In fact,

$$(LX, Y) = \operatorname{tr}((A_0 X - X A_0) Y^*) = \operatorname{tr}(Y^* A_0 X - Y^* X)$$
$$= \operatorname{tr}(X(Y^* A_0 - A_0 Y^*)) = \operatorname{tr}(X(A_0^* Y - Y A_0^*)^*)$$
$$= (X, A_0^* Y - Y A_0^*).$$

The preceding concepts make possible the application of the standard theorem of linear algebra which says that equations (5.2-12) are consistent if and only if the right members are orthogonal to the nullspace of L^*. The consistency conditions are, thus,

$$\text{tr}\left[\left(P'_{r-1} - \sum_{\substack{\mu+\nu=r \\ \mu>0}}(A_\mu P_\nu - P_\nu B_\mu)\right)A_0^k\right] = 0, \qquad k = 0, 1, \quad r \geqslant 1. \quad (5.2\text{-}17)$$

They will be satisfied by successive choices for the P_r and B_r, keeping in mind the aim of making B_r as simple as possible. It will be shown that three of the four entries of B_r can always be taken to be identically zero,

$$B_r(x) = \begin{bmatrix} 0 & 0 \\ b_{2r}(x) & 0 \end{bmatrix}, \qquad (5.2\text{-}18)$$

and that $b_{2r}(x)$ can be chosen as a polynomial of degree $\beta - 2$.

To do this for $r = 1$, note that the compatibility conditions (5.2-17) are then

$$\text{tr}\left[(P'_0 - A_1 P_0 + P_0 B_1)A_0^k\right] = 0, \qquad k = 0, 1. \quad (5.2\text{-}19)$$

From (5.2-15) it is known that P_0 must have the form

$$P_0(x) = C_{10}(x)I + C_{20}(x)A_0(x). \qquad (5.2\text{-}20)$$

Inserting (5.2-18) with $r = 0$ and (5.2-20) into (5.2-19) that condition becomes, after some calculations, the pair of differential equations

$$2c' - a_{11}c_{10} - (a_{21} - b_{21})c_{20} = 0,$$
$$2x^\beta c'_{20} - (a_{21} - b_{21})c_{10} - (x^\beta a_{11} - \beta x^{\beta-1})c_{20} = 0. \qquad (5.2\text{-}21)$$

Since the second of these equations has a singularity at $x = 0$, it has solutions that are only exceptionally holomorphic there. The further analysis depends so much on the value of β that different cases are best studied separately.

First Case: $\beta = 1$. It will be shown that one can choose

$$B_r \equiv 0, \qquad \text{i.e.,} \quad b_{2r} \equiv 0, \qquad \text{for all} \quad r > 0. \qquad (5.2\text{-}22)$$

The singularity of (5.2-21) is a regular one. The indicial exponents are 0, and $-\frac{1}{2}$ and, by the theory of such singularities, there is a fundamental matrix solution C of the system (5.2-21) that has the form

$$C(x) = \hat{C}(x)\begin{pmatrix} 1 & 0 \\ 0 & x^{-1/2} \end{pmatrix}, \qquad \hat{C}(0) = I \qquad (5.2\text{-}23)$$

with a holomorphic $\hat{C}$. This implies that c_{10}, c_{20} can, indeed, be chosen so as to be holomorphic at $x = 0$, namely, as the entries of the first column of $\hat{C}$. Then, in particular,

$$c_{10}(0) = 1, \qquad c_{20}(0) = 0. \qquad (5.2\text{-}24)$$

Second Case: $\beta > 1$. The form of equation (5.2-21) suggests that b_{21} be chosen as the partial sum of degree $\beta - 2$ of the McLaurin series for a_{21}, for then the system (5.2-21) has again a regular singular point at $x = 0$. With the notation

$$a_{21}(x) - b_{21}(x) = k(x)x^{\beta - 1},$$

k holomorphic at $x = 0$, equation (5.2-21) can now be written

$$x\begin{bmatrix} c_{10} \\ c_{20} \end{bmatrix}' = \begin{bmatrix} \frac{1}{2}xa_{11}(x) & \frac{1}{2}x^{\beta}k(x) \\ \frac{1}{2}k(x) & \frac{1}{2}xa_{11}(x) - \frac{1}{2}\beta \end{bmatrix}\begin{bmatrix} c_{10} \\ c_{20} \end{bmatrix}. \tag{5.2-25}$$

The indicial roots are 0 and $-\beta/2$. Such a differential equation has at least one independent solution $(c_{10}, c_{20})^{T}$ that is holomorphic at $x = 0$. One sees, by insertion into (5.2-25) that the value at $x = 0$ for such a vector solution is a constant multiple of

$$c_{10}(0) = \beta, \qquad c_{20}(0) = k(0).$$

Thus, the consistency condition for $r = 1$, i.e., (5.2-19) has been satisfied for all values of β by taking B_1 in the form (5.2-18) and so that b_{21} is zero when $\beta = 1$. When $\beta > 1$, then B_1 is a polynomial of degree $\beta - 2$. Formula (5.2-20) then yields a matrix P_0 satisfying $LP_0 = 0$ and $P(0) = I$.

The arguments now continue by induction with respect to r. Assume, therefore, that the consistency conditions for the calculation of P_1, $P_2 \ldots P_r$ from equation (5.2-17) have already been satisfied and that the matrices $P_1, P_2, \ldots, P_{r-1}$, as well as $B_1, B_2, \ldots, B_r$ have been chosen so that (5.2-18) is true and that the B_j's are polynomials of degree $\beta - 2$ for $\beta \geqslant 2$ and equal to zero for $\beta = 1$.

Let $\tilde{P}_r(x)$ be one of the holomorphic solutions of (5.2-12). Then

$$P_r(x) = c_{1r}(x)I + c_{2r}(x)A_0(x) + \tilde{P}_r(x),$$

with arbitrary scalar coefficients c_{1r}, c_{2r}, is the general solution of (5.2-12). The consistency condition (5.2-17) for the calculations of P_{r+1} from (5.2-12) then has the form

$$2c_{1r}' - a_{11}c_{1r} - (a_{21} - b_{21})c_{2r} = f_{1r},$$
$$2x^{\beta}c_{2r}' - (a_{21} - b_{21})c_{1r} - (x^{\beta}a_{11} - \beta x^{\beta-1})c_{2r} = f_{2r}, \tag{5.2-26}$$

with

$$f_{1r} = g_{1r} - \text{tr}(P_0 B_{r+1}),$$
$$f_{2r} = g_{2r} - \text{tr}(P_0 B_{r+1} A_0). \tag{5.2-27}$$

Here, the g_{jr}, $j = 1, 2$, are holomorphic functions that depend on B_1, $B_2, \ldots, B_r$, but not on B_{r+1}.

The left members of (5.2-26) are the same as for the homogeneous equations (5.2-21). It follows by the method of variation of parameters that

$$\frac{1}{2}C(x)\int^x C^{-1}(t)\begin{pmatrix} f_{1r}(t) \\ t^{-1}f_{2r}(t) \end{pmatrix} dt, \qquad (5.2\text{-}28)$$

in which C denotes a fundamental matrix solution of (5.2-21) or (5.2-25), is a vector solution of (5.2-26). The integral here may be any antiderivative. For $\beta = 1$ the integrand is a vector whose components have the order of magnitude $[O(1), O(t^{-1/2})]^T$ at $t = 0$ because of (5.2-21). There is, therefore, an antiderivative with orders of magnitude $[O(x), O(x^{1/2})]^T$, and the corresponding vector (5.2-28) is then holomorphic. This proves Theorem 5.2-1 for $n = 2$. If $\beta \geqslant 2$, observe that

$$\text{tr}(P_0 B_{r+1} A_0) = \text{tr}\left[(c_{10} A_0 + c_{20} A_0^2) B_{r+1} \right] = c_{10} b_{2,r+1}.$$

Therefore, if $b_{2,r+1}$ in formula (5.2-18) for B_{r+1} is taken as the partial sum of degree $\beta - 2$ of the McLaurin series for g_{2r}/c_{10}, then f_{2r} has a zero of order $x^{\beta-1}$ at $x = 0$. Here the fact that $c_{10}(0) \neq 0$ has been used essentially. With this choice of $b_{2,r+1}$ the argument becomes the same as for $\beta = 1$. The result is the next theorem.

Theorem 5.2-2. *If $n = 2$ and $A(x, \epsilon)$ is as in (5.2-3)–(5.2-7), then for $h = 1$, $\beta \geqslant 2$, the differential equations (5.2-1) and (5.2-3) are formally equivalent to*

$$\epsilon z' = \left[A_0(x) + \begin{bmatrix} 0 & 0 \\ \sum\limits_{r=1}^{\infty} b_{2r}(x)\epsilon^r & 0 \end{bmatrix} \right] z,$$

in a region containing the point $x = 0$. The b_{2r} are polynomials of degree $\beta - 2$ at most.

The most important case of this theorem is the one for $\beta = 2$, for then the b_{2r} are constants and the solution of the simplified differential equation can be expressed in terms of parabolic cylinder functions. For $\beta > 2$, no such explicit solution is available.

5.3. The Case $h > 1$

When $h > 1$, the method of the preceding section can be modified by an artifice, as follows. One rewrites the series $A(x, \epsilon)$ in the form

$$\sum_{r=0}^{\infty} A_r \epsilon^r = \sum_{s=0}^{\infty} \left(\sum_{j=0}^{h-1} A_{s+j} \epsilon^j \right) \epsilon^{hs}$$

and introduces an auxiliary parameter

$$\mu = \epsilon^h.$$

With

$$K_s(x, \epsilon) := \sum_{j=0}^{h-1} A_{s+j}(x)\epsilon^j$$

the original formal differential equation (5.1-1) takes on the form

$$\mu y' = \left(\sum_{s=0}^{\infty} K_s(x, \epsilon)\mu^s \right) y \qquad (5.3\text{-}1)$$

or

$$\mu y' = \mathbf{K}(x, \epsilon, \mu) y$$

with

$$\mathbf{K}(x, \epsilon, \mu) = \sum_{s=0}^{\infty} K_s(x, \epsilon)\mu^s$$

and

$$\mathbf{K}(x, \epsilon, \epsilon^h) = \mathbf{A}(x, \epsilon).$$

Many of the operations of the preceding section can now be carried out with μ playing the role of ϵ and the K_s replacing the A_r. I shall describe the procedure for $n = 2$, $\beta = 1$ only. The matrix K_0 in (5.3-1) corresponds to the A_0 of before. It has the form

$$K_0(x, \epsilon) = \begin{bmatrix} 0 & 1 \\ x + \sum_{j=1}^{h-1} a_{2j}(x)\epsilon^j & \sum_{j=1}^{h-1} a_{1j}(x)\epsilon^j \end{bmatrix}$$

$$=: \begin{bmatrix} 0 & 1 \\ k_{20}(x, \epsilon) & k_{10}(x, \epsilon) \end{bmatrix} \qquad (5.3\text{-}2)$$

The later coefficient matrices are

$$K_s(x, \epsilon) = \begin{bmatrix} 0 & 0 \\ k_{2s}(x, \epsilon) & k_{1s}(x, \epsilon) \end{bmatrix} \qquad (5.3\text{-}3)$$

with

$$k_{2s}(x, \epsilon) = \sum_{j=0}^{h-1} a_{2,s+j}(x)\epsilon^j, \qquad k_{1s}(x, \epsilon) = \sum_{j=0}^{h-1} a_{1,s+j}(x)\epsilon^j, \qquad s = 1, 2, \ldots \ . \qquad (5.3\text{-}4)$$

The matrix $K_0(x, \epsilon)$, which now plays the former role of $A_0(x)$ in the theory for $h = 1$, does not have the simple form of $A_0(x)$, which was important for the success of the reduction method. A few preliminary transformations are therefore necessary. First, set

$$y = \exp\left[\frac{1}{2\mu} \int_0^x k_{10}(t, \epsilon)\, dt \right] \begin{bmatrix} 1 & 0 \\ \frac{1}{2}k_{10}(x, \epsilon) & 1 \end{bmatrix} v. \qquad (5.3\text{-}5)$$

Since $\mu = \epsilon^h$, this is, of course, not an equivalence transformation in the sense of Section 4.1. A short calculation shows that the differential equation for v is of the same type as (5.3-1) with the leading coefficient

$$\begin{pmatrix} 0 & 1 \\ k_{20} - \frac{1}{2}k'_{10} + \frac{1}{4}k^2_{10} & 0 \end{pmatrix}$$

instead of K_0. On the assumption that the change of variable (5.3-5) has been performed beforehand, it can be stipulated without loss of generality that

$$k_{10}(x, \epsilon) \equiv 0 \tag{5.3-6}$$

in (5.3-2).

To simplify the coefficient

$$k_{20}(x, \epsilon) = x + \sum_{j=1}^{h-1} a_{2j}(x)\epsilon^j,$$

the technique of the proof of Lemma 5.1-2 can be applied. Let

$$\xi = \phi(x, \epsilon), \quad y = \Omega(x, \epsilon)w := \begin{pmatrix} 1 & 0 \\ 0 & \omega(x, \epsilon) \end{pmatrix}w. \tag{5.3-7}$$

The differential equation (5.3-1) then becomes, if (5.3-6) is assumed,

$$\mu \frac{dw}{d\xi} = L(\xi, \epsilon, \mu)w, \tag{5.3-8}$$

with

$$L(\xi, \epsilon, \mu) := \frac{dx}{d\xi} \begin{bmatrix} 0 & \omega(x, \epsilon) \\ \omega^{-1}(x, \epsilon)k_2(x, \epsilon, \mu) & k_1 \end{bmatrix}$$
$$- \mu \frac{dx}{d\xi} \Omega^{-1}(x, \epsilon)\Omega(x, \epsilon). \tag{5.3-9}$$

(The symbol L here has nothing to do with the operator L in Section 5.2.) Except for the changed notation, the series k_2, k_1 are the same as a_2, a_1 in (5.1-7), when $n = 2$. The functions ω and ϕ should satisfy the relations

$$\frac{dx}{d\xi}\omega = 1 \quad \text{and} \quad \omega^{-1}k_{20}\frac{dx}{d\xi} = \xi$$

if the analog to Lemma 5.1-2 for $n = 2$ is to hold. One finds

$$\left(\frac{d\xi}{dx}\right)^2 \xi = k_{20}$$

and hence

$$\xi = \phi(x, \epsilon) = \left[\frac{3}{2} \int_{x_1}^{x} (k_{20}(t, \epsilon))^{1/2}\, dt \right]^{2/3}. \tag{5.3-10}$$

The constant x_1 with $|x_1| \leqslant x_0$ is still arbitrary. Also,

$$\omega(x, \epsilon) = \left(\frac{k_2(x, \epsilon)}{\phi(x, \epsilon)} \right)^{1/2}. \tag{5.3-11}$$

To decide whether or not ϕ is holomorphic at $x = \epsilon = 0$, change the variable of integration to

$$\tau = k_{20}(t, \epsilon) = t + \sum_{j=1}^{h-1} a_{2j}(t)\epsilon^j.$$

This τ is a holomorphic function of t at $t = 0$, for all ϵ, with a holomorphic inverse for small t and ϵ. One has

$$\int_0^x (k_{20}(t, \ ))^{1/2} dt = \int_{\tau(x_1, \epsilon)}^{\tau(x, \epsilon)} \frac{\tau^{1/2}}{1 + \sum_{j=1}^{h-1} a'_{2j}(t)\epsilon^j} dt$$

$$= \int_{\tau(x_1, \epsilon)}^{\tau(x, \epsilon)} \tau^{1/2} [1 + \epsilon g(\tau, \epsilon)] dt, \tag{5.3-12}$$

with g holomorphic for small τ and ϵ. The equation

$$t + \sum_{j=1}^{h-1} a_{2j}(t)\epsilon^j = 0$$

for t has at least one solution that tends to zero with t, for all small ϵ. Let x_1 be that solution (which depends on ϵ). Then the last integral in (5.3-12) has the form

$$\tfrac{2}{3}\tau(x, \epsilon)^{3/2} [1 + \epsilon g^*(x, \epsilon)]$$

with $g^*(x, \epsilon)$ holomorphic near $x = \epsilon = 0$. Thus, (5.3-10) becomes

$$\xi = \phi(x, \epsilon) = \tau(x, \epsilon) [1 + \epsilon g^{**}(x, \epsilon)],$$

where g^{**} is holomorphic near $x = \epsilon = 0$. Finally, (5.3-11) implies that

$$\omega(x, \epsilon) = (1 + \epsilon g^{**}(x, \epsilon))^{-1/2},$$

which is holomorphic.

The result just proved is strictly analogous to Lemma 5.1-2 and is now also stated as a lemma.

Lemma 5.3-1. *If $k_{10}(x, \epsilon) \equiv 0$ and*

$$k_{20}(x, \epsilon) = x + \sum_{j=1}^{h-1} a_{2j}(x)\epsilon^j,$$

there exists a change of variables from (y, x) to (w, ξ) which transforms the *formal differential equation (5.3-1) into (5.3-8) with*

$$\mathbf{L}(\xi, \epsilon, \mu) = \sum_{s=0}^{\infty} L_s(\xi, \epsilon)\mu^s,$$

$$L_0(\xi, \epsilon) = \begin{bmatrix} 0 & 1 \\ \xi & 0 \end{bmatrix},$$

$$L_s(\xi, \epsilon) = \begin{bmatrix} 0 & 0 \\ l_{s2}(\xi, \epsilon) & l_{s1}(\xi, \epsilon) \end{bmatrix}, \qquad s > 0.$$

The functions l_{s1}, l_{s2} *are holomorphic in a neighborhood of* $\xi = \epsilon = 0$ *that is independent of* ϵ.

To facilitate the comparison with previous reasoning, the notation used before the last transformation will be restored in part. In precise terms, the differential equation is now

$$\mu y' = \mathbf{A}(x, \epsilon, \mu) y \tag{5.3-13}$$

with

$$\mathbf{A}(x, \epsilon, \mu) = \sum_{r=0}^{\infty} A_r(x, \epsilon)\mu^r, \tag{5.3-14}$$

$$A_0(x, \epsilon) = \begin{bmatrix} 0 & 1 \\ x & 0 \end{bmatrix}, \tag{5.3-15}$$

$$A_r(x, \epsilon) = \begin{bmatrix} 0 & 0 \\ a_{2r}(x, \epsilon) & a_{1r}(x, \epsilon) \end{bmatrix}, \qquad r > 0. \tag{5.3-16}$$

The a_{jr}, $r = 0, 1, 2, \ldots$, $j = 1, 2$, are holomorphic in x and ϵ in a neighborhood of $x = \epsilon = 0$, independent of r.

From here on, the reasoning is almost literally the same as in the proof of Theorem 5.2-1 for $n = 2$. The letter μ plays the role of ϵ, and ϵ is now an additional parameter. The rational operations and differentiations do not destroy the holomorphic dependence on ϵ, since all divisors appearing in the calculations are different from zero at $x = \epsilon = 0$. Thus, a series

$$\mathbf{P}(x, \epsilon, \mu) = \sum_{s=0}^{\infty} P_s(x, \epsilon)\mu^s$$

can be constructed so that the formal transformation

$$y = \mathbf{P}(x, \epsilon, \mu)z$$

changes the differential equation (5.3-13) into

$$\mu z' = A_0(x)z.$$

The P_s are holomorphic in x and ϵ at $x = \epsilon = 0$ and $P_0(0, 0) = I$. Finally, μ can be replaced by ϵ^h and the series $\mathbf{P}(x, \epsilon, \mu)$ reordered so as to become a series in powers of ϵ. This completes the proof of the theorem below.

Theorem 5.3-1. *Assume that in the formal differential equation*

$$\epsilon^h y' = A(x, \epsilon) y, \qquad h > 0, \tag{5.3-17}$$

the coefficient matrix has the form

$$A(x, \epsilon) = \begin{pmatrix} 0 & 1 \\ a_2(x, \epsilon) & a_1(x, \epsilon) \end{pmatrix} \tag{5.3-18}$$

with

$$a_j(x, \epsilon) = \sum_{r=0}^{\infty} a_{jr}(x) \epsilon^r, \qquad j = 1, 2, \tag{5.3-19}$$

where the a_{jr} are holomorphic in a disk $|x| \leqslant x_0$. If

$$\sum_{r=0}^{h-1} a_{2r}(x) \epsilon^r \equiv x, \qquad \sum_{r=0}^{h-1} a_{1r}(x) \epsilon^r \equiv 0, \tag{5.3-20}$$

then the differential equation is formally equivalent to

$$\epsilon^h y' = \begin{pmatrix} 0 & 1 \\ x & 0 \end{pmatrix}. \tag{5.3-21}$$

It must be emphasized that this theorem is much less special than appears at first, for as has been shown here, *any differential equation of the form* (5.3-17) *with*

$$A(x, \epsilon) = \sum_{r=0}^{\infty} A_r(x) \epsilon^r$$

can be explicitly transformed into one that satisfies conditions (5.3-18), (5.3-19), *and* (5.3-20) *by suitable changes of the dependent and independent variables followed by a return to the previous names for the variables, provided*

(i) the Jordan form of $A_0(0)$ is $\begin{pmatrix} 0 & 1 \\ 0 & 0 \end{pmatrix}$;

(ii) $(d/dx)\{\det[A_0(x) - \lambda I]\} \neq 0$ at $x = 0$.

Condition (ii) is tantamount to saying that $\beta = 1$ in Lemma 4.1-2.

5.4. The General Theory for $n = 2$

Throughout Chapter V so far, it has been assumed that $A_0(0)$ has only one elementary divisor. When $n = 2$, this is condition (i) formulated at the end of Section 5.3. The eigenvalue may, of course, be taken as zero, without loss of generality, because of the existence of transformation (5.1-4). Then, the only two possibilities for the Jordan form of the nilpotent matrix $A_0(0)$ are

$$\begin{pmatrix} 0 & 1 \\ 0 & 0 \end{pmatrix} \quad \text{and} \quad \begin{pmatrix} 0 & 0 \\ 0 & 0 \end{pmatrix}.$$

The first of these has been the subject of a thorough study in the literature —and in the present account. The only systematic account of the second case, to my knowledge, is in the work of Richard J. Hanson [18], [19] (see also [20]).

For a brief description of this theory consider again the formal differential equation

$$\epsilon^h y' = \mathbf{A}(x, \epsilon) y, \tag{5.4-1}$$

$$\mathbf{A}(x, \epsilon) = \sum_{r=0}^{\infty} A_r(x) \epsilon^r \tag{5.4-2}$$

in the case $n = 2$, without any restricting hypotheses on the $A_r(x)$ beyond the assumption that they are holomorphic in $|x| < x_0$. As in Section 5.1, the transformation

$$y = \exp\left\{ \frac{1}{2} \epsilon^{-h} \int_0^x \operatorname{tr} \mathbf{A}(t, \epsilon) \, dt \right\} z \tag{5.4-3}$$

annuls the traces of all matrices $A_r(x)$. The hypothesis

$$\operatorname{tr} A_r(x) = 0, \qquad r = 0, 1, \ldots, \tag{5.4-4}$$

can therefore be introduced without losing generality.

The formal transformation (5.4-3) has the expanded form

$$y = \left[\exp\left\{ \frac{1}{2} \sum_{r=0}^{h-1} \int_0^x A_r(t) \, dt \, \epsilon^{r-h} \right\} \sum_{r=0}^{\infty} T_r(x) \epsilon^r \right] z$$

with holomorphic matrices $T_r(x)$.

The next lemma is the basis of a classification of the differential equation under consideration.

Lemma 5.4-1. *If the two-by-two matrix valued function A_0 is holomorphic in x at $x = 0$, and if $\operatorname{tr} A_0(x) \equiv 0$, then A_0 is holomorphically similar, at $x = 0$, to a matrix of the form*

$$B_0(x) = k(x) \begin{pmatrix} 0 & x^\mu \\ x^\nu & 0 \end{pmatrix}, \tag{5.4-5}$$

where k is holomorphic and not zero at $x = 0$. The exponents μ and ν are integers in the interval $[0, \infty]$. Here, $x^\infty = 0$, by definition.

PROOF. Let

$$A_0(x) = \begin{bmatrix} a(x) & b(x) \\ c(x) & -a(x) \end{bmatrix} = \begin{bmatrix} \tilde{a}(x) x^\lambda & \tilde{b}(x) x^\mu \\ \tilde{c}(x) x^\nu & -\tilde{a}(x) x^\lambda \end{bmatrix},$$

where λ, μ, ν are integers in $[0, \infty]$, and $\tilde{a}, \tilde{b}, \tilde{c}$ are holomorphic and not zero at $x = 0$.

If $\lambda \geqslant \mu$, set

$$T(x) = \begin{pmatrix} 1 & 0 \\ \gamma(x) & 1 \end{pmatrix}.$$

One verifies directly that $T^{-1}A_0T$ has zero diagonal if $a + b\gamma = 0$, i.e., if

$$\gamma(x) = \frac{\tilde{a}(x)}{\tilde{b}(x)} x^{\lambda - \mu}.$$

If $\lambda \geqslant \nu$, set

$$T(x) = \begin{pmatrix} 1 & \beta(x) \\ 0 & 1 \end{pmatrix}$$

and find, similarly, that the diagonal of $T^{-1}A_0T$ is zero if

$$\beta(x) = \frac{\tilde{a}(x)}{\tilde{c}(x)} x^{\lambda - \nu}.$$

Finally, if $\lambda < \mu$ and $\lambda < \nu$, take

$$T(x) = \begin{pmatrix} \alpha(x) & 1 \\ 1 & 1 \end{pmatrix}.$$

The condition that $T^{-1}A_0T$ have zero diagonal becomes $\alpha(a - c) + b + a = 0$, or

$$\alpha(x) = \frac{b(x) + a(x)}{c(x) - a(x)} = \frac{\tilde{b}(x)x^{\mu - \lambda} + \tilde{a}(x)}{\tilde{c}(x)x^{\nu - \lambda} - \tilde{a}(x)},$$

which is holomorphic at $x = 0$, with $\alpha(0) = -1$. In each case $\det T(0) \neq 0$. Therefore, A_0 is holomorphically similar to a matrix with zero diagonal.

To complete the proof of the lemma it may now be assumed that

$$A_0 = \begin{pmatrix} 0 & b \\ c & 0 \end{pmatrix},$$

$$b(x) = \tilde{b}(x)x^{\mu}, \qquad c(x) = \tilde{c}(x)x^{\nu}.$$

The matrix

$$S = \begin{pmatrix} 1 & 0 \\ 0 & (\tilde{c}/\tilde{b})^{1/2} \end{pmatrix}$$

is holomorphic and invertible at $x = 0$, and

$$S^{-1}A_0S = (\tilde{b}\tilde{c})^{1/2} \begin{bmatrix} 0 & x^{\mu} \\ x^{\nu} & 0 \end{bmatrix}.$$

This proves the lemma.

Lemma 5.4-2. *For holomorphic matrices A_0 with trace zero the pair of integers μ, ν defined in Lemma 5.4-1 is invariant under holomorphic similarity transformations. If μ and ν are finite, the function k^2 is also invariant.*

PROOF. Let

$$\tilde{B}(x) = \tilde{k}(x)\begin{pmatrix} 0 & x^{\tilde{\mu}} \\ x^{\tilde{\nu}} & 0 \end{pmatrix}.$$

If $\tilde{B}_0$ is similar to the B_0 of (5.4-5) then $\det B_0 = \det \tilde{B}_0$, i.e.,

$$k^2 x^{\mu+\nu} = \tilde{k}^2 x^{\tilde{\mu}+\tilde{\nu}}. \tag{5.4-6}$$

Consider first the case that μ and ν are finite. Since

$$(k(x)/\tilde{k}(x))^2 = x^{\tilde{\mu}+\tilde{\nu}-\mu-\nu}$$

and $k(0) \neq 0$, $\tilde{k}(0) \neq 0$, it must be true that

$$\tilde{\mu} + \tilde{\nu} = \mu + \nu \quad \text{and} \quad k^2 \equiv \tilde{k}^2.$$

Thus, $\tilde{\mu}$ and $\tilde{\nu}$ are also finite. Now assume that

$$TB_0 = \tilde{B}_0 T, \tag{5.4-7}$$

where

$$T = \begin{bmatrix} \alpha & \beta \\ \gamma & \delta \end{bmatrix}.$$

The order of the exponents in B_0 and $\tilde{B}_0$ is arbitrary, for

$$\begin{pmatrix} 0 & 1 \\ 1 & 0 \end{pmatrix}\begin{pmatrix} 0 & x^{\mu} \\ x^{\nu} & 0 \end{pmatrix}\begin{pmatrix} 0 & 1 \\ 1 & 0 \end{pmatrix} = \begin{pmatrix} 0 & x^{\nu} \\ x^{\mu} & 0 \end{pmatrix}.$$

Therefore, it can be stipulated that

$$\nu \geqslant \mu, \qquad \tilde{\nu} \geqslant \tilde{\mu}.$$

Now assume that $\tilde{\mu} > \mu$ (if $\tilde{\mu} < \mu$, change the labeling). Then (5.4-7) implies that

$$T(x)\begin{pmatrix} 0 & 1 \\ x^{\nu-\mu} & 0 \end{pmatrix} = \pm x^{\tilde{\mu}-\mu}\begin{pmatrix} 0 & 1 \\ x^{\tilde{\nu}-\tilde{\mu}} & 0 \end{pmatrix}T(x).$$

The right-hand side is zero for $x = 0$. Hence, $\alpha(0) = 0$ and $\gamma(0) = 0$, i.e., $\det T(0) = 0$, so that B_0 and $\tilde{B}_0$ cannot be holomorphically similar in $|x| \leqslant x_0$. When μ or ν is infinite, the proof is almost the same but simpler.

Observe that while k^2 is invariant, k itself need not be invariant. For instance,

$$\begin{pmatrix} 0 & 1 \\ 1 & 0 \end{pmatrix}\begin{pmatrix} 0 & 1 \\ -1 & 0 \end{pmatrix}\begin{pmatrix} 0 & 1 \\ 1 & 0 \end{pmatrix} = (-1)\begin{pmatrix} 0 & 1 \\ -1 & 0 \end{pmatrix}.$$

From now on it will be assumed that the leading coefficient matrix in (5.4-2) already has the form

$$A_0(x) = k(x)\begin{pmatrix} 0 & x^{\mu} \\ x^{\nu} & 0 \end{pmatrix}, \qquad \nu \geqslant \mu. \tag{5.4-8}$$

The factor $k(x)$ can be removed by changes of the dependent and independent variables in the differential equation (5.4-1), more or less as in Lemma

5.1-2. The description of this simplification will be kept briefer than in the proof of that lemma:

Set

$$\xi = \phi(x), \qquad y = \begin{pmatrix} 1 & 0 \\ 0 & \omega \end{pmatrix} v \tag{5.4-9}$$

in differential equation (5.4-2). The leading term of the series for the coefficient matrix in the transformed differential equation is then

$$\frac{dx}{d\xi} \begin{pmatrix} 0 & k\omega x^{\mu} \\ k\omega^{-1} x^{\nu} & 0 \end{pmatrix}.$$

This will reduce to

$$\begin{pmatrix} 0 & \xi^{\mu} \\ \xi^{\nu} & 0 \end{pmatrix}$$

if, and only if,

$$\frac{dx}{d\xi} k\omega x^{\mu} = \xi^{\mu},$$

$$\frac{dx}{d\xi} k\omega^{-1} x^{\nu} = \xi^{\nu},$$

or, equivalently, if

$$\frac{dx}{d\xi} k x^{(\mu+\nu)/2} = \xi^{(\mu+\nu)/2}$$

and

$$\omega x^{(\mu-\nu)/2} = \xi^{(\mu-\nu)/2}.$$

The first of the last two relations yields

$$\xi = \phi(x) = \left[\frac{\mu + \nu + 2}{2} \int_0^x k(t) t^{(\mu+\nu)/2} \, dt \right]^{2/(\mu+\nu+2)},$$

and the second leads to

$$\omega(x) = [\phi(x)]^{(\mu-\nu)/2} x^{(\nu-\mu)/2}.$$

As in Lemma 5.1-2 one verifies that ϕ and ω can be taken as holomorphic functions without branch point at $x = 0$.

The result of Section 5.4, so far, can be summarized as follows. *No generality is lost in the local asymptotic theory of differential equations of order two of the form (5.4-1) and (5.4-2) if $A_0(x)$ is assumed to have the form*

$$A_0(x) = \begin{pmatrix} 0 & x^{\mu} \\ x^{\nu} & 0 \end{pmatrix}, \qquad 0 \leqslant \mu, \quad \nu < \infty. \tag{5.4-10}$$

The case $\mu = 0$, $\nu > 0$ was studied in some detail in Sections 5.2 and 5.3. In Ref. 19, Hanson has extended these methods to the case of any μ and ν. Here, his main result will be stated, but the proof is omitted.

Theorem 5.4-1. *Every formal second-order system of differential equations of the form (5.4-1) and (5.4-2) can be transformed into one,*

$$\epsilon^h \frac{dz}{d\xi} = \left(\sum_{r=0}^{\infty} B_r(\xi)\epsilon^r \right) z =: \mathbf{B}(\xi, \epsilon)z,$$

in which the B_r are polynomials whose degrees have a bound independent of r. In the transformation

$$x = \psi(\xi, \epsilon), \qquad y = \left(\sum_{r=0}^{\infty} P_r(\xi)\epsilon^r \right) z$$

by which this simplification can be achieved, the functions $\psi(\xi, \epsilon), P_r(\xi)$ are holomorphic in a disk $|\xi| \leqslant \xi_0$ for $|\epsilon| \leqslant \epsilon_0$. Also, $(\partial\psi(\xi, 0)/\partial\xi)|_{\xi=0} \neq 0$ and P_0^{-1} is holomorphic at $\xi = 0$. The functions ψ and P_r can be calculated by rational operations, differentiations, and quadratures. In particular,

$$B_0(\xi) = \begin{pmatrix} 0 & \xi^\mu \\ \xi^\nu & 0 \end{pmatrix}, \qquad 0 \leqslant \mu, \quad \nu \leqslant \infty.$$

Hanson's method allows him to give more detailed information on the polynomials B_r, as stated in the next theorem.

Theorem 5.4-2. *In the notation of Theorem 5.4-1, let $\nu \geqslant \mu$ and*

$$B_r = \begin{bmatrix} b_r^{11} & b_r^{12} \\ b_r^{21} & b_r^{22} \end{bmatrix}, \qquad r > 0.$$

Then the degrees $d(b_r^{jk})$ of the polynomials b_r^{jk} satisfy the inequalities

$$d(b_r^{11}) < \mu - 1, \qquad d(b_r^{12}) < \mu, \qquad d(b_r^{21}) < \nu, \qquad d(b_r^{22}) < \mu - 1.$$

(Here, a polynomial of negative degree is, by definition, identically zero.) If $\mu = 0$, one has, even, $d(b_r^{21}) < \nu - 1$.

CHAPTER VI
Uniform Transformations at Turning Points: Analytic Theory

6.1. Preliminary General Results

The results of Chapter V are of limited interest by themselves. Only if the formal series constructed there are asymptotic representations of asymptotic transformations in sets of the x-plane that contain a turning point has really useful information been attained.

The first task in this direction is to supplement the formal theorems 2.3-1 and 5.1-1 by a corresponding analytic statement. This will be done by proving the following theorem whose formulation resembles that of Theorem 2.3-1.

Theorem 6.1-1. *Let* $A(\cdot,\epsilon) \in H_{nn}(D)$ *for all* $\epsilon \in E := \{\epsilon \,|\, 0 < |\epsilon| \leqslant \epsilon_0,$ $|\arg\epsilon| \leqslant \delta\}$ *and assume that*

$$A(x,\epsilon) \sim \mathbf{A}(x,\epsilon) := \sum_{r=0}^{\infty} A_r(x)\epsilon^r \tag{6.1-1}$$

with $A_r \in H_{nn}(D)$; *as* $\epsilon \to 0$ *in* E, *uniformly for* $x \in D$. *Assume also that* A_0 *is block diagonal*:

$$A_0 = A_0^{11} \oplus A_0^{22}, \tag{6.1-2}$$

and that $A_0^{11}(x), A_0^{22}(x)$ *have no common eigenvalues for any* $x \in D$. *Then there exists in every sufficiently small subregion* $D^* \subset D$, *and in every sufficiently narrow sector* E *a matrix-valued function* $P(\cdot,\epsilon) \in H_{nn}(D^*)$ *with an asymptotic expansion*

$$P(x,\epsilon) \sim \mathbf{P}(x,\epsilon) = \sum_{r=0}^{\infty} P_r(x)\epsilon^r, \qquad P_0 = I, \qquad P_r \in H_{nn}(D), \tag{6.1-3}$$

as $\epsilon \to 0$ in E, uniformly for $x \in D^*$, such that the transformation

$$y = Pv \tag{6.1-4}$$

takes the differential equation

$$\epsilon^h \frac{dy}{dx} = Ay \tag{6.1-5}$$

into an equation

$$\epsilon^h \frac{dv}{dx} = Bv \tag{6.1-6}$$

with the property

$$B = B^{11} \oplus B^{22}, \tag{6.1-7}$$

where B^{11} has the same dimension as A_0^{11}.

PROOF. By the Borel–Ritt theorem there exist matrix-valued functions $\tilde{P}$ holomorphic in x for $x \in D$ and with the asymptotic expansion

$$\tilde{P}(x,\epsilon) \sim \sum_{r=0}^{\infty} P_r(x)\epsilon^r \qquad \text{as} \quad \epsilon \to 0, \quad x \in D, \quad \epsilon \in E. \tag{6.1-8}$$

Here, the series in the right-hand member is the one introduced in (2.3-6). With $\tilde{P}$ in the place of P, the transformation (6.1-4) takes equation (6.1-5) into one of the form (6.1-8) with the property that if B is written in the partitioned form

$$B = \begin{bmatrix} B^{11} & B^{12} \\ B^{21} & B^{22} \end{bmatrix}$$

induced by the decomposition of A_0 in (6.1-2), then

$$B^{12} \sim 0, \qquad B^{21} \sim 0, \qquad x \in D.$$

For the remainder of the proof it is convenient to stipulate that this transformation has been performed already and to return to the previous notation. In other words, it will be assumed, without loss of generality, that

$$A^{12} \sim 0, \qquad A^{21} \sim 0 \qquad \text{as} \quad \epsilon \to 0, \quad \epsilon \in E, \quad x \in D. \tag{6.1-9}$$

Transformation (6.1-4) takes (6.1-5) into (6.1-6) if P satisfies the differential equation

$$\epsilon^h P' = AP - PB. \tag{6.1-10}$$

The matrix B is to have the form (6.1-7). The procedure in the proof of Theorem 2.3-1 makes it plausible to try finding a matrix P with the required property which has the form

$$P = \begin{bmatrix} I & P^{21} \\ P^{12} & I \end{bmatrix}. \tag{6.1-11}$$

Then (6.1-10) is equivalent to the four relations

$$0 = A^{11} + A^{12}P^{21} - B^{11}, \qquad\qquad \epsilon^h(P^{12})' = A^{11}P^{12} + A^{12} - P^{12}B^{22},$$

$$\epsilon^h(P^{21})' = A^{21} + A^{22}P^{21} - P^{21}B^{11}, \qquad 0 = A^{21}P^{12} + A^{22} - B^{22}.$$

$$(6.1\text{-}12)$$

By elimination of B^{11} and B^{22} one finds the nonlinear differential equations

$$\epsilon^h(P^{12})' = A^{11}P^{12} - P^{12}A^{22} + A^{12} - P^{12}A^{21}P^{12},$$
$$\epsilon^h(P^{21})' = A^{22}P^{21} - P^{21}A^{11} + A^{21} - P^{21}A^{12}P^{21}, \qquad (6.1\text{-}13)$$

each of which contains only one unknown matrix. Each of the two equations (6.1-13) is a system of $N := m(n - m)$ scalar differential equations, if m is the order of the matrix A_0^{11}. Assembling these entries into a vector w of dimension $m(n - m)$, each differential equation in (6.1-13)—for instance the one for P^{12}—can be rewritten in a new, self-explanatory notation as

$$\epsilon^h w' = Kw + g. \qquad (6.1\text{-}14)$$

Here, K is an $N \times N$ matrix holomorphic in x for $x \in D$, having an asymptotic series in powers of ϵ:

$$K(x, \epsilon) \sim \sum_{r=0} K_r(x)\epsilon^r. \qquad (6.1\text{-}15)$$

The assumption that A_0^{11}, A_0^{22} have no common eigenvalues in D translates itself into the statement that $K_0(x)$ is holomorphically invertible for all $x \in D$. The entries of the vector g in (6.1-14) are quadratic functions of the entries of w with coefficients that—thanks to hypothesis (6.1-9)—are asymptotic to zero for all $x \in D$, as $\epsilon \to 0$ in E.

If it can be shown that equation (6.1-14) has solutions that are asymptotic to zero, then the corresponding transformation matrix P in (6.1-11) is asymptotic to I, which proves the theorem under the nonrestrictive assumption (6.1-9). To construct such a solution of (6.1-14), at least in sufficiently small regions, consider the neighborhood of a point in d, say the point $x = 0$, and write (6.1-14) in the form

$$\epsilon^h w' = K_0(0)w + \tilde{g}(w, x, \epsilon) \qquad (6.1\text{-}16)$$

with

$$\tilde{g}(w, x, \epsilon) = g(w, x, \epsilon) + (K(x, \epsilon) - K_0(0))w. \qquad (6.1\text{-}17)$$

Since any change of variable $w = Tw^*$ with a constant invertible matrix T does not change the structure of this differential equation, it may be assumed, from now on, that $K_0(0)$ is in Jordan form

$$K_0(0) = J.$$

The differential equation (6.1-16) is then equivalent to the integral equation

$$w(x,\epsilon) = e^{\epsilon^{-h}Jx}c(\epsilon) + \int_{\Gamma_x} e^{\epsilon^{-h}J(x-t)}\tilde{g}(w(t,\epsilon),t,\epsilon)\epsilon^{-h}\,dt, \quad (6.1\text{-}18)$$

where $c(\epsilon)$ is an arbitrary vector independent of x and Γ_x is a path in the t-plane from a fixed point to x.

From here on the reasoning resembles so much the arguments in the proof of Theorems 3.4-1 and 3.4-2 that a concise description is justified. In one respect the situation is simpler, here, in as much as the exponent of the exponential factor in the integrand is linear in $x - t$. On the other hand, the nonlinearity of $\tilde{g}$ in w causes technical complications, which are, however, minor thanks to the properties introduced in formula (6.1-9). More burdensome is the presence of the term $(K(x,\epsilon) - K_0(0))w$ in (6.1-17). The term has the form $(O(x) + O(\epsilon))w$, as x and ϵ tend to zero. Therefore, to make the usual proof of existence of a solution of (6.1-18) by the contraction mapping theorem (or, equivalently, by Picard type iterations) applicable, x must be restricted to a possibly small neighborhood of the origin. The fact that $A_0^{11}(x)$ and $A_0^{22}(x)$ have nowhere in D common eigenvalues implies that none of the N eigenvalues of $K(0) = J$ is zero. Therefore, one can choose for D^* a rhombus, as in the proof of Theorem 3.4-1, and for E a sufficiently narrow sector, so that a certain set of straight paths γ_{xj}, $j = 1, 2, \ldots, N$, in D^* is strictly progressive. Actually the γ_{xj} are all one or the other of two segments ending at $t = x$. (See Fig. 3.1 accompanying the proof of Theorem 3.4-1.) The path γ_{xj} is used in the jth component of the integral equation (6.1-18). This change of paths amounts to a change in the constant $c(\epsilon)$ in (6.1-18). Now, that constant will be chosen as zero.

Once it has been established that the differential equation has a solution w bounded in $D^* \times E$, one more estimate of the right-hand member of (6.1-18) shows that $w \sim 0$, because the coefficients of the quadratic function g are small—even asymptotic to zero. Therefore, $P \sim I$ in (6.1-11). Thus, P has the form (6.1-3) with all $P_r = 0$, $r > 0$. This very special property of P is, of course, due to the relation (6.1-9), which was a consequence of a preliminary transformation $y = \tilde{P}v$ with $\tilde{P}$ as in (6.1-8), where the $P_r, r > 0$, need not be all zero, yet. This completes the proof of Theorem 6.1-1.

Theorem 6.1-2. *As in Theorem 6.1-1, let $A(\cdot,\epsilon) \in H_{nn}(D)$ for all $\epsilon \in E$, and let (6.1-1) be true. Assume that the characteristic polynomial $\phi = \det(\lambda I - A_0(x))$ of the matrix A_0 can be factored into p monic polynomials ϕ_j of positive degrees m_j with coefficients holomorphic in D:*

$$\phi = \phi_1\phi_2 \ldots \phi_p,$$

so that for no $x \in D$ any two of these polynomials have a common zero. Then the differential equation (6.1-5) is, in a neighborhood of each point of D equivalent to a set of p systems of first order differential equations in m_j dependent variables, respectively, $j = 1, 2, \ldots, p$, and with ϕ_j as the characteristic polynomial of the leading coefficient matrix.

PROOF.   Under the assumption made on the holomorphic factorization of ϕ, Theorem 12.3-1 can be applied to the effect that a holomorphic similarity transformation with a coefficient matrix independent of ϵ changes the differential equation into one whose leading coefficient matrix is block-diagonal as required in formula (6.1-2). By virtue of Theorem 6.1-1 the differential equation is then holomorphically equivalent to a pair of completely uncoupled differential equations of the same type. After $p - 2$ consecutive such reductions the statement of the theorem is at hand.

REMARK.   Theorems 6.1-1 and 6.1-2 are strictly local, although the decomposition described in Theorem 12.3-1 of the Appendix is globally valid in all of D. A global uncoupling of the given differential equation by one and the same transformation with an asymptotic series in powers of ϵ, valid in large regions would be a boon to the theory. On the other hand, it is quite possible that such a theorem does not exist, and then one would like to see counterexamples.

The idea of using the Borel–Ritt theorem to pass from a formal equivalence transformation to an analytic one that almost leads to the desired simple features, as it was described in the preceding proof of Theorem 6.1-1, is of such general applicability that it is worthwhile to state it as a lemma. Two formal differential equations

$$\epsilon^h y' = \mathbf{A} y, \qquad \epsilon^h v' = \mathbf{B} v \qquad (6.1\text{-}19)$$

with

$$\mathbf{A}(x,\epsilon) = \sum_{r=0}^{\infty} A_r(x)\epsilon^r, \qquad (6.1\text{-}20)$$

$$\mathbf{B}(x,\epsilon) = \sum_{r=0}^{\infty} B_r(x)\epsilon^r, \qquad (6.1\text{-}21)$$

and $A_r \in H_{nn}(D)$, $B_r \in H_{nn}(D)$ were called formally equivalent in D when there existed a formal matrix series

$$\mathbf{P}(x,\epsilon) = \sum_{r=0}^{\infty} P_r(x)\epsilon^r, \qquad P_r \in H_{nn}(D), \qquad P_0(x) \text{ invertible in } D,$$

$$(6.1\text{-}22)$$

such that the formal transformation

$$y = \mathbf{P} v$$

took the first differential equation in (6.1-19) into the second one. This was tantamount to the formal relation

$$\epsilon^h \mathbf{P}' - \mathbf{AP} + \mathbf{PB} = 0. \qquad (6.1\text{-}23)$$

In an obvious extension of that terminology, the series in (6.1-20) and (6.1-21) may be called formally equivalent if the two differential equations in (6.1-19) are formally equivalent, i.e., if formal series P in (6.1-22) exist for which (6.1-23) is true. By the Borel–Ritt theorem there exist then functions $A, B, \tilde{P}$ such that

$$\epsilon^h \tilde{P}' - A\tilde{P} + \tilde{P}B \sim 0, \qquad \text{in } D,$$

i.e.,

$$\tilde{P}^{-1}A\tilde{P} - \epsilon^h \tilde{P}^{-1}\tilde{P}' = B - \Omega,$$

where $\Omega \sim 0$. Stated as a lemma:

Lemma 6.1-1. *If A and B are formally equivalent for $x \in D$ then there exists a matrix $A + \Omega$ with*

$$\Omega \sim 0 \qquad \text{in } D, \tag{6.1-24}$$

such that B and $A + \Omega$ are asymptotically equivalent in D, for $\epsilon \to 0$ in any prescribed sector.

The remaining problem is to show that A and $A + \Omega$ are asymptotically equivalent, at least in some part of D that contains a turning point. This difficult general problem has been only very partially carried out. Theorem 6.1-2 is one useful result, and Lemma 6.1-1 makes the arguments slightly more transparent.

Once the transformation problem has been changed into one with $B = A + \Omega$, a formal transformation $y = (\sum_{r=0}^{\infty} P_r \epsilon^r)v$ which takes the differential equation with coefficient matrix $A + \Omega$ into one with coefficient A is, of course, the one with coefficients

$$P_0 = I, \qquad P_r = 0, \qquad r > 0. \tag{6.1-25}$$

It is not the only one, but since in this theory nothing is known of Ω except that it is asymptotic to zero, choosing the P_r different from (6.1-3) cannot lead to better results.

It must be shown that the differential equation

$$\epsilon^h P' = (A + \Omega)P - PA$$

for P has a solution $P \sim I$. Equivalently, with $R := P - I$, the differential equation

$$\epsilon^h R' = (A + \Omega)R - RA + \Omega \tag{6.1-26}$$

for R must be shown to possess a solution for which

$$R \sim 0$$

in a set with a turning point.

In this formulation, A designates the matrix of the simplified differential equation. In much of the existing work along these lines, that differential equation is one whose asymptotic theory is already known, independently of the general theory. Often it is solvable by Bessel, Parabolic Cylinder, or Whittaker functions. The more general theory is, as yet, very incomplete and rather difficult. The basic approach is always to represent the solution of the differential equation (6.1-26) as an integral by means of some variant of the variation of parameter method and to appraise that integral.

In general terms, if a matrix differential equation of the form

$$X' = BX - XA + F \qquad (6.1\text{-}27)$$

for X is given and U, V are fundamental matrix solutions of

$$U' = AU, \qquad V' = BV, \qquad (6.1\text{-}28)$$

then the matrix

$$X(x) = V(x)\left\{ C + \int^x V^{-1}(t) F(t) U(t)\, dt \right\} U^{-1}(x) \qquad (6.1\text{-}29)$$

with an arbitrary constant matrix C is the general solution of (6.1-26). The fact that (6.1-29) is a solution can be verified by differentiation. That it is the *general* solution is a consequence of the uniqueness theorem for differential equations.

In the present application of formula (6.1-29) the matrix X is the R of (6.1-26), and A, B in (6.1-28) become $\epsilon^{-h} A$, $\epsilon^{-h}(A + \Omega)$, respectively, in (6.1-26), while F is $\epsilon^{-h}\Omega$.

Formula (6.1-29) is different from, but related to, (6.1-18). There, the fact that the simplified coefficient matrix was the direct sum of two matrices whose leading terms had no common eigenvalues dominated the details of the procedure. Here, no such special information is available.

For $n > 2$ the difficulties in appraising the matrix R in (6.1-26) have been overcome only for certain classes of differential equations (see [76], [92]). In view of the complexity of that theory, only two very simple special—but important—second-order cases will be presented here, as examples of more general arguments.

For $n = 2$ the most general results for differential equations of the type (5.2-1) satisfying conditions (5.2-2)–(5.2-7) are due to Sibuya [77] who analyzed them for any positive value of the exponent β in (5.2-5). For the considerably simpler cases $\beta = 1$, $\beta = 2$, the theory had been completed earlier by a number of authors, including Langer [38], McKelvey [47], Kazarinoff [31], Wasow [90] (Russian edition), Lee [40], and others. Only these two cases will be dealt with in this chapter.

6.2. Differential Equations Reducible to Airy's Equation

Consider the class of differential equations of order two of the form

$$\epsilon y' = [A + \Omega] y \qquad (6.2\text{-}1)$$

with

$$A(x) = \begin{pmatrix} 0 & 1 \\ x & 0 \end{pmatrix} \qquad (6.2\text{-}2)$$

and

$$\Omega(x,\epsilon)\sim 0, \qquad \text{as} \quad \epsilon\to 0 \quad \text{in} \quad |\arg\epsilon| \leqslant \delta_0, \qquad (6.2\text{-}3)$$

uniformly for $x \in D$. The problem is to find a matrix P such that

$$P(x,\epsilon)\sim I \qquad \text{as} \quad \epsilon\to 0 \quad \text{in} \quad |\arg\epsilon| \leqslant \delta < \delta_0$$

for x in a set that contains the turning point at $x = 0$ as a boundary point or—better still—as an interior point and such that the change of variables

$$y = Pz \qquad (6.2\text{-}4)$$

transforms (6.2-1) into

$$\epsilon z' = Az. \qquad (6.2\text{-}5)$$

The fact that (6.2-5) can easily be transformed into Airy's equation whose solution is thoroughly known will be used sparingly, since many aspects of the problem to be solved carry over to more difficult differential equations that cannot be reduced to such well explored equations.

For instance, the formal theory of Chapter II leads to *formal* solutions of the differential equations (6.2-1) and (6.2-5). They must be the same, of course. At the turning point $x = 0$ the coefficients in the series that constitute the formal solution have singularities. The order of these singularities grows with the power of ϵ those coefficients multiply, and that rate of growth is important for the further analysis. It can be read off from the asymptotic series for Airy's equation, but it will be found here by carrying out explicitly the block-diagonalization described in Section 2.3.

That process begins with the transformation

$$y = \begin{bmatrix} 1 & 1 \\ x^{1/2} & -x^{1/2} \end{bmatrix} v$$

which changes

$$\epsilon y' = Ay$$

into

$$\epsilon v' = \left[\begin{pmatrix} x^{1/2} & 0 \\ 0 & -x^{1/2} \end{pmatrix} - \frac{1}{4x} \begin{pmatrix} 1 & -1 \\ -1 & 1 \end{pmatrix} \epsilon \right] v$$

(compare Section 2.3). In the notation of Theorem 2.3-1,

$$A_0(x) = J(x) = x^{1/2} \begin{pmatrix} 1 & 0 \\ 0 & -1 \end{pmatrix},$$

$$A_1(x) = x^{-1}\frac{1}{4} \begin{pmatrix} 1 & -1 \\ -1 & 1 \end{pmatrix} \qquad (6.2\text{-}6)$$

$$A_r(x) = 0, \qquad r > 1.$$

Next, one applies equations (2.3-14) and (2.3-16) and finds, by a simple induction argument that, in the notation used there,

$$B_r(X) = O(x^{-3r/2}), \qquad P_r = O(x^{-(3r+1)/2}), \qquad (6.2\text{-}7)$$

Incidentally, it is always true that the order of the singularities of the diagonalized matrices B_r and P_r at a turning point grow linearly with r, but the rate of growth is not always so easy to determine (see Chapter VII).

After the complete formal diagonalization, each of the two resulting scalar formal differential equations can be integrated, as in Chapter II, and by noting the order of the singularities at each step one finds the lemma below. The somewhat tedious calculations are omitted.

Lemma 6.2-1. *The differential equation (6.2-1), (6.2-2) has a formal BFM-solution ("Basic Formal Matrix Solution")*

$$Y(x,\epsilon) = \hat{Y}(x,\epsilon)\exp\left\{ \epsilon^{-1}\frac{2}{3}x^{3/2}\begin{pmatrix} 1 & 0 \\ 0 & -1 \end{pmatrix}\right\}, \tag{6.2-8}$$

where

$$\hat{Y}(x,\epsilon) = \sum_{r=0}^{\infty} Y_r(x)\epsilon^r \tag{6.2-9}$$

with

$$Y_0(x) = x^{-1/4}\begin{pmatrix} 1 & 1 \\ x^{1/2} & -x^{1/2} \end{pmatrix} \tag{6.2-10}$$

and

$$Y_r(x) = O(x^{-3r/2-1/4}) \quad \text{as} \quad x \to 0.$$

The information contained in this lemma is the basis of a significant generalization of the basic existence theorem 3.3-1 to domains whose distance from the turning point $x = 0$ tends to zero with ϵ, provided the shrinkage is not faster than $O(\epsilon^{2/3})$. To avoid a lengthy interruption of the main argument, and also because analogous theorems exist for a much wider class of problems, the property needed will be stated here but proved later, in Chapter VII. For simplicity, ϵ will be restricted to positive values. The extension of the theorem to sufficiently narrow open sectors of the ϵ-plane makes the formulation more cumbersome but changes the arguments only trivially.

Let

$$q(x,\epsilon) = \epsilon^{-1}\frac{2}{3}x^{3/2} \tag{6.2-11}$$

and denote by

$$y(x,\epsilon) = \hat{y}(x,\epsilon)e^{q(x,\epsilon)} \tag{6.2-12}$$

one (either one) of the two columns of the formal matrix solution (6.2-8). The function q has, of course, two branches, so that (6.2-12) represents two formal vector solutions. The sign of $\operatorname{Re} q$ is, as always, essential for the asymptotic analysis. This sign is constant in each of the three sectors $S_j = \{-\frac{1}{3}\pi + \frac{2}{3}\pi j < \arg x < \frac{1}{3}\pi + \frac{2}{3}\pi j\}$, $j \bmod 3$. The boundaries of these sectors are the rays

$$l_j := \{x/\arg x = \pi + \tfrac{2}{3}\pi j\}, \qquad j \bmod 3. \tag{6.2-13}$$

Theorem 6.2-1. *In (6.2-12) choose one of the two formal vector solutions of the differential equation (6.2-1), (6.2-2), (6.2-3) described by formula (6.2-12). Let S_j be the sector in which $\operatorname{Re} q(x,\epsilon) < 0$ for that solution. Then there exists a solution $y(x,\epsilon)$ which is asymptotically represented by (6.2-12) in the part of the annulus*

$$\epsilon^{2/3} x_1 < |x| \leqslant x_0$$

obtained by removing l_j. Here x_0, x_1 are certain positive constants. The asymptotic representation is such that

$$\left| \sum_{r=0}^{N} y_r(x)\epsilon^r - y(x,\epsilon)e^{-q(x,\epsilon)} \right| \leqslant k_N |\epsilon x^{-3/2}|^N$$

in every compact subset of the domain described. The constant k_N does not depend on x or ϵ, provided that compact subset is bounded away from l_j.

Inside the shrinking disk $|x| \leqslant x_1 \epsilon^{2/3}$ the analytic continuations of the solutions called y in Theorem 6.2-1 must remain holomorphic in x, but they may well have singularities with respect to ϵ at $\epsilon = 0$. The lemma below—which is also a special case of a more general theorem—shows that the solutions under consideration remain bounded in that disk, as $\epsilon \to 0 +$.

Lemma 6.2-2. *The vector solutions of (6.2-1), (6.2-2) called y in Theorem 6.2-1 are bounded for*

$$|x| \leqslant \epsilon^{2/3} x_1$$

as $\epsilon \to 0 +$.

PROOF. The change of variables

$$x = \xi \epsilon^{2/3}, \qquad y = \begin{pmatrix} 1 & 0 \\ 0 & \epsilon^{1/3} \end{pmatrix} w$$

in (6.2-1) and (6.2-2) leads to the differential equation

$$\frac{dw}{d\xi} = \left[\begin{pmatrix} 0 & 1 \\ \xi & 0 \end{pmatrix} + \begin{pmatrix} 1 & 0 \\ 0 & \epsilon^{-1/3} \end{pmatrix} \Omega(\xi\epsilon^{2/3}, \epsilon) \begin{pmatrix} \epsilon^{-1/3} & 0 \\ 0 & 1 \end{pmatrix} \right] y. \tag{6.2-14}$$

Since $\Omega \sim 0$ by (6.2-3) this is a regular perturbation of the equation

$$\frac{dz}{d\xi} = \begin{pmatrix} 0 & 1 \\ \xi & 0 \end{pmatrix} z.$$

Therefore, the particular matrix solution W of (6.2-14) that is characterized by the initial condition $W(0,\epsilon) = I$ is bounded for all $|\xi| \leqslant \xi_1$ with ξ_1 arbitrary, $|\arg \epsilon| \leqslant \delta$, provided $\xi\epsilon^{2/3} \in D$. This last condition can be satisfied by limiting the size of ϵ by the inequality $|\epsilon| \leqslant \epsilon_0$ with a suitable value of ϵ_0 that depends on D and on ξ_1. Now set $\xi_1 = x_1$, then the corresponding solution

$$Y(x,\epsilon) = \begin{pmatrix} 1 & 0 \\ 0 & \epsilon^{1/3} \end{pmatrix} W(x\epsilon^{-2/3},\epsilon)$$

of (6.2-1) is also bounded, as long as $|x| \leqslant x_1\epsilon^{2/3}$. The continuation of the particular solutions called y in Theorem 6.2-1 can be written

$$y(x,\epsilon) = Y(x,\epsilon)Y^{-1}(x_1\epsilon^{-2/3},\epsilon)y(x_1,\epsilon).$$

By Theorem 6.2-1 the last right-hand factor remains bounded, as $\epsilon \to 0+$. The other two factors in the right-hand member are bounded for $|x| \leqslant \epsilon^{2/3}x_1$, by what has just been proved, and, thus, the lemma has been established.

After these preparations, the solution of the differential equation (6.2-1) can be tackled for the problem at hand. The matrices U and V of formulas (6.1-28) are here solutions of

$$U' = \epsilon^{-1}AU, \qquad V' = \epsilon^{-1}(A + \Omega)V \qquad (6.2\text{-}15)$$

with A as in (6.2-2), and $\Omega \sim 0$ for $|x| \leqslant x_0$, $\epsilon \to 0$. U and V will be chosen as matrix solutions that have the asymptotic representations described in Theorem 6.2-1. Remember that U and V have the same expansions in the regions where that theorem applies, because the *formal* solutions of the two differential equations are the same. To fix the ideas, assume that x is in the sector

$$\Sigma := \left\{ -\frac{\pi}{3} < \arg x < \pi \right\}.$$

Let

$$S := \left\{ x \mid |x| \leqslant \epsilon^{2/3}x_1 \right\}.$$

Then

$$U(x,\epsilon) = \hat{U}(x,\epsilon)\exp\left\{ \epsilon^{-1}\frac{2}{3}x^{3/2}\begin{pmatrix} 1 & 0 \\ 0 & -1 \end{pmatrix} \right\},$$
$$V(x,\epsilon) = \hat{V}(x,\epsilon)\exp\left\{ \epsilon^{-1}\frac{2}{3}x^{3/2}\begin{pmatrix} 1 & 0 \\ 0 & -1 \end{pmatrix} \right\} \qquad (6.2\text{-}16)$$

with

$$U(x,\epsilon) \sim V(x,\epsilon) \sim \hat{Y}(x,\epsilon) \qquad (6.2\text{-}17)$$

uniformly in $\Sigma - S$.

Now consider the representation

$$R(x,\epsilon) = \int^x V(x,\epsilon)V^{-1}(t,\epsilon)\Omega(t,\epsilon)\epsilon^{-1}U(t,\epsilon)U^{-1}(x,\epsilon)\,dt \qquad (6.2\text{-}18)$$

for the matrix R in (6.1-26). By formula (6.1-29) it is a solution of the differential equation (6.1-26), no matter how the path of integration is chosen, provided its initial point is independent of x. It is not even necessary to choose the same initial point in each of the four integrals that are the entries of the matrix in (6.2-18). These four paths of integration will be denoted by γ_x^{jk}, $j, k = 1, 2$. As always in the asymptotic theory of differential equations, the proper choice of the paths is crucial and must be described with care.

At $x = 0$ the series $\hat{Y}$ in (6.2-17) becomes meaningless, but if $\hat{U}, \hat{V}$ are *defined* by (6.2-16), then these functions are bounded in S, as well, thanks to Lemma 6.2-1; and (6.2-16) is then valid in all of Σ, including $x = 0$, with functions $\hat{U}, \hat{V}$ that are bounded, as $\epsilon \to 0$, together with their inverses. To rewrite (6.2-18) in a more explicit but not excessively long form, the abbreviations

$$g(x, t, \epsilon) := \exp\left\{ \frac{2}{3\epsilon} (x^{3/2} - t^{3/2}) \right\}, \qquad G = \begin{pmatrix} g & 0 \\ 0 & -g \end{pmatrix}$$

and

$$F = \hat{V}^{-1} \epsilon^{-1} \Omega \hat{U}, \qquad F = \begin{pmatrix} f_{11} & f_{12} \\ f_{21} & f_{22} \end{pmatrix} \qquad (6.2\text{-}19)$$

are introduced. Then (6.2-18) becomes

$$R(x, \epsilon) = \hat{V}(x, \epsilon) \int^x G(x, t, \epsilon) F(t, \epsilon) G^{-1}(x, t, \epsilon) \, dt \, \hat{U}^{-1}(x, t)$$

$$= \hat{V}(x, \epsilon) \begin{bmatrix} \int_{\gamma_x^{11}} f_{11}(t, \epsilon) \, dt & \int_{\gamma_x^{12}} g^2(x, t, \epsilon) f_{12}(t, \epsilon) \, dt \\ \int_{\gamma_x^{21}} g^{-2}(x, t, \epsilon) f_{21}(t, \epsilon) \, dt & \int_{\gamma_x^{22}} f_{22}(t, \epsilon) \, dt \end{bmatrix}$$

$$\times \hat{U}^{-1}(x, \epsilon).$$

$$(6.2\text{-}20)$$

The paths γ_x^{jk} in the t-plane will be chosen as follows: γ_x^{11} and γ_x^{22} are straight segments from the origin to x. To describe γ_x^{jk}, $j \neq k$, consider the mapping $\mathcal{M}$ of x to s defined by $s = x^{3/2}$. Then

$$\mathcal{M}(\Sigma) = \left\{ s \mid 0 \leqslant |s| \leqslant x_0^{3/2}, \; -\tfrac{1}{2}\pi < \arg x < \tfrac{3}{2}\pi \right\}. \qquad (6.2\text{-}21)$$

If the variable of the integrals in (6.2-20) is changed to $\sigma = t^{3/2}$ the integrals in the off-diagonal entries become

$$\int_{\delta_s^{12}} e^{4(s-\sigma)/3\epsilon} f_{12}(\sigma^{2/3}, \epsilon) \frac{2}{3} \sigma^{-1/3} \, d\sigma,$$

$$\int_{\delta_s^{21}} e^{4(\sigma-s)/3\epsilon} f_{21}(\sigma^{2/3}, \epsilon) \frac{2}{3} \sigma^{-1/3} \, d\sigma$$

with paths of integration δ_s^{jk} which are the images of γ_x^{jk}. Every point $\sigma = s$ of the sector $\mathcal{M}(\Sigma)$ on the Riemann surface over the σ-plane can be reached from $\sigma = x_0^{3/2}$ by a path in $\mathcal{M}(\Sigma)$ along which $\operatorname{Re}\sigma$ decreases. Let δ_s^{12} be such a path and define γ_x^{12} as the preimage of δ_s^{12} in the t-plane. Also, every such point $\sigma = s$ can be reached from $\sigma = -x_0^{3/2}$ by a path in $\mathcal{M}(\Sigma)$ along which $\operatorname{Re}\sigma$ increases. Take γ_x^{21} as the preimage of that path. As F in (6.2-19) is asymptotic to zero, uniformly in Σ, and g^2 as well as g^{-2} are bounded along the paths γ_x^{12} and γ_x^{21}, respectively, the matrix R in (6.2-20) is also asymptotic to zero.

This completes the proof that a differential equation which is *formally* equivalent to

$$\epsilon y' = \begin{pmatrix} 0 & 1 \\ x & 0 \end{pmatrix} y \tag{6.2-22}$$

in a disk $|x| \leqslant x_0$ is asymptotically equivalent to it in the part of the disk that lies in the sector $-\frac{1}{3}\pi < \arg x < \pi$. Instead of this sector each of the two sectors $-\frac{1}{3}\pi + j\frac{2}{3}\pi < \arg x < \pi + j\frac{2}{3}\pi, j = 1, 2$, could have been made the basis of the reasoning. Therefore, asymptotic equivalence exists in these sectors, as well. The *formal* transformation into (6.2-22) is the same in each of these three sectors, but the three transformation matrices asymptotically represented by the formal series may well be different from each other, although, of course, two of them whose domains of asymptotic representation by the series have a sector in common must be asymptotically equal in that sector. The results of Chapter V and VI, so far, concerning asymptotic reduction of differential equations to Airy's equation are summarized below.

Consider a vectorial differential of the form

$$\epsilon^h y' = A(x, \epsilon) y \tag{6.2-23}$$

of order two whose coefficient matrix A is holomorphic in both variables at $x = \epsilon = 0$. There exists then a transformation of the variables of the form

$$y = e^{1/2\epsilon^h} \left[\int_0^x \operatorname{tr} A(t, 0) \, dt \right] T(x) u,$$

$$x = \phi(\xi),$$

which takes the differential equation (6.2-23) into one of the same form with a leading coefficient matrix

$$\begin{pmatrix} 0 & \xi^\mu \\ \xi^\nu & 0 \end{pmatrix}$$

with non-negative integers μ, ν. The matrix T and the function ϕ can be easily constructed. If, in particular, $h = 1$, $\mu = 0$, $\nu = 1$, then the equation is asymptotically reducible to Airy's equation, as described below.

Theorem 6.2-2. *The formal differential equation*

$$\epsilon y' = \left(\sum_{r=0}^\infty A_r(x)\epsilon^r \right) y \tag{6.2-24}$$

with

$$A_0(x) = \begin{pmatrix} 0 & 1 \\ x & 0 \end{pmatrix} \tag{6.2-25}$$

is formally equivalent to

$$\epsilon z' = \begin{pmatrix} 0 & 1 \\ x & 0 \end{pmatrix} z \tag{6.2-26}$$

in $|x| \leq x_1$ $(x_1 > 0$ *a constant). If*

$$A(x,\epsilon) \sim \sum_{r=0}^{\infty} A_r(x)\epsilon^r \qquad \text{as} \quad \epsilon \to 0+, \tag{6.2-27}$$

uniformly in $|x| \leq x_1$, *then in each of the three sectorial domains*

$$S_j = \{ x \,|\, |x| \leq x_0, \; -\tfrac{1}{3}\pi + j\tfrac{2}{3}\pi < \arg x < \pi + j\tfrac{2}{3}\pi \}, \tag{6.2-28}$$

$(j \bmod 3$, *and* x_0 *a positive constant) the two differential equations* (6.2-24), (6.2-26) *are asymptotically equivalent, by means of transformations*

$$y = P^j(x,\epsilon)z \qquad (j \bmod 3),$$

with

$$P^j(x,\epsilon) \sim \sum_{r=0}^{\infty} P_r(x)\epsilon^r \qquad \text{as} \quad \epsilon \to 0+,$$

uniformly in every compact subset of S_j *(which may contain the origin). The* P_r *are holomorphic in* $|x| \leq x_1, \, 0 < x \leq x_1$.

6.3. Differential Equations Reducible to Weber's Equation

In Theorem 5.2-2 it was stated that a fairly large class of differential equations is formally equivalent to

$$\epsilon y' = \begin{bmatrix} 0 & 1 \\ x^2 + \epsilon \sum_{r=0}^{\infty} b_r \epsilon^r & 0 \end{bmatrix} y \tag{6.3-1}$$

with b_r independent of x and ϵ. To pass from formal equivalence to a statement on asymptotic equivalence, let b denote one of the infinitely many scalar functions of ϵ alone which are asymptotically represented by $\sum_{r=0}^{\infty} b_r \epsilon^r$, i.e.,

$$b(\epsilon) \sim \mathbf{b}(\epsilon) := \sum_{r=0}^{\infty} b_r \epsilon^r, \qquad \epsilon \to 0 \quad \text{in} \quad |\arg\epsilon| \leq \epsilon_0. \tag{6.3-2}$$

The question then is whether the originally given differential equation, which was *formally* equivalent to (6.3-1), is asymptotically equivalent to

$$\epsilon z' = \begin{pmatrix} 0 & 1 \\ x^2 + \epsilon b(\epsilon) & 0 \end{pmatrix} z = : A(x,\epsilon)z \tag{6.3-3}$$

in a domain containing $x = 0$, at least as a boundary point. The method used to answer this question is the same as in Section 6.2; the explanations will therefore be kept shorter. No use will be made of the fact that equation (6.3-1) is equivalent to a form of the so-called Weber equation and that the solutions are the well-known parabolic cylinder functions.

As was shown in Section 6.1, any differential equation formally equivalent to (6.3-1) is asymptotically equivalent to a differential equation of the form

$$\epsilon y' = (A(x,\epsilon) + \Omega(x,\epsilon)) y, \tag{6.3-4}$$

where A is as in (6.3-3), and $\Omega \sim 0$, as in (6.2-3).

The analog of Lemma 6.2-1 is

Lemma 6.3-1. *The differential equation* (6.3-1) *(and therefore* (6.3-4)*) has a formal BFM-solution*

$$Y(x,\epsilon) = \hat{Y}(x,\epsilon)\exp\left\{ \frac{1}{2\epsilon} x^2 \begin{pmatrix} 1 & 0 \\ 0 & -1 \end{pmatrix} \right\} \tag{6.3-5}$$

where

$$\hat{Y}(x,\epsilon) = \sum_{r=0}^{\infty} Y_r(x)\epsilon^r,$$

with

$$Y_0(x) = \begin{bmatrix} x^{(b_0-1)/2} & x^{-(b_0+1)/2} \\ x^{(b_0+1)/2} & -x^{-(b_0-1)/2} \end{bmatrix},$$

$$Y_r(x) = O(x^{-2r}), \quad r > 0, \quad as \quad x \to 0. \tag{6.3-6}$$

PROOF. The transformation

$$y = \begin{pmatrix} 1 & 1 \\ x & -x \end{pmatrix} v$$

diagonalizes the leading coefficient in (6.3-1). One finds the new formal differential equations

$$\epsilon v' = \left[\begin{pmatrix} x & 0 \\ 0 & -x \end{pmatrix} + \frac{\epsilon}{2x} \begin{pmatrix} b_0 + 1 & b_0 - 1 \\ -b_0 + 1 & -b_0 + 1 \end{pmatrix} \right.$$
$$\left. + \frac{\epsilon}{2x} \begin{pmatrix} 1 & 1 \\ -1 & -1 \end{pmatrix} \sum_{r=1}^{\infty} b_r \epsilon^r \right] v.$$

In the notation of Theorem 2.3-1 this means that

$$A_0(x) = J(x) = x\begin{pmatrix} 1 & 0 \\ 0 & -1 \end{pmatrix}, \qquad A_1(x) = \frac{1}{2x}\begin{pmatrix} b_0 - 1 & b_0 + 1 \\ -b_0 + 1 & -b_0 - 1 \end{pmatrix},$$

$$A_r(x) = \frac{b_r}{2x}\begin{pmatrix} 1 & 1 \\ -1 & -1 \end{pmatrix}, \qquad r > 1.$$

Now one proceeds exactly as in the proof of (6.2-7) and finds, for the present differential equation, that

$$B_r(x) = O(x^{-2r+1}), \qquad P_r(x) = O(x^{-2r}), \qquad r > 0, \quad x \to 0.$$

Integration of the completely diagonalized differential equation then leads to the formulas (6.3-6), and the lemma is proved.

Let the function q be defined [differently from its meaning in (6.2-11)] by

$$q(x, \epsilon) = x^2/2\epsilon. \tag{6.3-7}$$

The two columns of the formal matrix solution (6.3-5) may be written, in self-explanatory notation, as

$$\mathbf{y}_\pm(x, \epsilon) := \mathbf{y}_\pm(x, \epsilon)e^{\pm x^2/2}. \tag{6.3-8}$$

The general asymptotic existence theorem in regions that expand to a turning point, as $\epsilon \to 0$—to be proved in Section 7.7—which is the basis of Theorem 6.2-1 also implies the theorem below.

Theorem 6.3-1. *Let*

$$\Sigma_j := \left\{ x \mid -\frac{\pi}{4} + j\frac{\pi}{2} < \arg x < \frac{\pi}{4} + j\frac{\pi}{2} \right\}, \qquad j \bmod 4. \tag{6.3-9}$$

For each j, the one of the two formal solutions (6.3-8) of equation (6.3-1) for which the exponent $\pm x^2/2\epsilon$ has negative real part in Σ_j is an asymptotic representation of a vector solution y of (6.3-3), as $\epsilon \to 0+$, valid in that part of the annulus

$$\epsilon^{1/2}x_1 \leqslant |x| \leqslant x_0 \tag{6.3-10}$$

that is outside $\overline{\Sigma}_j$. Here x_0, x_1 are certain constants. The asymptotic representation is such that

$$\left| \sum_{r=0}^{} y_r(x)\epsilon^r - y(x, \epsilon)e^{\pm q(x, \epsilon)} \right| \leqslant k_N |x|^{-2}\epsilon|^N$$

for all $N \geqslant 0$ in every compact subset of the annulus (6.3-10) that omits a sector containing the closure of Σ_j. The constant k_N is independent of x and ϵ.

The analog of Lemma 6.2-2 holds in the present context. The vector solutions of equation (6.3-4) called y in Theorem 6.3-1 are bounded for $|x| \leqslant \epsilon^{1/2}x_1$, as $\epsilon \to 0+$. The proof is omitted, because it is almost literally the same as for Lemma 6.2-2.

Theorem 6.2-2 corresponds to a very similar theorem for the differential equation of this section:

Theorem 6.3-2.   *The formal differential equation*

$$\epsilon \mathbf{y}' = \left\{ \begin{pmatrix} 0 & 1 \\ x^2 & 0 \end{pmatrix} + \epsilon \sum_{r=0}^{\infty} A_r(x)\epsilon^r \right\} \mathbf{y}$$

is formally equivalent to

$$\epsilon \mathbf{z}' = \begin{bmatrix} 0 & 1 \\ x^2 + \epsilon \sum_{r=0}^{\infty} b_r \epsilon^r & 0 \end{bmatrix} \mathbf{z}$$

in $|x| \leqslant x_1$, where x_1 and b_r are certain constants. If

$$A(x,\epsilon) \sim \sum_{r=0}^{\infty} A_r(x)\epsilon^r \qquad as \quad \epsilon \to 0+$$

and

$$b(\epsilon) \sim \sum_{r=0}^{\infty} b_r \epsilon^r \qquad as \quad \epsilon \to 0+,$$

uniformly in $|x| \leqslant x_1$, then in each of the four sectorial domains

$$\Sigma_j = \{ x/|x| \leqslant x_0, \ -\tfrac{1}{4}\pi + j\tfrac{1}{2}\pi < \arg x < \tfrac{3}{4}\pi + j\tfrac{1}{2}\pi \} \qquad (j \bmod 4)$$

the two differential equations

$$\epsilon y' = \left[\begin{pmatrix} 0 & 1 \\ x^2 & 0 \end{pmatrix} + \epsilon A(x,\epsilon) \right] y,$$

$$\epsilon z' = \begin{pmatrix} 0 & 1 \\ x^2 + \epsilon b(\epsilon) & 0 \end{pmatrix} z$$

are asymptotically equivalent by means of transformations $y = P^j(x,\epsilon)z$ ($j \bmod 4$), with

$$P^j(x,\epsilon) \sim \sum_{r=0}^{\infty} P_r(x)\epsilon^r \qquad in \ D_j \quad as \quad \epsilon \to 0+.$$

The P_r are holomorphic in $|x| \leqslant x_2$ (x_2 a constant), and $P_0(0)$ is invertible. The asymptotic relation is uniform in every compact subset of D_j (which may contain the origin).

The only differences in the proofs of Theorems 6.2-2 and 6.3-2 are caused by the replacement of the factor $x^{3/2}$ in the exponent in formula (6.2-8) by the factor x^2 in the exponent of formula (6.3-8). As a consequence of this difference the sectors of asymptotic validity claimed in the two theorems are not the same. Otherwise, the reasoning is exactly as before, and a repetition of the details is superfluous.

6.4. Uniform Transformations in a Full Neighborhood of a Turning Point

So far, the results stated in this section deserve the designation "uniform transformations" with only partial justification, because they are valid only in certain sectors with vertices at the turning point, not in full neighborhoods of that point. Therefore, the complete asymptotic analysis near the turning point still requires the solution of lateral connection problems. This is particularly awkward when dealing with boundary value problems in which the solution to be found is characterized by values at points in more than one such sector.

In a small number of cases it has been possible to find transformations to decisively simpler problems so that the asymptotic expansions of these transformations are valid in a *full* neighborhood of the turning point. This includes, in particular, the second-order differential equations that are formally reducible to Airy's equation or to Weber's equation. However, as Sibuya has shown, the existence of such simplifications having asymptotic expansions in a full disk about the turning point cannot be taken for granted and may even be exceptional.

Here is a simple example due to Sibuya which illustrates well the inherent difficulties of this problem. Let

$$A(x,\epsilon) = \begin{pmatrix} x & \alpha(\epsilon) \\ 0 & -x \end{pmatrix} \tag{6.4-1}$$

with

$$\alpha(\epsilon) \sim 0 \quad \text{as} \quad \epsilon \to 0+, \quad \alpha(0) = 0, \tag{6.4-2}$$

and consider the differential equation

$$\epsilon y' = A(x,\epsilon). \tag{6.4-3}$$

If there exists a transformation

$$y = P(x,\epsilon)z$$

which takes (6.4-3) into

$$\epsilon z' = A(x,0)z \tag{6.4-4}$$

in $|x| \leq x_0$, then P must satisfy the differential equation

$$\epsilon P' = A(x,\epsilon)P - PA(x,0). \tag{6.4-5}$$

Only solutions P that have an asymptotic series in powers of ϵ are of interest here. If such a series $\sum_{r=0}^{\infty} P_r(x)\epsilon^r$ exists one must have

$$P_0(x) = \begin{bmatrix} P_{11}^0(x) & 0 \\ 0 & P_{22}^0(x) \end{bmatrix}, \quad P_r(x) = 0, \quad r > 0, \tag{6.4-6}$$

as one sees by substituting the series into (6.4-5). The functions P_{11}^0, P_{22}^0

must be holomorphic and nowhere zero in $|x| \leqslant x_0$, but are otherwise arbitrary. An elementary calculation shows that the general solution

$$P = \begin{pmatrix} P_{11} & P_{12} \\ P_{21} & P_{22} \end{pmatrix}$$

of (6.4-5) is

$$P_{11}(x,\epsilon) = \alpha(\epsilon) \int_{\gamma_1(x)} e^{-\xi^2/\epsilon}\, d\xi + k_2(\epsilon),$$

$$P_{12}(x,\epsilon) = \epsilon^{-1}\alpha(\epsilon)k_3(\epsilon) \int_{\gamma_2(x)} e^{(x^2-\xi^2)/\epsilon}\, d\xi + k_4(\epsilon)e^{x^2/\epsilon}, \qquad (6.4\text{-}7)$$

$$P_{21}(x,\epsilon) = k_1(\epsilon)e^{-x^2/\epsilon},$$

$$P_{22}(x,\epsilon) = k_3(\epsilon),$$

with k_1, k_2, k_3, k_4 arbitrary functions of ϵ and $\gamma_1(x), \gamma_2(x)$ paths in the ξ-plane that end at $\xi = x$.

If this expression is to have an asymptotic expansion that satisfies (6.4-6) in a full neighborhood of $x = 0$, and for which $P_0(0)$ is invertible, it is necessary that

$$k_1(\epsilon) \equiv 0, \qquad k_2(\epsilon) \equiv k_2(0), \qquad k_3(\epsilon) \equiv k_3(0) \neq 0,$$

and that P_{12}, which is equal to

$$P_{12}(x,\epsilon) = e^{x^2/\epsilon} \left\{ \epsilon^{-1}\alpha(\epsilon)k_3(0) \int_{\gamma_2(x)} e^{-\xi^2/\epsilon}\, d\xi + k_4(\epsilon) \right\}$$

be asymptotic to zero for all small x. The same must be true for

$$P_{12}(x,\epsilon) - P_{12}(-x,\epsilon) = e^{x^2/\epsilon}\epsilon^{-1}\alpha(\epsilon)k_3(0) \int_{-x}^{x} e^{-\xi^2/\epsilon}\, d\xi$$

$$= e^{x^2/\epsilon}\epsilon^{-1}\alpha(\epsilon)k_3(0)2 \int_0^x e^{-\xi^2/\epsilon}\, d\xi.$$

Now,

$$\int_0^x e^{-\xi^2/\epsilon}\, d\xi = \sqrt{\epsilon} \int_0^{x/\sqrt{\epsilon}} e^{-s^2}\, ds,$$

so that

$$\lim_{\epsilon \to 0+} \frac{1}{\sqrt{\epsilon}} \int_0^x e^{-\xi^2/\epsilon}\, d\xi = \sqrt{\pi}$$

and, therefore, $P_{12}(x,\epsilon) - P_{12}(-x,\epsilon)$ is asymptotic to zero for $x > 0$ if, and only if

$$\alpha(\epsilon) = O(e^{-\beta/\epsilon}) \qquad \text{as} \quad \epsilon \to 0+$$

for some $\beta > 0$. In this example nothing was assumed of $\alpha(\epsilon)$ beyond the property of being asymptotic to zero. Therefore, an asymptotic theory that applies to differential equations whose coefficient matrix is known only through its asymptotic expansion cannot always admit simplifications with asymptotic expansion uniformly valid in a full neighborhood of a turning point. The problem is, however, not completely settled by this remark. For, uniform simplifications *do* exist in some important cases, as will be shown here. Also, in most applications one starts from differential equations whose coefficient matrix is *holomorphic* in ϵ at $\epsilon = 0$, so that its series in powers of ϵ converge, instead of being only an asymptotic representation. It is thinkable that the divergent series which enter the formulas for asymptotic solution in this case have special properties which allow estimates of the asymptotic error term better than what can be said when it is only known that $A(x,\epsilon) \sim \sum_{r=0}^{\infty} A_r(x)\epsilon^r$. As A is not holomorphic in ϵ at $\epsilon = 0$ in the preceding counterexample, it sheds no light on this question.

6.5. Complete Reduction to Airy's Equation

In this section the given differential equation of order two is assumed to have the form

$$\epsilon y' = A(x,\epsilon)y,$$

where

$$A(x,\epsilon) \sim \sum_{r=0}^{\infty} A_r(x)\epsilon^r \qquad \text{as} \quad \epsilon \to 0+,$$

uniformly in $|x| \leqslant x_0$, and

$$A_0(x) = \begin{pmatrix} 0 & 1 \\ x & 0 \end{pmatrix}.$$

The parameter ϵ is taken positive, but the arguments that follow are not essentially more difficult if ϵ is allowed to tend to zero in a narrow sector of the ϵ-plane and results in such a sector are desired.

Thanks to Theorem 5.2-1 and Lemma 6.1-1 the given differential equation may be taken, without losing generality, to have the form

$$\epsilon y' = \left[\begin{pmatrix} 0 & 1 \\ x & 0 \end{pmatrix} + \Omega(x,\epsilon) \right] y \tag{6.5-1}$$

with

$$\Omega \sim 0 \qquad \text{as} \quad \epsilon \to 0+, \quad \text{uniformly in} \quad |x| \leqslant x_0. \tag{6.5-2}$$

The aim of this section is to find a transformation

$$y = Pz \tag{6.5-3}$$

which takes equation (6.5-1) into

$$\epsilon z' = \begin{pmatrix} 0 & 1 \\ x & 0 \end{pmatrix} z \tag{6.5-4}$$

and such that P possesses an asymptotic series in powers of ϵ valid, as $\epsilon \to 0+$, uniformly in a *full* neighborhood of $x = 0$.

The abbreviations below are useful for the description of the arguments that follow. The letter j is taken modulo 3.

$$S = \{x/|x| \leqslant x_0\},$$
$$l_j = \{x/|x| \leqslant x_0, \arg x = -\pi + \tfrac{2}{3}\pi j\},$$
$$S_j = \{x/|x| \leqslant x_0, \arg l_j < \arg x < \arg l_{j+1}\},$$
$$\Sigma_j = \{x/|x| \leqslant x_0, \arg l_{j-1} < \arg x < \arg l_{j+1}\}.$$

(6.5-5)

Figure 6.1 may be helpful.

The asymptotic properties of the simple equation (6.5-4) are known in great detail. Let u_1 be the so-called Airy function. It is a particular solution to Airy's equation $d^2u/dx^2 = xu$. By a simple symmetry consideration one finds two more solutions, u_2 and u_3, and, as is proved in the theory of Airy's equation (see, e.g., [90] or [61]), these three solutions are connected by the identity

$$u_1 + u_2 + u_3 = 0.$$

The three vector-valued functions z_j ($j \bmod 3$) defined by

$$z_j(x,\ ) = \begin{bmatrix} u_j(x\epsilon^{-2/3}) \\ \epsilon^{1/3}u_j'(x\epsilon^{-2/3}) \end{bmatrix}$$

(6.5-6)

are then solutions of (6.5-4), as is easily verified. They, too, are related by the formula

$$z_1 + z_2 + z_3 = 0.$$

(6.5-7)

According to Theorem 6.2-2, there exist three matrix-valued functions P^j of x and ϵ such that

$$y = P^j z \qquad (j \bmod 3)$$

(6.5-8)

takes equation (6.5-1) into (6.5-4) and that

$$P^j \sim I \qquad \text{for} \quad x \in \Sigma_j.$$

(6.5-9)

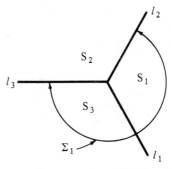

Figure 6.1. Sectors of asymptotic validity for Airy's equation.

The construction of a matrix P with the desired asymptotic properties depends on asymptotic properties of certain ones of the solutions $P^j z_k$ of (6.5-1). These properties will be stated and explained now, but the proofs will be postponed to the end of this section.

Lemma 6.5-1. *There exist three scalar functions k_j of ϵ such that $k_j(\epsilon) \sim 1$, as $\epsilon \to 0+$, and that the solutions*

$$y_j := k_j P^j z_j \qquad (6.5\text{-}10)$$

of (6.5-1) are linked by the same connection formula as the three z_j, i.e., that

$$y_1 + y_2 + y_3 = 0. \qquad (6.5\text{-}11)$$

Let

$$Z_{j,j+1} := (z_j, z_{j+1}) \qquad (6.5\text{-}12)$$

be the matrix solution of (6.5-4) with columns z_j, z_{j+1}. The solutions u_j, u_{j+1} of Airy's equation are known to be independent. Therefore, by (6.5-6), the matrix $Z_{j,j+1}$ is a *fundamental* solution. Formula (6.5-7) implies that

$$Z_{j+1,j+2} = Z_{j,j+1} K_j, \qquad K = \begin{pmatrix} 0 & -1 \\ 1 & -1 \end{pmatrix}. \qquad (6.5\text{-}13)$$

Thanks to (6.5-7) and (6.5-11), the relation (6.5-13) is the same as the analogous one for (6.5-1), namely,

$$Y_{j+1,j+2} = Y_{j,j+1} K \qquad (6.5\text{-}14)$$

with

$$Y_{j,j+1} := (y_j, y_{j+1}). \qquad (6.5\text{-}15)$$

If the matrix Q^j is defined by

$$Q^j := Y_{j,j+1} Z_{j,j+1}^{-1}, \qquad (6.5\text{-}16)$$

formulas (6.5-13) and (6.5-14) imply that

$$Q^{j+1} = Q^j,$$

so that Q^j does not depend on j, and one can write

$$Q := Q^j. \qquad (6.5\text{-}17)$$

The transformation

$$y = Qz \qquad (6.5\text{-}18)$$

takes every solution of (6.5-4) into one of (6.5-1). It will now be proved that Q is a matrix with the property desired in this section, namely, $Q \sim I$ in all of S. To do this, the asymptotic properties of the z_j have to be used. These are simple consequences of the theory of Airy's function and formula (6.5-6), or they can be derived from the general theory. Let

$$q_j(x,\epsilon) := \epsilon^{-1} \tfrac{2}{3} x^{3/2}, \qquad x \in S - l_{j-1}. \qquad (6.5\text{-}19)$$

with the understanding that the determination of the two-valued function in the right-side member of (6.5-19) is to be the one that has *negative* real part in S_j and is continued analytically into $S - l_{j-1}$. Observe that q_j is not defined for $x \in l_{j-1}$. Now, define $\hat{z}_j$ by

$$z_j = \hat{z}_j e^{q_j}, \qquad x \in S - l_{j-1}. \tag{6.5-20}$$

The only important asymptotic property of $\hat{z}_j$ in the present context is that it is uniformly bounded in every proper subsector of $S - l_{j-1}$. Let $\hat{y}_j$ be, analogously, defined by

$$y_j = \hat{y}_j e^{q_j}, \qquad x \in S - l_{j-1} \tag{6.5-21}$$

then the lemma below is true, which will be proved at the end of this section.

Lemma 6.5-2

$$\hat{y}_j \sim \hat{z}_j \qquad for \quad x \in S - l_{j-1}.$$

By insertion of (6.5-20) and (6.5-21) into (6.5-12) and (6.5-15), and then into (6.5-16), it is seen that Lemma 6.5-2 implies

$$Q^j \sim I \qquad \text{in } \Sigma_j.$$

As $Q = Q^j$ for all j and $\Sigma_1 \cup \Sigma_2 \cup \Sigma_3 = S$, the aim of this section has been reached, and the theorem below is proved.

Theorem 6.5-1. *Let the coefficient matrix A of the differential equation of order two,*

$$\epsilon y' = A(x, \epsilon) y,$$

be holomorphic for x in a neighborhood of $x = 0$ and for $0 < \epsilon \leqslant \epsilon_0$, and let it have there the uniformly asymptotic expansion

$$A(x, \epsilon) \sim \sum_{r=0}^{\infty} A_r(x) \epsilon^r \qquad as \quad \epsilon \to 0 +$$

with

$$A_0(x) = \begin{pmatrix} 0 & 1 \\ x & 0 \end{pmatrix}.$$

Then there exists a matrix P holomorphic for $|x| \leqslant x_0, 0 < \epsilon \leqslant \epsilon$, which has the asymptotic expansion

$$P(x, \epsilon) \sim \sum_{r=0}^{\infty} P_r(x) \epsilon^r \qquad as \quad \epsilon \to 0 +,$$

uniformly valid in $|x| \leqslant x_1$ for all $x < x_0$, and such that the transformation takes the given differential equation into

$$\epsilon z' = A_0(x) z.$$

PROOF OF LEMMA 6.5-1. Since the three solutions $P^j z_j$ of the second-order differential equation (6.5-1) must be dependent, the existence of three numbers k_j independent of x such that

$$k_1 P^1 z_1 + k_2 P^2 z_2 + k_3 P^3 z_3 = 0 \qquad (6.5\text{-}22)$$

is clear. Regarding this relation as a set of two homogeneous linear algebraic equations for the scalars k_j one sees that their general solution is

$$k_1 = r \det(P^2 z_2, P^3 z_3), \qquad k_2 = r \det(P^3 z_3, P^1 z_1),$$
$$k_3 = r \det(P^1 z_1, P^2 z_2), \qquad (6.5\text{-}23)$$

where r is an arbitrary scalar. The k_j do not depend on x, therefore, the right members can be evaluated at $x = 0$. This point belongs to all Σ_j so that $P^j|_{x=0} \sim I$, for all j. The vectors z_j remain bounded at $x = 0$, as $\epsilon \to 0+$, according to (6.5-6), and therefore

$$\det(P^j z_j, P^{j+1} z_{j+1})|_{x=0} \sim \det(z_j, z_{j+1})|_{x=0}.$$

The right member of this relation is independent of j, because of (6.5-7), and is different from zero for $\epsilon > 0$. With the choice

$$r = 1/\det(z_j, z_{j+1})|_{x=0},$$

the k_j in (6.5-23) have the required property

$$k_j(\epsilon) \sim 1 \qquad \text{as} \quad \epsilon \to 0+. \qquad (6.5\text{-}24)$$

PROOF OF LEMMA 6.5-2. By (6.5-10), (6.5-20), and (6.5-21),

$$\hat{y}_j = k_j P^j \hat{z}_j.$$

Since $k_j \sim 1$ and $P^j \sim I$ in Σ_j, one has $\hat{y}_j \sim \hat{z}_j$ in Σ_j. To prove this relation in Σ_{j+1}, as well, represent y_j as a linear combination of the two solutions $P^{j+1} z_j$, $P^{j+1} z_{j+1}$ of equation (6.5-1). Then there exist two scalars, f, g, depending on ϵ only, such that

$$y_j = f P^{j+1} z_j + g P^{j+1} z_{j+1}. \qquad (6.5\text{-}25)$$

This identity can be used to calculate f and g, and the choice of x in the calculation is arbitrary. Let $x = x_j$ be the point on the bisector of S_j at distance x_0 from 0. At that point $P^{j+1} \sim I$, and $\hat{z}_j$ as well as $\hat{z}_{j+1}$ are bounded. A short calculation using Cramer's rule, formulas (6.5-10), (6.5-20), and (6.5-24) shows that

$$f(\epsilon) \sim 1, \qquad g(\epsilon) = O(e^{2q(x_j, \epsilon)}) \qquad \text{as} \quad \epsilon \to 0+. \qquad (6.5\text{-}26)$$

If this result is inserted in (6.5-25) that formula becomes an asymptotic representation of y_j valid in all of Σ_{j+1}. Division by e^q takes it into an expansion of the form

$$\hat{y}_j = f P^{j+1} \hat{z}_j + O\{\exp(2q_j(x_j, \epsilon) - 2q_j(x, \epsilon))\}, \qquad (6.5\text{-}27)$$

valid for all $x \in \Sigma_{j+1}$. Here, the fact that $q_{j+1} = -q_j$ in Σ_{j+1} has been used. Thanks to the choice of the point x_j, one has $\operatorname{Re} q_j(x_j, \epsilon) < \operatorname{Re} q_j(x, \epsilon)$ for all x with $|x| < x_0$. This means that the last term in (6.5-27) is asymptotic to zero for $x \in \Sigma_{j+1}$. Finally, remembering that $f \sim 1$ and $P^{j+1} \sim I$ in Σ_{j+1}, (6.5-27) is seen to reduce to $\hat{y}_j \sim \hat{z}_j$ in Σ_{j+1}. But $\Sigma_j \cup \Sigma_{j+1} = S - l_{j-1}$, and, thus, the proof of the lemma is complete.

6.6. Reduction to Weber's Equation in Wider Sectors

Again, the differential equation is of order two and of the form

$$\epsilon y' = A(x, \epsilon) y$$

with

$$A(x, \epsilon) \sim \sum_{r=0}^{\infty} A_r(x)\epsilon^r \qquad \text{as} \quad \epsilon \to 0+,$$

uniformly in $|x| \leq x_0$. Now, however, the leading coefficient is quadratic in x:

$$A_0(x) = \begin{pmatrix} 0 & 1 \\ x^2/4 & 0 \end{pmatrix}. \tag{6.6-1}$$

(The factor $\frac{1}{4}$ is a normalization convenient for comparison with Weber's equation.) The analysis below is largely analogous to that in the preceding section, and the presentation will try to exhibit this analogy clearly.

Here is the notation for the sectors and separation lines of importance in this section.

$$
\left.
\begin{aligned}
S &= \{x/|x| \leq x_0\} \\
l_j &= \left\{x/|x| \leq x_0, \arg x = -\frac{\pi}{4} + (j-1)\frac{\pi}{2}\right\} \\
S_j &= \{x/|x| \leq x_0, \arg l_j < \arg x < \arg l_{j+1}\} \\
\Sigma_j &= \{x/|x| \leq x_0, \arg l_{j-1} < \arg x < \arg l_{j+1}\} \\
T_j &= \{x/|x| \leq x_0, \arg l_{j-1} < \arg x < \arg l_{j+2}\}
\end{aligned}
\right\} \quad (j \bmod 4)
$$

(see Fig. 6.2).

Theorem 5.2-2 states that the given differential equation is *formally* equivalent to one in which A_0 is as in (6.6-1) and

$$A_r = \begin{pmatrix} 0 & 0 \\ \mu_{r-1} & 0 \end{pmatrix}, \qquad r > 0$$

for certain uniquely determined constants μ_{r-1}. It then follows from

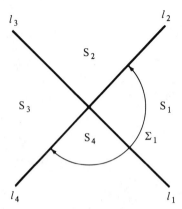

Figure 6.2. Sectors of asymptotic validity for Weber's equation.

Lemma 6.1-1 that the given differential equation may, without loss of generality, be assumed to have already the form

$$\epsilon y' = \left[\begin{pmatrix} 0 & 1 \\ (x^2/4) + \epsilon\mu(\epsilon) & 0 \end{pmatrix} + \Omega(x,\epsilon) \right] y \tag{6.6-2}$$

where

$$\Omega(x,\epsilon) \sim 0 \qquad \text{as} \quad \epsilon \to 0+, \tag{6.6-3}$$

uniformly in S, and μ is one of the infinitely many functions with the asymptotic expansion

$$\mu(\epsilon) \sim \sum_{r=0}^{\infty} \mu_r \epsilon^r \qquad \text{as} \quad \epsilon \to 0+. \tag{6.6-4}$$

Of course, Ω depends on the choice of μ.

The aim of this section is to transform the differential equation (6.6-2) into

$$\epsilon z' = \begin{pmatrix} 0 & 1 \\ (x^2/4) + \epsilon\mu(\epsilon) & 0 \end{pmatrix} z \tag{6.6-5}$$

by means of a transformation

$$y = P(x,\epsilon)z \tag{6.6-6}$$

in which P has an expansion in powers of ϵ, as $\epsilon \to 0+$ valid in sectors larger than Σ_j, for which it is already known to be true by Theorem 6.3-2. Of course, the most desirable result would be to find such a matrix P with an expansion valid in all of S, as was done in the preceding section. However, the reasoning in that section when carried over to the present problem yields only matrices P with an asymptotic expansion in T_j. The difficult problem of replacing T_j by S in this statement has been studied by Roy Lee in Ref. 40. Here, only the weaker result will be proved. A proof of Lee's theorem can be found in the next section.

The several functions P mentioned above have the same series expansion, but with different sectors of validity. Since $\Omega \sim 0$ in S, the condition that

$$P(x, \epsilon) \sim I, \qquad \epsilon \to 0+, \tag{6.6-7}$$

for x in the sector where the particular function P has an asymptotic power series, is not a restriction.

The differential equation (6.6-5) is equivalent to the scalar equation

$$\frac{d^2w}{dt^2} = \left(\frac{t^2}{4} + \mu(\epsilon) \right) w \tag{6.6-8}$$

by means of the transformations

$$z = \left(\frac{w}{\epsilon^{1/2} \, dw/dt} \right), \qquad x = t\epsilon^{1/2}. \tag{6.6-9}$$

Except for its dependence on ϵ, equation (6.6-8) is a form of Weber's equation. Its solutions, the parabolic cylinder functions, are well known (see, e.g., [1]). Some of their properties will be stated now because they are needed later.

If Weber's equation is written, as usual, in the form

$$\frac{d^2u}{dt^2} = \left(\frac{t^2}{4} + p \right) u$$

with a constant p, then there is one particular solution $u = U(t, p)$ that has been studied in great detail. From it one obtains three more solutions by symmetry considerations so that

$$u = u_j(t, p) = U\big(i^{j-1}t, (-1)^{j-1}p\big) \qquad (j \bmod 4) \tag{6.6-10}$$

are four known solutions. By (6.6-9) they correspond to the four vector solutions of (6.6-5),

$$z_j(x, \epsilon) = \begin{bmatrix} U\big(\epsilon^{-1/2}i^{j-1}x, (-1)^{j-1}\mu(\epsilon)\big) \\ \epsilon^{1/2}i^{j-1}U'\big(\epsilon^{-1/2}i^{j-1}x, (-1)^{j-1}\mu(\epsilon)\big) \end{bmatrix} \qquad (j \bmod 4). \tag{6.6-11}$$

To describe the asymptotic properties of z_j, as $\epsilon \to 0+$, define $q_j(x, \epsilon)$ as

$$q_j(x, \epsilon) = \pm x^2/4\epsilon \qquad \text{for} \quad x \in T_j,$$

with the sign chosen so that

$$\operatorname{Re} q_j(x, \epsilon) < 0 \qquad \text{for} \quad x \in S_j. \tag{6.6-12}$$

Recall that $T_j = S_{j-1} \cup l_j \cup S_j \cup l_{j+1} \cup S_{j+1}$, so that S_j is the middle one of the three sectors that make up T_j. Define $\hat{z}_j$ in T_j by

$$z_j(x, \epsilon) = \hat{z}_j(x, \epsilon)e^{q_j(x, \epsilon)}. \tag{6.6-13}$$

Then the well known asymptotic properties of the parabolic cylinder functions imply that $\hat{z}_j$ is bounded in T_j.

Any three of the four functions u_j in (6.6-10) are linearly dependent, since they are solutions of the same linear homogeneous differential equations. The same relations must hold for the z_j. Let a_j, c_j be constants such that

$$a_j z_{j-1} + b_j z_j + c_j z_{j+1} = 0 \qquad (j \bmod 4). \qquad (6.6\text{-}14)$$

The ratios of the a_j, b_j, c_j are well-known simple functions of $\mu(\epsilon)$. They are different from zero if the following condition is satisfied.

Hypothesis VI-1. *Let μ_0 be the leading coefficient in the expansion* (6.6-4). *Then*

$$\mu_0 \neq \tfrac{1}{2}(2m + 1), \qquad m \text{ arc integer.}$$

This is a consequence of formulas (6.7-5) and (6.7-6), below. Under this hypothesis one may choose $b_j = 1$.

It follows from Theorem 6.3-2 that there exist four matrix-valued functions P^j of x and ϵ such that the transformation

$$y = P^j z \qquad (j \bmod 4) \qquad (6.6\text{-}15)$$

takes (6.6-2) into (6.6-5) and that $P^j \sim I$ in Σ_j. The solutions $P^j z_j$ of (6.6-2) also must be linearly related, and so is any set of more than two scalar multiples of these solutions. The lemma to be stated now is the exact analog of Lemma 6.5-1.

Lemma 6.6-1. *Assume Hypothesis VI-1 to be true. For every j (mod 4) there exist three scalar functions, $\alpha_j, \beta_j, \gamma_j$ of ϵ such that the three solutions*

$$\alpha_j P^{j-1} z_{j-1}, \quad \beta_j P^j z_j, \quad \gamma_j P^{j+1} z_{j+1} \qquad (6.6\text{-}16)$$

of (6.6-2) *are linked by the same connection formula as z_{j-1}, z_j, z_{j+1}, i.e., that*

$$a_j \alpha_j P^{j-1} z_{j-1} + \beta_j P^j z_j + c_j \gamma_j P^{j+1} z_{j+1} = 0. \qquad (6.6\text{-}17)$$

Moreover,

$$\alpha_j \sim 1, \qquad \beta_j \sim 1, \qquad \gamma_j \sim 1. \qquad (6.6\text{-}18)$$

The proof of this lemma is postponed to the end of this section. By virtue of Lemma 6.6-1 the solutions (6.6-18)

$$y_{j-1} := \alpha_j P^{j-1} z_{j-1}, \qquad y_j := \beta_j P^j z_j, \qquad y_{j+1} := \gamma_j P^{j+1} z_{j+1} \quad (6.6\text{-}19)$$

satisfy the same linear relation as z_{j-1}, z_j, z_{j+1}, namely

$$a_j y_{j-1} + y_j + c_j y_{j+1} = 0. \qquad (6.6\text{-}20)$$

It must be emphasized that in using the notation of (6.6-19) it has been tacitly assumed that j is kept fixed throughout the remainder of this proof. A completely precise notation would require double indices. To give an example, if $j = 2$, then $y_1 = \alpha_1 P^1 z_1$, but if $j = 1$, then $y_1 = \beta_1 P^1 z_1$. The less precise notation used here has the advantage of emphasizing the close relation between equations (6.6-2) and (6.6-5). Because of Hypothesis VI-1 the matrices

$$Z_{j-1,j} := (z_{j-1}, z_j), \qquad Z_{j,j+1} := (z_j, z_{j+1}) \qquad (6.6\text{-}21)$$

are two fundamental matrix solutions of equation (6.6-5). The constant matrix K_j in the relation

$$Z_{j,j+1} = Z_{j-1,j} K_j \qquad (6.6\text{-}22)$$

which must connect them can easily be calculated from (6.6-14). Since the linear relation (6.6-20) is identical with (6.6-14), one has also

$$Y_{j,j+1} = Y_{j-1,j} K_j. \qquad (6.6\text{-}23)$$

Here,

$$Y_{j,j+1} := (y_j, y_{j+1}), \qquad Y_{j-1,j} := (y_{j-1}, y_j). \qquad (6.6\text{-}24)$$

Formulas (6.6-20) and (6.6-21) imply that

$$Y_{j-1,j} Z_{j-1,j}^{-1} = Y_{j,j+1} Z_{j,j+1}^{-1}. \qquad (6.6\text{-}25)$$

Denote this matrix by Q_j. It has the property that the transformation

$$y = Q_j z$$

takes the differential equation (6.6-2) into (6.6-5).

It will now be proved that the matrix Q_j has the property which is the aim of this section, namely,

$$Q_j \sim I \qquad \text{in } T_j. \qquad (6.6\text{-}26)$$

To this end the asymptotic properties of the solutions y_{j-1}, y_j, y_{j+1} are needed. Let the functions $\hat{y}_{j-1}, \hat{y}_j, \hat{y}_{j+1}$ be defined by

$$y_{j+\nu}(x, \epsilon) = \hat{y}_{j+\nu}(x, \epsilon) e^{q_{j+\nu}(x,\epsilon)}, \qquad x \in T_{j+\nu}, \quad \nu = -1, 0, 1. \qquad (6.6\text{-}27)$$

Then the lemma below (whose proof will also be found at the end of this section) holds.

Lemma 6.6-2

$$\hat{y}_{j+\nu} \sim \hat{z}_{j+\nu} \qquad \text{in } T_{j+\nu}, \quad \nu = -1, 0, 1 \qquad (6.6\text{-}28)$$

Now insert (6.6-13) and (6.6-27) into (6.6-21) and (6.6-24), obtaining

$$Z_{j,j+1} = (\hat{z}_j, \hat{z}_{j+1}) \exp \begin{bmatrix} q_j & 0 \\ 0 & q_{j+1} \end{bmatrix} \qquad \text{in } T_j \cap T_{j+1} \quad (j \bmod 4)$$

$$Y_{j,j+1} = (\hat{y}_{j-1}, \hat{y}_j) \exp \begin{bmatrix} q_{j-1} & 0 \\ 0 & q_j \end{bmatrix} \qquad \text{in } T_{j-1} \cap T_j,$$

$$Y_{j,j+1} = (\hat{y}_j, \hat{y}_{j+1}) \exp \begin{bmatrix} q_j & 0 \\ 0 & q_{j+1} \end{bmatrix} \qquad \text{in } T_j \cap T_{j+1}.$$

Hence, using (6.6-25) and the definition of Q_j,

$$Q_j = Y_{j-1,j} Z_{j-1,j}^{-1} = (\hat{y}_{j-1}, \hat{y}_j)(\hat{z}_{j-1}, \hat{z}_j)^{-1}$$

and also,

$$Q_j = Y_{j,j+1} Z_{j,j+1}^{-1} = (\hat{y}_j, y_{j+1})(\hat{z}_j, z_{j+1})^{-1}.$$

From (6.6-28) and the last two identities it follows that $Q_j \sim I$, both in $T_{j-1} \cap T_j$ and $T_j \cap T_{j+1}$, i.e.,

$$Q_j \sim I \qquad \text{in } T_j.$$

Thus, the aim of this section has been reached and can be stated as a theorem.

Theorem 6.6-1. *Let the coefficient matrix of the differential equation of order two*

$$\epsilon y' = A(x, \epsilon) y, \tag{6.6-29}$$

be holomorphic for x in a neighborhood of $x = 0$ and for $0 < \epsilon \leqslant \epsilon_0$, and let it have there the uniformly asymptotic expansion

$$A(x, \epsilon) \sim \sum_{r=0}^{\infty} A_r(x) \epsilon^r \qquad as \quad \epsilon \to 0+. \tag{6.6-30}$$

with leading term

$$A_0(x) = \begin{pmatrix} 0 & 1 \\ \frac{1}{4} x^2 & 0 \end{pmatrix}. \tag{6.6-31}$$

Let μ_r, $r \geqslant 0$ be the constants introduced in Theorem 5.2-2. Assume Hypothesis VI-1 to be true. Then, corresponding to every scalar function μ of ϵ with the expansion

$$\mu(\epsilon) \sim \sum_{r=0}^{\infty} \mu_r \epsilon^r \qquad as \quad \epsilon \to 0+ \tag{6.6-32}$$

and to every one of the four sectors

$$T_j = \left\{ |x| \leqslant x_0 / - \frac{3}{4} \pi + j \frac{\pi}{2} < \arg x < \frac{3}{4} \pi + j \frac{\pi}{2} \right\} \qquad (j \bmod 4)$$

there exists a matrix-valued function Q_j of x and ϵ, holomorphic for $|x| \leqslant x_0$, $0 < \epsilon \leqslant \epsilon_0$ which has an asymptotic expansion

$$Q_j(x, \epsilon) \sim \sum_{r=0}^{\infty} P_r(x) \epsilon^r \qquad as \quad \epsilon \to 0+, \quad x \in T_j$$

such that the transformation

$$y = Q_j x \tag{6.6-33}$$

takes the differential equation into

$$\epsilon z' = \begin{pmatrix} 0 & 1 \\ \frac{1}{4} x^2 + \epsilon \mu(\epsilon) & 0 \end{pmatrix} z, \tag{6.6-34}$$

PROOF OF LEMMA 6.6-1.   If the relations (6.6-17) are true for one value of x, they are identically valid, since the three vectors in (6.6-16) are solutions of the same differential equation. *From now on it is understood that $x = 0$ in this proof.* The equations (6.6-17) constitute a system of two linear homogeneous scalar equations for the three numbers $\alpha_j, \beta_j, \gamma_j$. Its general solution is

$$\alpha_j = r \det\left(P^j z_j, c_j P^{j+1} z_{j+1}\right)$$
$$\beta_j = r \det\left(c_j P^{j+1} z_{j+1}, a_j P^{j-1} z_{j-1}\right) \qquad (6.6\text{-}35)$$
$$\gamma_j = r \det\left(a_j P^{j-1} z_{j-1}, P^j z_j\right)$$

with an arbitrary constant r. The point $x = 0$ chosen in the above relation belongs to all sectors Σ_j ($j \bmod 4$), hence, $P^j(0,\epsilon) \sim I$, as in (6.6-7), for all j. The a_j, c_j are well known bounded functions of μ, and the z_j ($j \bmod 4$) are bounded for all small ϵ at $x = 0$, by (6.6-11). Hence, the three determinants in (6.6-35) are asymptotic to their values obtained when the matrices P^{j-1}, P^j, P^{j+1} are replaced by I. But then the right members of (6.6-35) reduce to solutions of the system

$$\alpha_j a_j z_{j-1} + \beta_j z_j + \gamma_j c_j z_{j+1} = 0$$

which, by (6.6-14), with $b_j = 1$, are $\alpha_j = \beta_j = \gamma_j = r$. Hence, the solutions (6.6-35) have the required property (6.6-18) if r is taken equal to 1. This proves Lemma 6.6-1.

PROOF OF LEMMA 6.6-2.   Each P^j ($j \bmod 4$) is already known to be asymptotic to I in Σ_j by Theorem 6.3-1. Hence, (6.6-19) and (6.6-18), together with (6.6-13) and (6.6-27) imply that relation (6.6-28) is true in $\Sigma_{j+\nu}$, $\nu = -1, 0, 1$. To prove it in $\Sigma_{j+\nu+1}$ as well, represent $y_{j+\nu}$ as a linear combination of the two solutions $P^{j+\nu+1} z_{j+\nu}, P^{j+\nu+1} z_{j+\nu+1}$ of equation (6.6-2). There exist two scalars, f, g, depending on ϵ only such that

$$y_{j+\nu} = f P^{j+\nu+1} z_{j+\nu} + g P^{j+\nu+1} z_{j+\nu+1}. \qquad (6.6\text{-}36)$$

(The dependence of f and g on $j + \nu$ need not be put in evidence.) The quantities f, g will now be asymptotically calculated from (6.6-30) taken at the point $x = x_{j+\nu}$ on the bisector of $S_{j+\nu}$ at distance x_0 from 0. As $S_{j+\nu} \subset \Sigma_{j+\nu+1}$, one has $P^{j+\nu+1} \sim I$ at $x = x_{j+\nu}$ and $\hat{z}_{j+\nu}$ as well as $\hat{z}_{j+\nu+1}$ are there bounded. By Cramer's rule one finds from (6.6-3), with the help of (6.6-13), (6.6-19), (6.6-27), and Lemma 6.6-1, that

$$f(\epsilon) \sim 1, \qquad g(\epsilon) = O\big(\exp\big[2q_{j+\nu}(x_{j+\nu},\epsilon)\big]\big), \qquad (6.6\text{-}37)$$

as $\epsilon \to 0 +$. The proof now continues literally as in the proof of Lemma 6.5-2, except that j has to be replaced by $j + \nu$, $\nu = -1, 0, 1$, and that $\Sigma_{j+\nu} \cup \Sigma_{j+\nu+1} = T_{j+\nu}$, which is a sector of angle $3\pi/2$, not 2π. The obvious details are omitted.

6.7. Reduction to Weber's Equation in a Full Disk

In Theorem 6.6-1 the differential equation $\epsilon y' = A(x, \epsilon)y$ with the properties described there was transformed into the simpler equation (6.6-34) by means of the transformations (6.6-33), whose coefficient matrices were asymptotic to I in the sectors T_j of angle $3\pi/2$, but not necessarily in a full neighborhood of the turning point at $x = 0$. Now, following R. Lee's method [40], a transformation matrix P with $P \sim I$ for $|x| \leq x_0$ will be constructed. In one important respect the new result is, however, weaker than Theorem 6.6-1: there, μ was allowed to be *any* function with the expansion $\mu(\epsilon) \sim \sum_{r=0}^{\infty} \mu_r \epsilon^r$, as $\epsilon \to 0 +$. The result of this section is true only for a particular choice of this function. On the other hand it is true even without Hypothesis VI-1. As in Section 6.6, the given differential equation may be assumed to already have the form (6.6-2) with *some* function μ satisfying (6.6-4). As this function will not be changed throughout this section it need not be exhibited explicitly in the notation. The function γ introduced in the theorem now to be stated is not, in general, equal to μ.

Theorem 6.7-1. *Consider the differential equation (6.6-29) with the properties described in Theorem 6.6-1. Then there exists a function γ of ϵ with the expansion*

$$\gamma(\epsilon) \sim \sum_{r=0}^{\infty} \mu_r \epsilon^r, \qquad \epsilon \to 0 + \tag{6.7-1}$$

and a matrix P with the asymptotic expansion

$$P(x, \epsilon) \sim I \quad as \quad \epsilon \to 0 + \quad for \quad |x| \leq x_0 \tag{6.7-2}$$

such that the transformation

$$y = Pz \tag{6.7-3}$$

takes the differential equation (6.6-29) into

$$\epsilon z' = \begin{bmatrix} 0 & 1 \\ \dfrac{x^2}{4} + \epsilon\gamma(\epsilon) & 0 \end{bmatrix} z. \tag{6.7-4}$$

The principal effort in the proof of the theorem lies in establishing a lemma analogous to, but stronger than, Lemma 6.6-1. As in Section 6.6, one starts with the four vectors z_j of (6.6-11). They are solutions of the differential equation (6.6-5) but not, in general, of (6.7-4).

In this section the explicit values of the coefficients in the linear relations among the four z_j are needed. In Section 6.6 it was sufficient to know that such relations exist. Two of these relations are

$$z_3 = az_1 + bz_2$$
$$z_4 = cz_1 + dz_2 \tag{6.7-5}$$

with

$$a(\mu) = ie^{\pi i \mu}, \qquad b(\mu) = \frac{\sqrt{2\pi}}{\Gamma(\frac{1}{2} + \mu)} \exp\left\{\pi i\left(\frac{\mu}{2} - \frac{1}{4}\right)\right\}$$

$$c(\mu) = \frac{\sqrt{2\pi}}{\Gamma(\frac{1}{2} - \mu)} \exp\left\{\pi i\left(\frac{\mu}{2} + \frac{1}{4}\right)\right\}, \qquad d(\mu) = -ie^{\pi i \mu} \tag{6.7-6}$$

(see [1]). All the other relations among any three of the four z_j are immediate consequences of the two in (6.7-5) and (6.7-6).

Let P_j be the matrices described in formula (6.6-15). Then

$$w_j = P_j z_j \qquad (j \bmod 4) \tag{6.7-7}$$

are vector solutions of equations (6.6-2), and so are the functions

$$y_j = k_j w_j \tag{6.7-8}$$

for any choice of the four constants k_j ($j \bmod 4$). These k_j may depend on ϵ. By analogy with Lemma 6.6-1 one would like to determine the k_j so that the y_j satisfy the same connection formulas as the z_j in (6.7-5) and (6.7-6). It turns out that this is not possible in this precise sense. The best that can be achieved is that the connection formulas for the y_j are

$$y_3 = a(\gamma)y_1 + b(\gamma)y_2$$
$$y_4 = c(\gamma)y_1 + d(\gamma)y_3 \tag{6.7-9}$$

with a particular function γ of ϵ for which

$$\gamma(\epsilon) \sim \mu(\epsilon) \tag{6.7-10}$$

but not necessarily $\gamma \equiv \mu$.

To prove the fact just claimed, insert (6.7-8) into (6.7-9) and interpret the resulting formulas,

$$k_1 a(\gamma)w_1 + k_2 b(\gamma)w_2 - k_3 w_3 = 0$$
$$k_1 c(\gamma)w_1 + k_2 d(\gamma)w_2 - k_4 w_4 = 0, \tag{6.7-11}$$

as four linear homogeneous scalar equations for the four unknowns k_j. Nontrivial solutions exist if and only if the determinant of that system vanishes. The essential part of the present section is the proof of the following lemma.

Lemma 6.7-1. *There exists a function γ of ϵ with the expansion*

$$\gamma(\epsilon) \sim \sum_{r=0}^{\infty} \mu_r \epsilon^r \tag{6.7-12}$$

such that for $\gamma = \gamma(\epsilon)$ the system (6.7-9) possesses nontrivial solutions k_j ($j \bmod 4$). They have the asymptotic form

$$k_j(\epsilon) \sim 1, \qquad \epsilon \to 0 + . \tag{6.7-13}$$

PROOF. The y_j in (6.7-9) depend on x. However, if equations (6.7-9) are true for one value of x they are true identically, since the y_j are solutions of the same differential equation. Throughout this proof x will be taken as equal to zero. This fact will not be indicated in the notation unless misunderstanding is to be feared. Since $x = 0$ belongs to all sectors Σ_j,

$$P_j|_{x=0} \sim I. \tag{6.7-14}$$

Write

$$z_j = \begin{pmatrix} z_{j1} \\ z_{j2} \end{pmatrix}, \quad w_j = \begin{pmatrix} w_{j1} \\ w_{j2} \end{pmatrix}, \quad \text{etc.}$$

Then formula (6.6-11) shows that $z_{j2} = O(\epsilon^{1/2})$. This means that the coefficients of two out of the four scalar equations (6.7-11) for the k_j vanish as $\epsilon \to 0+$. To avoid this inconvenience one can multiply those two equations by $\epsilon^{-1/2}$, which does not alter the solutions k_j. The determinant of the system is then

$$\Delta(\gamma,\epsilon) := \epsilon^{-1}\det \begin{bmatrix} a(\gamma)w_{11} & b(\gamma)w_{21} & -w_{31} & 0 \\ a(\gamma)w_{12} & b(\gamma)w_{22} & -w_{31} & 0 \\ c(\gamma)w_{11} & d(\gamma)w_{21} & 0 & -w_{41} \\ c(\gamma)w_{12} & d(\gamma)w_{22} & 0 & -w_{42} \end{bmatrix} \tag{6.7-15}$$

For $\gamma = \mu_0$, $\epsilon \to 0+$, $k_j = 1$, the system (6.7-9) reduces to (6.7-5) taken at $\mu = \mu_0$, $\epsilon = 0+$. This proves that $\Delta(\mu_0, 0) = 0$ and that the corresponding solutions of (6.7-11) are $k_1 = k_2 = k_3 = k_4 = \text{const.}$, the constant being arbitrary.

To show that the equation $\Delta(\gamma, \epsilon) = 0$ can be solved for γ in terms of ϵ, at least for all small non-negative ϵ, it suffices to prove that

$$\frac{\partial \Delta(\gamma,\epsilon)}{\partial \gamma} \bigg|_{\substack{\gamma=\mu \\ \epsilon=0+}} \neq 0 \tag{6.7-16}$$

and to invoke the implicit function theorem. The determinant in (6.7-15) can be developed in terms of its first two rows by Laplace's formula. With the help of the notation $|v, w|$ for the determinant whose columns are the two-dimensional vectors v, w, the resulting expression can be written

$$\Delta(\gamma,\epsilon) = a(\gamma)d(\gamma)\epsilon^{-1}|w_1, w_3||w_2, w_4| - b(\gamma)c(\gamma)\epsilon^{-1}|w_2, w_3||w_1, w_4|. \tag{6.7-17}$$

In this formula only a, b, c, d depend on γ. Formulas (6.7-6) and the well-known identity

$$\pi/\Gamma(\tfrac{1}{2} - \gamma)\Gamma(\tfrac{1}{2} + \gamma) = \cos \pi\gamma$$

imply that

$$a(\gamma)d(\gamma) = e^{2\pi i\gamma}, \qquad b(\gamma)c(\gamma) = e^{2\pi i\gamma} + 1. \tag{6.7-18}$$

Hence,

$$\frac{\partial \Delta(\gamma, \epsilon)}{\partial \gamma} = 2\pi i e^{2\pi i \gamma} \left\{ \epsilon^{-1} \left[|w_1, w_3| |w_2, w_4| - |w_1, w_4| |w_2, w_3| \right] \right\}. \quad (6.7\text{-}19)$$

As $\epsilon \to 0+$, the expression in braces in the right-hand member of (6.7-19) becomes

$$\lim_{\epsilon \to 0+} \left\{ \epsilon^{-1} \left[|z_1, z_3| |z_2, z_4| - |z_1, z_4| |z_2, z_4| \right] \right\}_{\mu = \mu_0}.$$

To simplify this expression observe that (6.7-5) implies the identities

$$0 = a|z_1, z_3| + b|z_2, z_3|,$$
$$0 = c|z_1, z_4| + d|z_2, z_4|.$$

With their help and one more reference to (6.7-18) one sees that at $\gamma = \mu_0$ formula (6.7-19) reduces to

$$\lim_{\epsilon \to 0+} \left(\frac{\partial \Delta(\gamma, \epsilon)}{\partial \gamma} \right) \Bigg|_{\gamma = \mu_0} = 2\pi i \lim_{\epsilon \to 0+} \left\{ \epsilon^{-1/2} |z_1, z_4| \cdot \epsilon^{-1/2} |z_2, z_3| \right\}$$

Here, as throughout this proof, z_j is taken at $x = 0$. One verifies by means of (6.6-11) and reference 1 that the right member of the last formula is not zero. Thus, (6.7-16) has been proved.

The implicit function theorem now guarantees the existence of a unique function γ of ϵ which reduces to μ_0, as $\epsilon \to 0+$, and such that

$$\Delta(\gamma(\epsilon), \epsilon) \equiv 0. \quad (6.7\text{-}20)$$

To prove also the stronger claim that $\gamma(\epsilon) \sim \mu(\epsilon)$, return to (6.7-15) and recall that the determinant obtained by replacing the w_j with z_j and γ with $\mu(\epsilon)$ is zero identically in ϵ, because (6.7-11) is true for $k_j = 1$, if those substitutions are made. This follows from (6.7-5). As $P_j|_{x=0} \sim I$, it is therefore true that

$$\Delta(\mu(\epsilon), \epsilon) \sim 0.$$

From this relation and (6.7-20) the successive derivatives of the functions μ and γ at $\epsilon = 0 + 1$ can be calculated by repeated differentiations with respect to ϵ. The inequality (6.7-16) is essential for this procedure. Since it is already known that $\mu(0+) = \mu_0 = \gamma(0+)$, it follows successively that all derivatives of $\mu(\epsilon)$ and $\gamma(\epsilon)$ are equal at $\epsilon = 0+$. This means that $\gamma(\epsilon) \sim \mu(\epsilon)$.

It must still be shown that the k_j which solve (6.7-11) when γ is the solution of $\Delta(\gamma, \epsilon) = 0$ are asymptotically equal and may therefore be chosen so that (6.7-13) is true. As before, recall that replacing γ by $\mu(\epsilon)$ and P_j by I in (6.7-11) produces a system for the k_j which is satisfied whenever $k_1 = k_2 = k_3 = k_4$. As $\gamma(\epsilon) \sim \mu(\epsilon)$, $P_j|_{x=0} \sim I$, this remains at least asymptotically true for (6.7-11) itself. This completes the proof of Lemma 6.7-1.

The proof of Theorem 6.7-1 is now relatively easy. Denote by z_j^γ the vectors obtained instead of z_j if $\mu(\epsilon)$ in (6.6-11) is replaced by $\gamma(\epsilon)$. The z_j^γ solve equation (6.7-4) and, thanks to Lemma 6.7-1 just proved, the solutions y_j of (6.6-2) introduced in (6.7-7) and (6.7-8) are linked by exactly the same connection formulas as the z_j^γ, provided the k_j are the quantities so called in Lemma 6.7-1. Let

$$Y_{j,j+1} := (y_j, y_{j+1}), \qquad Z_{j,j+1}^\gamma := (z_j^\gamma, z_{j+1}^\gamma),$$

be the matrix solutions of equation (6.6-2) and of (6.7-4), respectively, with the column vectors indicated by this notation as was done in Section 6.6. The 4 matrices $Y_{j,j+1}$ are obtained from each other by right-hand multiplication with invertible constant matrices. The corresponding matrices for the $Z_{j,j+1}^\gamma$ are exactly the same. Formulas (6.6-11), (6.7-5), and (6.7-6) imply that they are fundamental matrices. Therefore, the matrix

$$P := Y_{j,j+1}(Z_{j,j+1}^\gamma)^{-1}$$

is independent of j. The transformation

$$y = Pz$$

takes the differential equation (6.6-29) into (6.7-4).

To complete the proof of Theorem 6.7-1 only formula (6.7-2) remains to be established. Now, in T_j

$$z_j^\gamma = \hat{z}_j^\gamma e^{q_j}$$

and $\hat{z}_j^\gamma$ is bounded there. By Lemma 6.6-2, one also has

$$y_j = \hat{y}_j e^{q_j},$$

with $\hat{y}_j^\gamma \sim \hat{z}_j$ in T_j. Hence

$$Y_{j,j+1} = \hat{Y}_{j,j+1}\exp\begin{pmatrix} q_j & 0 \\ 0 & -q_j \end{pmatrix}$$

with

$$\hat{Y}_{j,j+1} \sim \hat{Z}_{j,j+1}^\gamma \qquad \text{in} \quad T_j \cap T_{j+1}.$$

It follows that

$$P = \hat{Y}_{j,j+1}(\hat{Z}_{j,j+1}^\gamma)^{-1} \sim I \qquad \text{in} \quad T_j \cap T_{j+1}.$$

This being true for all j, formula (6.7-2) has been proved.

CHAPTER VII

Extensions of the Regions of Validity of the Asymptotic Solutions

7.1. Introduction

Except for the formal theory of Chapter II and Theorem 3.3-1, all results proved so far have been local. They are valid in unspecified, possibly small, neighborhoods of a point. Several techniques exist for the study of the asymptotic properties of the solutions in larger domains. Most of those methods have to be adapted to specific examples, because general theories are either too difficult or so cumbersome as to yield little insight. Here is a short list of the various approaches available.

Theorem 3.3-1 is global in character, but it applies only to regions called "accessible" in Chapter III and the verification of that property of accessibility has been, so far, carried out only in the proof of the local Theorems 3.4-2 and 3.4-3. In many more or less specialized situations, accessibility can be proved in larger domains. Particularly important for the applications are results valid in unbounded regions, as in Chapters IX and X to follow.

A very small number of differential equations with especially simple coefficients can be solved by representing the solutions as integrals whose asymptotic evaluation is possible in large regions. The uniform theory of Chapter VI then makes those formulas available for full neighborhoods of the turning points of more general differential equations. In that chapter no attempt was made to carry this uniform transformation out in large domains, but that could be done, at least for differential equations whose coefficients have appropriately simple properties, such as being polynomials (see Chapters IX and X).

At points "very close" to turning points (the meaning of this phrase will become clearer below) the asymptotic expansions derived in Chapters II and III are not valid. There, the solutions can be expanded in series of a different form and there arises what I call the "central connection problem." To decide which asymptotic "outer" expansions, i.e., expansions valid away from the turning point, represent a solution given by its "inner" expansion—one valid at, or, at least, closer to, the turning points—one has to "match" the two types of expansions in a region where both are valid.

Actually, in all but relatively simple problems more than two types of expansions are needed to represent a solution asymptotically all the way to a turning point, and the central connection problem becomes so complicated that it takes a very important potential application to justify such calculations.

As one circles a turning point, the validity of the asymptotic expansion of a solution ceases abruptly, where certain curves are crossed. To find an asymptotic expansion for the analytic continuation of the solution on the other side of such a curve may be called the "lateral connection" problem. It can be tackled either by a direct matching procedure or by combining two central connection formulas.

The purpose of this and the following chapter is to explain these several methods, with emphasis on the analytic theory. In many applications matching methods are customarily used without complete proofs that the formalisms are logically sound. On the other hand, there exist many special examples requiring ingenious modifications of the general theory. Those applications are best studied separately, rather than being collected in a monograph as this one.

7.2. Regions of Asymptotic Validity Bounded by Separation Curves: The Problem

In this section the small regions of validity for the asymptotic expansions of solutions described in Theorems 3.4-2 and 3.4-3 will be extended "sideways" so as to fill certain—often wide—sectors with vertex at a turning point. The regions may, however, have to lie in a small neighborhood of that turning point.

As in previous chapters the differential equation will have the general form (2.1-3), but with the simplifying hypothesis that $h = 1$:

$$\epsilon y' = A(x, \epsilon) y \qquad (7.2\text{-}1)$$

with

$$A(x, \epsilon) \sim \sum_{r=0}^{\infty} A_r(x) \epsilon^r \qquad \text{as} \quad \epsilon \to 0+, \quad x \in D, \quad |\epsilon| \leqslant \epsilon_0. \quad (7.2\text{-}2)$$

One more, rather strong, restriction will have to be imposed, in order to deal with a manageable problem.

Hypothesis VII-1. *There is a point $x_0 \in D$ such that in $D - \{x_0\}$ all eigenvalues of $A_0(x)$ are simple.*

This is a strengthened form of Hypothesis V-1. There, the eigenvalues were of constant multiplicity in $D - \{x_0\}$, but not necessarily simple.

The point x_0 may be assumed to be at the origin: $x_0 = 0$. For investigations in the neighborhood of that point, Theorem 6.1-2 can be invoked to the effect that $A_0(0)$ may be stipulated to have only one distinct eigenvalue. As previously, it can even be assumed that $A_0(0)$ *is nilpotent and in Jordan form.* In this situation the construction of the formal solution of (7.2-1) in $D - \{x_0\}$ involves no shearing transformations. The first formal block-diagonalization described in Section 2.3 suffices to generate full diagonalization of the system and, hence, its formal solution as described in Theorem 2.8-1. That solution has the form

$$\mathbf{Y}(x, \epsilon) := \hat{\mathbf{Y}}(x, \epsilon) e^{(1/\epsilon) Q(x)} \tag{7.2-3}$$

with

$$\hat{\mathbf{Y}}(x, \epsilon) := \sum_{r=0}^{\infty} Y_r(x) \epsilon^r, \qquad \det Y_0(x) \neq 0 \qquad \text{in } D, \tag{7.2-4}$$

$$Q(x) = \int_0^x \Lambda(t)\, dt, \tag{7.2-5}$$

$$\Lambda(t) = \mathrm{diag}(\lambda_1(t), \lambda_2(t), \ldots, \lambda_n(t)). \tag{7.2-6}$$

Strictly speaking, the existence of a formal solution of the form described above was proved in Chapter II only for subregions of $D \subset \mathbb{C}$ in which the function λ_j are holomorphic. However, the domain of existence can be enlarged to Riemann domains D as shown below.

The functions λ_j, $j = 1, 2, \ldots, n$, have a branch point at $x = 0$. In the present context the nature of this branch point is of importance. As the λ_j are the zeros of the polynomial

$$\det(\lambda I - A_0(x)),$$

they are, in the neighborhood of each point of $D - \{0\}$, the branches of s holomorphic functions $\phi_\nu(x^{1/m_\nu})$, $\nu = 1, 2, \ldots, s$, $1 \leqslant \nu \leqslant n$, of a fractional power x^{1/m_ν}. If m_ν is chosen as the smallest such integer, then $\phi_\nu(x^{1/m_\nu})$ represents m_ν distinct functions of x in the neighborhood of every point of $D - \{0\}$. Therefore $\sum_{\nu=1}^{s} m_\nu = n$. The m_ν eigenvalues corresponding to each function ϕ_ν undergo a cyclic permutation as x circles the origin. This set of

m_ν eigenvalues will be called a "cycle" of eigenvalues. Let

$$\phi_\nu(x^{1/m_\nu}) = \sum_{r=r_\nu}^{\infty} \phi_\nu x^{r/m_\nu}, \qquad \nu = 1, 2, \ldots, s. \qquad (7.2\text{-}7)$$

By Hypothesis VII-1 at most one of these expansions is identically zero. For the others it may be assumed that $\phi_{\nu,r_\nu} \neq 0$. Then $r_\nu \geqslant 1$, because $A_0(0)$ is nilpotent, i.e., $\lambda_j(0) = 0$, $j = 1, 2, \ldots, n$. One may partially order the eigenvalues by the stipulation that the eigenvalues of the μth cycle precede those of the νth cycle whenever $r_\mu/m_\mu < r_\nu/m_\nu$. Within each cycle a natural ordering is one in which one counterclockwise revolution around $x = 0$ changes each eigenvalue into the next one modulo m_ν.

If p is the least common multiple of the m_ν, $\nu = 1, 2, \ldots, s$, then the eigenvalues are all holomorphic functions on the p-sheeted Riemann domain over D. This domain will be denoted by D^p. Let the eigenvalues be $\lambda_1(x), \lambda_2(x), \ldots, \lambda_n(x)$ in an ordering that agrees with the partial ordering described above. The functions

$$q_j(x) = \int_0^x \lambda_j(t) \, dt \qquad (7.2\text{-}8)$$

are also holomorphic on D^p. Also, if λ_j belongs to the νth cycle, then

$$q_j(x) = O(x^{1+r_\nu/m_\nu}) \qquad \text{as} \quad x \to 0. \qquad (7.2\text{-}9)$$

The matrix A_0 in the original form (7.2-1), (7.2-2) of the differential equation can be changed into the matrix Λ of formula (7.2-6) by a similarity transformation with a matrix which is the Vandermonde matrix of the n eigenvalues λ_j. This matrix is holomorphic on the Riemann domain $D^p - \{0\}$. The diagonalization of the later terms A_r in (7.2-2) by the method of Chapter II involves only rational operations with matrices, including the taking of inverses of holomorphic matrices invertible in $D^p - \{0\}$. Therefore, the coefficients Y_r in (7.2-4) are holomorphic on $D^p - \{0\}$.

The n columns of the formal matrix solution (7.2-3) of the differential equation (7.2-1) have the form

$$y_j(x, \epsilon) = \hat{y}_j(x, \epsilon) e^{q_j(x)/\epsilon}, \qquad j = 1, 2, \ldots, n, \qquad (7.2\text{-}10)$$

where the quantities

$$\hat{y}_j(x, \epsilon) := \sum_{r=0}^{\infty} y_{jr}(x) \epsilon^r \qquad (7.2\text{-}11)$$

are certain formal series. It is known from Theorem 3.4-4 that there exist fundamental systems of n vector solutions $y_j(x, \epsilon)$, $j = 1, 2, \ldots, n$, for which

$$\hat{y}_j(x, \epsilon) := y_j(x, \epsilon) e^{-q_j(x)/\epsilon} \sim \hat{y}_j(x, \epsilon) \qquad \text{as} \quad \epsilon \to 0 + \qquad (7.2\text{-}12)$$

uniformly in sufficiently small neighborhoods of every point of $D^p - \{0\}$.

These solutions depend on the point chosen. The main aim of this section is to strengthen that statement by finding solutions that are asymptotically represented by (7.2-10) in regions that are not small with respect to $\arg x$, although $|x|$ may still have to be restricted to sufficiently small values.

As is known from Chapter III, the sign of the harmonic functions

$$p_{jk}(x) := \mathrm{Re}[\, q_j(x) - q_k(x)\,], \qquad j, k = k, 2, \ldots, n, \qquad (7.2-13)$$

plays a fundamental role in the asymptotic theory. The curves in D^p defined by the equations

$$p_{jk}(x) = 0, \qquad j \neq k,$$

will be called the *separation curves* at the turning point $x = 0$ for the differential equation (7.2-1). The term "Stokes curves" or "Antistokes curves" are frequently used for these curves, but for the sake of clarity it is preferable to employ those names in a more restricted sense to be explained presently. Observe that it is quite possible for two different pairs (j_1, k_1) and (j_2, k_2) to define the same separation curve.

One has

$$q_{jk}(x) := q_j(x) - q_k(x) = c_{jk} x^{r_{jk}/p}\big[1 + O(x^{1/p})\big] \qquad \text{as} \quad x \to 0, \quad (7.2-14)$$

with certain constants c_{jk}, r_{jk}. As a consequence of Hypothesis VII-1 the inequalities $c_{jk} \neq 0$ hold, and $r_{jk} > 0$, because $A_0(0)$ may be assumed to be nilpotent. Therefore, each separation curve is an analytic arc in D^p with one end point at $x = 0$ and the other one on the boundary of D^p, provided D is small enough. The tangents at $x = 0$ have the directions

$$\arg x = \left(\frac{2m + 1}{2} \pi - \arg c_{jk} \right) \frac{p}{r_{jk}}, \qquad m = 0, 1, \ldots, 2r_{jk}^{-1}. \quad (7.2-15)$$

In the well-known asymptotic theory as $x \to \infty$ for analytic linear differential equations without a parameter, the curves analogous to the separation arcs defined above are straight rays, and their importance lies in the fact that the sectors of asymptotic validity of formal solutions are *always* bounded by separations rays. It is important to realize that *this is no longer true* for asymptotic expansions of solutions with respect to a *parameter*.

To show that, let y_j and u_k be vector solutions of equation (7.2-1) having the asymptotic representation (7.2-12) in some region $S \in D^p$, and let $\tilde{x} \in S$ be a point not on a separation curve. Let $k \neq j$, and consider the branch of the curve $p_{jk}(x) = p_{jk}(\tilde{x})$ that contains $\tilde{x}$. The vector

$$y^*(x, \epsilon) = y_j(x, \epsilon) + e^{q_{jk}(\tilde{x})/\epsilon} y_k(x, \epsilon)$$

is then also a solution of (7.2-1). It can be written

$$y^*(x, \epsilon) = \big\{ \hat{y}_j(x, \epsilon) + \hat{y}_k(x, \epsilon) \exp([\, q_{jk}(\tilde{x}) - q_{jk}(x)\,]/\epsilon) \big\} e^{q_j(x)/\epsilon}.$$

This shows that in the subset S_1 of S where $p_{jk}(\tilde{x}) < p_{jk}(x)$ the solution y^*

has the same asymptotic expansion as y_j, while this is false in $S - S_1$. Thus, the region of asymptotic validity of the expansion

$$y^*(x, \epsilon) e^{-q_j(x)/\epsilon} \sim \hat{\mathbf{y}}(x, \epsilon)$$

is not bounded by separation curves.

It is a problem of some interest to construct solutions that maintain the same asymptotic representation in regions—preferably large ones—that *are* bounded by separation curves. For then some—though not all—of the methods developed for the analysis of lateral connection problems for the parameterless asymptotic theory can be extended to expansions with respect to a parameter. Such solutions will be found in the next section.

7.3. Solutions Asymptotically Known in Sectors Bounded by Separation Curves

An inspection of the proof of Theorem 3.3-1 shows that it remains valid for the differential equation (7.2-1), (7.2-2) if the region $D \subset \mathbb{C}$ of that theorem is replaced by the Riemann domain $D^p - \{0\}$. For convenience this fact is restated below in the present notation.

Theorem 7.3-1. *Let the differential equation* (7.2-1), (7.2-2) *satisfy Hypothesis VII-1 with* $x_0 = 0$, *and let* $\mathbf{y}_j(x, \epsilon), j = 1, 2, \ldots, n$, *in* (7.2-10) *be n formal solutions. If* $D^* \subset D^p - \{0\}$ *is compact and accessible with respect to* q_j *for* $\epsilon > 0$, *then the asymptotic representation* (7.2-12) *for the solution* y_j *in that formula is uniformly valid in* D^*.

The term "accessible" was introduced in Definitions 3.3-1 and 3.3-2. In the present application, $E = \{\epsilon / 0 < \epsilon \leqslant \epsilon_0\}$.

There remains the task of finding domains—preferably large ones—in D^p that are accessible with respect to a given q_j. From now on in this section j will be arbitrary but fixed.

For a fixed j the separation curves $\rho_{jk}(x) = 0$, $k = 1, 2, \ldots, n$, $k \neq j$, form a subset of the set of all separation curves. They divide D^p into a set of sectors, none of which contains curves $\rho_{jk}(x) = 0$ in its interior, although other separation curves may well be present. These sectors will be called "q_j-sectors." A "*double q_j-sector*" is a sector consisting of two adjacent q_j-sectors and their common boundary arc. The other two bounding separation curves are not part of that double q_j-sector, nor is the origin.

It appears to be difficult to establish the existence of large sectors of accessibility without subjecting the function q_{jk} defined in (7.2-14) to restrictive conditions. Only the simplest situation is going to be completely analyzed here, namely, the case that for some j the term designated by

$O(x^{1/p})$ in (7.2-14) is absent for all $k = 1, 2, \ldots, n$. This type of differential equations is of importance in itself, and the arguments to be given illustrate the procedure to be followed in other applications.

It will be convenient to assume that D is the disk

$$D = \{x \in C / |x| \leqslant a\}.$$

Let D_d^p consist of the points of D^p at distance $d > 0$ or more from the origin.

Lemma 7.3-1. *Assume that for some fixed j ($1 \leqslant j \leqslant n$)*

$$q_{jk}(x) = c_{jk} x^{r_{jk}/p}, \qquad c_{jk} \neq 0, \quad k = 1, 2, \ldots, n, \quad k \neq j, \qquad (7.3\text{-}1)$$

and let S^* be a compact subset of a double q_j-sector S. Then, for d sufficiently small, there exists a domain $S' \subset D_d^p$ with

$$S^* \subset S' \subset S \cap D_d^p$$

so that S' is accessible for q_j.

PROOF. For each $k \neq j$, $k = 1, 2, \ldots, n$, the conformal mapping

$$\xi = q_{jk}(x)$$

takes S into a domain $\xi_k(S)$ on the logarithmic Riemann surface over the complex ξ-plane. Thanks to assumption (7.3-1) the domains S and $\xi_k(S)$ are each bounded by a circular arc and two radial segments. If $\xi_k(S)$ contains a segment on the imaginary axis of the ξ-plane, it must be the image of the one separation curve belonging to q_j which lies in the interior of the double q_j-sector S. Hence, $\xi_k(S)$ is a sector of central angle not exceeding 2π and it contains at most one segment of the imaginary axis in its interior. For a given compact $S^* \subset S$ choose a point α on the bisector of S so close to the origin that the parallels through α to the bounding rays of S do not meet S^*. These parallels together with an arc of the boundary of D bound a subregion $S' \subset S$. Then choose d so small that $S' \subset S \cap D_d^p$. Inspection of the image $\xi_k(S')$ shows that the rightmost point (or one of them) of the boundary $\partial \xi_k(S')$ can be connected with every point of $\xi_k(S')$ by a path along which $\operatorname{Re} \xi$ does not increase. This means that along the preimage in the x-plane of such a path the function $\operatorname{Re} q_{jk}$ is a nonincreasing function of x. Therefore, G is accessible for q_j, as was to be proved.

Lemma 7.3-1, when combined with Theorem 7.3-1, establishes the following result.

Theorem 7.3-2. *If the differential equation (7.2-1), (7.2-2) satisfies Hypothesis VII-1 (with $x_0 = 0$) and condition (7.3-1) for a particular j and all k, then, corresponding to every compact subset S^* of a double q_j-sector there exists a vector solution y_j of the differential equation for which the asymptotic representation (7.2-12) is valid, uniformly in S^*.*

It is easy to modify the construction in the preceding proof of Theorem 7.3-2 so that $\operatorname{Re} q_{jk}$ is *strictly* decreasing along the paths in S'. If the term $O(x^{1/p})$ in (7.2-14) is not absent, the path will then still be progressive (in the sense of Definition 3.3-1) provided the radius a of the disk D is sufficiently small. Therefore, if a domain S^* in a double q_j-sector is given, the method of proof used for Theorem 7.3-2 may still apply to all (or to a part) of S^*. The matter is best analyzed for specific cases rather than to attempt the formulation of a general theorem.

A stronger result can be established if there are regions in D^p where

$$\operatorname{Re} q_j(x) \leqslant \operatorname{Re} q_k(x), \qquad k = 1, 2, \ldots, n. \qquad (7.3\text{-}3)$$

Such regions are, of course, q_j-sectors if they are maximal. They will be called *recessive q_j-sectors*, and a solution y_j with the asymptotic representation (7.2-12) in all or part of such a sector is to be called recessive, there. A triple q_j-sector is defined as a subdomain of D^p consisting of the union of three contiguous q_j-sectors and the two separation curves between them.

Theorem 7.3-3. *If the middle q_j-sector of a triple q_j-sector is recessive for q_j, then the statement of Theorem 7.3-2 is valid in that triple q_j-sector.*

PROOF. Repeat the constructions in the proofs of Lemma 7.3-1 with the sector S now being the triple q_j-sector just described. Each region $\xi_j(S)$ is now a sector over the ξ-plane which contains at most two rays on which $\operatorname{Re} \xi = 0$. Therefore its central angle does not exceed 3π. The image of the middle one of the three q_j-sectors lies over the left ξ-half-plane. One can now verify directly that every compact subset S^* in S is contained in an accessible subregion of S. The construction is essentially the same as before, although $\xi_k(S)$ may now be a region on more than one sheet of the logarithmic Riemann surface over the ξ-plane. Again, the proof is completed by appealing to Theorem 7.3-1.

7.4. Singularities of Formal Solutions at a Turning Point

The asymptotic expansions for the solutions of differential equations constructed in Chapters II and III have been shown to be valid in certain regions at positive distance from turning points. If the regions are allowed to depend on the parameter ϵ, it is possible to improve those results so that they remain valid in some regions whose distance from a turning point tends to zero, as $\epsilon \to 0$. Two such facts have already been quoted and applied in Theorems 6.2-1 and 6.3-1, but no proofs were given there.

Actually, properties of that nature were already tacitly implied in the very first study of turning point problems, the one by Gans [14]. In Section

1.2 it was explained that Gans "matched" solutions written in terms of the variable t defined in (1.2-4) with solutions given in terms of x in regions bounded away from the turning point at $x = 0$. The image in the x-plane of any bounded region in the t-plane that does not depend on ϵ shrinks to the origin, as $\epsilon \to 0$ and, therefore, to justify the matching procedure, the validity of the "outer" expansion must be extended toward the turning point, or else the validity of the "inner" expansion in terms of t must be proved in domains that become unbounded, as $\epsilon \to 0$. Often both extensions are necessary.

As a first step in the investigation of these matters, the singularities of the coefficients Y_r and Q_j in the formal solution described in Theorem 2.8-1 will be analyzed in this section. As usual, the turning point will be placed at $x = 0$.

It will turn out that, while the Y_r and Q_j are, in general unbounded at $x = 0$, their growth, as $x \to 0$, is not faster than that of some negative power of x and, moreover, that the growth rate of Y_r depends linearly on r. The two definitions below serve to facilitate the description.

Definition 7.4-1. On the Riemann surface C_L of $\log x$ let D_L be the domain over the region $D \subset C$. A holomorphic function f from D_L into C is said to have growth rate σ_0 (σ_0 real) at $x = 0$, in a finite sector of D_L, if in that sector

$$\lim_{x \to 0} f(x) x^{\sigma} = 0 \qquad \text{for all} \quad \sigma > \sigma_0,$$

but not for $\sigma < \sigma_0$.

Note that $f(x) x^{\sigma_0}$ is not necessarily bounded at $x = 0$. The growth rate of a matrix-valued function is, by definition, the largest of the growth rates of its entries. Of course, not all functions have a finite growth rate.

Definition 7.4-2. Let f_r, $r = 0, 1, \ldots$, be a sequence of functions with finite growth rates σ_r at $x = 0$. The sequence is said to have restraint index κ at $x = 0$, if

$$\inf_{r \geq 1} \frac{r}{\sigma_r} = \kappa, \qquad \kappa \neq 0. \tag{7.4-1}$$

Relation (7.4-1) is tantamount to the inequalities

$$\frac{\sigma_r}{r} \leq \frac{1}{\kappa} \qquad \text{for all} \quad r \geq 1 \tag{7.4-2}$$

together with the statement that (7.4-2) is not true for any larger κ. If $\kappa > 0$, then σ_r increases at most linearly with r and the larger κ, the more the inequalities (7.4-2) "restrain" the worsening of the singularity of f_r at $x = 0$ with increasing r. The extension of these definitions from scalar to matrix-valued functions is obvious.

Theorem 7.4-1. *The sequence* $\{Y_r\}$ *in formula* (2.8-12) *for the formal solution of differential equations has positive restraint index at turning points, and the* Q_j *in that formula have finite growth rates there.*

PROOF. Around the turning point, say at $x = 0$, the eigenvalues of A_0 are holomorphic functions on a p-sheeted Riemann domain. If one sets $x = z^p$, this region is mapped onto a neighborhood of $z = 0$ in the z-plane. The neighborhood can be chosen so small that the multiplicity of the eigenvalues does not change there, except possibly at $z = 0$. By Theorem 12.2-2 there exists then a holomorphic matrix-valued function T of z a similarity transformation with which takes $A_0(z^p)$ into its Jordan form at every point of a deleted neighborhood of $z = 0$. At $z = 0$, T or T^{-1} may, of course, have a singularity as a function of z. A scrutiny of the proof of Theorem 12.2-2 and of the theorems on which it depends, in particular of Theorems 12.1-1, 12.1-2, and 12.2-1, shows that the singularity is, at worst, a pole with respect to z^p. The details of these verifications are omitted here.

If the first step of formal the block-diagonalization described in Section 2.3, namely, the transformation (2.3-1), is carried out in the neighborhood of $x = 0$, formula (2.3-3) shows that the sequence $\{B_r\}$ in (2.3-4) consists of matrices of finite growth rate at $x = 0$, and that (7.4-2) is true. The values of σ_r in (7.4-2) depends on $A_0(x)$ in a way not easy to analyze in general terms.

Next, it must be shown that the result of the block-diagonalization process described in Section 2.3 leads to a new differential equation and a formal transformation matrix, as described in Theorem 2.3-1 such that the new matrices B_r, P_r form sequences with positive restraint indices. Now, solution of the equation (2.3-16) yields the entries of the matrix P_1^{12} as functions of x by means of Cramer's rule. The determinant in the denominator of the calculation may be zero at $x = 0$. This zero has a certain integral order in terms of $z = x^{1/p}$. The matrix H_1 equals A_1, and P_1 has at worst a finite growth rate at $x = 0$. An easy induction with respect to r shows then that successive calculation of the B_r, H_r, and P_r leads to sequences of matrices with positive restraint index.

The shearing transformations of Section 2.4 may change the value of the restraint index of the coefficient matrix but not the fact that a positive restraint index exists.

The next simplification of the differential equation, the one based on Arnold's theory and described in Section 2.5, requires the successive solution of the equations in the sequence (2.5-7). For each r this is a consistent system of linear nonhomogeneous algebraic equations whose homogeneous part does not depend on r. The rate of growth of F_r in (2.5-8) increases linearly with r and, therefore, the solutions P_r of (2.5-7) can be chosen so as to have the same property.

Finally, as was shown in Section 2.8, one arrives at a system of differential equations consisting of a set of uncoupled systems each of which either is of order one, or has no factor ϵ^h in the left member. An inspection of the way these formal differential equations were solved in Section 2.8 shows that the coefficients in the power series for the results have finite growth rates and that the series have positive restraint index. The functions Y_r and Q_j in the final formula (2.8-12) may, however, fail to be defined on the Riemann domain of *finitely* many sheets, since the required integrations may introduce logarithmic terms. Thus, the proof of Theorem 7.4-1 is completed.

7.5. Asymptotic Expansions in Growing Domains

In the preceding section it was shown that the coefficients Y_r in the series $\sum_{r=0}^{\infty} Y_r \epsilon^{r/m}$ occurring in the formal solution (2.8-12) grow with r at most so that $Y_r(x)x^{\sigma+r/\kappa}$ tends to zero with x, for all $\sigma > \sigma_0$. Here σ_0 is a certain real constant, positive, zero, or negative, and κ is a positive constant. Accordingly, if x and ϵ both tend to zero, the terms of that series shrink at a rate that increases with r, as long as $x^{-1/\kappa}\epsilon^{1/m}$ tends to zero. Now define

$$\check{Y}_r(x) := Y_r(x)x^{\sigma+r/\kappa}, \tag{7.5-1}$$

and write

$$\sum_{r=0}^{\infty} Y_r(x)\epsilon^{r/m} = x^{-\sigma} \sum_{r=0}^{\infty} \check{Y}_r(x)(x^{-1/\kappa}\epsilon^{1/m})^r. \tag{7.5-2}$$

Then it can be expected that the formal summation in the right member is an asymptotic representation of a function of x and ϵ even in regions that depend on ϵ and whose distance from the origin tends to zero with ϵ so slowly that $x^{-1/\kappa}\epsilon^{1/m}$ shrinks to zero in that region. The functions $\hat{Y}_r$ depend on σ, of course. In many applications $\hat{Y}_r$ remains bounded even for $\sigma = \sigma_0$, but the distinction between σ and σ_0 has been retained in the definitions, because of the possible presence of logarithmic terms.

The extension of the Borel–Ritt theorem to series of the form occurring in (7.5-2) can be established by suitable modifications of the proof in [90], as will now be shown.

If the sequence $\{f_r\}$ has restraint index κ at $x = 0$ (cf. Definition (7.4-2)), then the formal series $\sum_{r=0}^{\infty} f_r(x)\epsilon^r$ will also be said to have restraint index κ at $x = 0$.

It is a simple matter to verify that if two formal series have positive restraint indices then this is also true of their formal sum and product, and likewise of their quotient, provided the denominator is not identically zero. The formal derivative with respect to x of such a series also has positive restraint index. (Details of these verifications were given in [96].)

Observe that the statements above contain no information as to the *size* of the restraint index of the sum or product, etc, in terms of the restraint indices of the two series. Such rules can be formulated but they are not always simple, except when the series to be combined have the same restraint index. Then the resulting series also have the same restraint index. (See [96].)

Theorem 7.5-1. (Generalized Borel–Ritt Theorem).   *Let* $\{f_r\}$ *be a sequence of functions of* x *holomorphic in*

$$D_L := \{x \in \mathbb{C}_L / 0 < |x| \leqslant x_0 < \infty\} \qquad (7.5\text{-}3)$$

and with the restraint index $\kappa > 0$ *at* $x = 0$, *so that*

$$\lim_{x \to 0} f_r(x) x^{\sigma + r/\kappa} = 0, \qquad r = 0, 1, \ldots,$$

for all numbers σ *greater than a certain real number* σ_0. *Let* S *and* E *be given sectors*:

$$S := \{x \in D_L / \vartheta_1 < \arg x < \vartheta_2\}, \qquad (7.5\text{-}4)$$

$$E := \{E \in \mathbb{C}_L / 0 < |\epsilon| \leqslant \epsilon_0, \delta_1 \leqslant \arg \epsilon \leqslant \delta_2\} \qquad (7.5\text{-}5)$$

with $\vartheta_1, \vartheta_2, \delta_1, \delta_2$ *arbitrary but finite. Then there exists, for any given* $c > 0$, *a function* f *of* x *and* ϵ *holomorphic in both variables in the set*

$$G_\kappa := \{(x,\epsilon)/(x,\epsilon) \in S \times E, |x^{-1/\kappa}\epsilon| < c\} \qquad (7.5\text{-}6)$$

so that f *has in* G_κ *the asymptotic expansion*

$$f(x,\epsilon) \sim \sum_{r=0}^{\infty} f_r(x)\epsilon^r \qquad as \quad \epsilon \to 0 \quad in \ E$$

in the sense that

$$\left| f(x,\epsilon) - \sum_{r=0}^{N} f_r(x)\epsilon^r \right| \leqslant c_N |x|^{-\sigma} |x^{-1/\kappa}\epsilon|^{N+1} \qquad (7.5\text{-}7)$$

for $(x,\epsilon) \in G_\kappa$, $N = 0, 1, \ldots$. *The function* f *and the constants* c_N *depend on* G_κ *and* c.

PROOF.   Let

$$a_r := \sup_{x \in S} |f_r(x) x^{\sigma + r/\kappa} c^r| \qquad (7.5\text{-}8)$$

and define

$$\alpha_r(x,\epsilon) := 1 - \exp\{-1/a_r(x^{-1/\kappa}\epsilon)^\beta\}, \qquad (7.5\text{-}9)$$

whenever $a_r \neq 0$. If $a_r = 0$, set $\alpha_r = 1$, say. Choose β in the interval $0 < \beta < 1$ so small that

$$\text{Re}[x^{-1/\kappa}\epsilon)^\beta] > 0 \qquad \text{for all} \quad (x,\epsilon) \in G_\kappa, \qquad (7.5\text{-}10)$$

which can be done because the constants $\vartheta_1, \vartheta_2, \delta_1, \delta_2$ are finite. Then

$$|\alpha_r(x,\epsilon)| \leqslant 1/a_r |x^{-1/\kappa}\epsilon|^\beta. \qquad (7.5\text{-}11)$$

This follows from the simple estimate

$$|1 - e^{-z}| = \left|\int_0^z e^{-t}\,dt\right| \le |z| \qquad \text{for} \quad \text{Re}\,z > 0.$$

Now, define f by the series

$$f(x, \epsilon) := \sum_{r=0}^{\infty} f_r(x)\alpha_r(x, \epsilon)\epsilon^r. \tag{7.5-12}$$

This series converges for $(x, \epsilon) \in G_\kappa$, as can be seen from the inequalities

$$|f_r(x)\alpha_r(x, \epsilon)\epsilon^r|$$
$$< |f_r(x)x^{\sigma + r/\kappa}x^{-\sigma - r/\kappa}a_r^{-1}|\,|x^{-1/\kappa}\epsilon|^{-\beta}|\epsilon^r|$$
$$< |x|^{-\sigma}|x^{-1/\kappa}\epsilon|^{-\beta}|c^{-1}x^{-1/\kappa}\epsilon|^r, \tag{7.5-13}$$

which are consequences of (7.5-8) and (7.5-11) and of the definition (7.5-6) of G_κ. For $|x^{-1/\kappa}\epsilon| \le c - \gamma$ ($\gamma > 0$, but otherwise arbitrary) the convergence is uniform. Therefore, f is holomorphic in G_κ.

To prove that $f(x, \epsilon) \sim \sum_{r=0}^{\infty} f_r(x)\epsilon^r$ in the sense of inequality (7.5-7), write

$$f(x, \epsilon) - \sum_{r=0}^{N} f_r(x)\epsilon^r = \Sigma_1 + \Sigma_2 + \Sigma_3, \tag{7.5-14}$$

where

$$\Sigma_1 = \sum_{r=0}^{N} f_r(s)(\alpha_r(x, \epsilon) - 1)\epsilon^r,$$

$$\Sigma_2 = f_{N+1}(x)\epsilon^{N+1},$$

$$\Sigma_3^{\infty} = \sum_{r=N+2} f_r(x)\alpha_r(x, \epsilon)\epsilon^r.$$

The function Σ_1 can be appraised with the help of (7.5-8). The abbreviation

$$\zeta := x^{-1/\kappa}\epsilon$$

will shorten the formulas. One has, from (7.5-8) and (7.5-9)

$$|f_r(x)(\alpha_r(x, \epsilon) - 1)\epsilon^r| \le |a_r|\,|x|^{-\sigma - r/\kappa}c^{-r}e^{-1/a_r\zeta^\beta}\epsilon^r|$$
$$= a_r c^{-r}|x|^{-\sigma}|\zeta^r e^{-1/a_r\zeta^\beta}\epsilon^r|$$
$$= a_r c^{-r}|x^{-\sigma}\zeta^{N+1}|\,|\zeta^{r-N-1}e^{-1/a_r\zeta^\beta}|.$$

The function of ζ alone between absolute bars is bounded in G, because $\text{Re}\,\zeta^\beta > 0$ there. It follows that Σ_1 has the bound required for the right member of (7.5-7). Again using (7.5-8) it is seen that

$$|\Sigma_2| \le |x|^{-\sigma}a_{N+1}c^{-N-1}|\zeta|^{N+1}.$$

Finally, the summation Σ_3 is appraised with the help of (7.5-13). The summation is dominated by

$$|\Sigma_3| \leqslant |x|^{-\sigma}|\zeta|^{-\beta} \sum_{r=N+2}^{\infty} \left|\frac{\zeta}{c}\right|^r$$

$$= |x|^{-\sigma}c^{-N-2}|\zeta|^{N+2-\beta}\frac{1}{1-|\zeta/c|}.$$

Since $\beta < 1$, this bound also has the required form, and the proof of the theorem is completed.

It is quite easy to show that if f and g are functions that have asymptotic expansions in G_κ with the same restraint index κ, then $f + g$, fg, have the same property. If $f_0(x) \neq 0$, the same is true of f^{-1}. Analogous statements hold for matrices. The proofs are omitted. They can, e.g., be found in Ref. 96. The value of σ_0 for these expansions varies in general under these operations. The next theorem states that the series for f may be termwise differentiated with respect to x.

Theorem 7.5-2. *Under the hypotheses of Theorem 7.5-1 one has*

$$\frac{df(x,\epsilon)}{dx} \sim \sum_{r=0}^{\infty} f_r'(x)\epsilon^r \qquad as \quad \epsilon \to 0 \quad in \ E,$$

or, more precisely,

$$\left|\frac{df(x,\epsilon)}{dx} - \sum_{r=0}^{N} f_r'(x)\epsilon^r\right| \leqslant c_N^*|x|^{-\sigma-1}|x^{-1/\kappa}\epsilon|^{N+1},$$

for $(x,\epsilon) \in G_\kappa^*$, $N = 0,1,\ldots$. *Here,* $G_\kappa^* \subset G_\kappa$ *is defined by the same inequalities as* G_κ, *but with constants* $\vartheta_1^* > \vartheta_1$, $\vartheta_2^* < \vartheta_2$, $\delta_1^* > \delta_1$, $\delta_2^* < \delta_2$, $c^* < c$. *The constant* c_N^* *depends on* G_κ^*.

PROOF.   Rewrite (7.5-7) in the form

$$x^{\sigma+(N+1)/\kappa}\left(f(x,\epsilon) - \sum_{r=0}^{N} f_r(x)\epsilon^r\right) =: \phi_N(x,\epsilon), \qquad (7.5\text{-}15)$$

$$|\phi_N(x,\epsilon)| \leqslant c_N|\epsilon|^{N+1} \qquad for \quad (x,\epsilon) \in G_\kappa. \qquad (7.5\text{-}16)$$

The result of differentiating (7.5-15) with respect to x can be written

$$x^{\sigma+1+(N+1)/\kappa}\left(f'(x,\epsilon) - \sum_{r=0}^{N} f_r'(x)\epsilon^r\right) = \psi_N(x,\epsilon), \qquad (7.5\text{-}17)$$

$$\psi_N(x,\epsilon) := x\phi_N'(x,\epsilon) - (\sigma + (N+1)/\kappa)x^{\sigma+(N+1)/\kappa}$$

$$\times \left(f(x,\epsilon) - \sum_{r=0}^{N} f_r(x)\epsilon^r\right). \qquad (7.5\text{-}18)$$

Here the dash denotes differentiation with respect to x. For $(x,\epsilon) \in G_\kappa^*$ the distance of x from the boundary of the projection of G_κ onto the x-plane is larger than $k|\epsilon|^\kappa$, k a certain positive constant. Therefore, the relation (7.5-16) and Cauchy's inequality for the derivatives of holomorphic functions imply that $|\phi'_N(x,\epsilon)| < c_N k^{-1}|\epsilon|^{N+1-\kappa}$, for $(x,\epsilon) \in G_\kappa^*$. Using once more (7.5-15) and (7.5-16) it follows from (7.5-18) that

$$|\psi_N(x,\epsilon)| \leqslant x_0 c_N k^{-1}|\epsilon|^{N+1-\kappa} + (\sigma + (N+1)/\kappa)c_N|\epsilon|^{N+1}.$$

As (7.5-15) is true for all $N > 0$, the last inequality remains true if N is replaced by an integer at least equal to $N + \kappa$. Therefore, the statement of the theorem follows from (7.5-17) for all N.

7.6. Asymptotic Solutions in Expanding Regions: A General Theorem

With the help of the results of the preceding two sections the methods of Chapter III will now be extended so as to apply in regions whose boundary depends on ϵ in such a way that it approaches the turning point at $x = 0$, as $\epsilon \to 0$. The descriptions in the proof—but not in the statement of the final theorem—will be simplified by restricting ϵ to the positive real axis and by replacing m with 1 in the formal solutions (2.8-12). The arguments are essentially the same without these restrictions, but the formulas would be longer.

The reasoning in Sections 3.2 and 3.3 can be adapted to regions in which Theorem 7.5-1 instead of the more restricted version of the Borel–Ritt theorem in Section 3.1 must be invoked. The arguments of Section 3.2 remain almost literally valid if the symbol "$\sim$" is now interpreted as in Theorem 7.5-1. Only in the statement and proof of a theorem analogous to Theorem 3.3-1 do nontrivial differences appear.

By Theorem 7.4-1 the matrices Y_r in formula (3.2-1) have positive restraint index at $x = 0$, i.e., for some real σ_0 and some $\kappa_0 > 0$ the product

$$Y_r(x)x^{\sigma+r/\kappa} \tag{7.6-1}$$

is bounded in a neighborhood of $x = 0$ whenever $\sigma > \sigma_0$, $0 < \kappa \leqslant \kappa_0$. Application of Theorem 7.5-1 then proves the existence of a matrix $\hat{U}$ with the expansion (3.2-1) in the sense of Theorem 7.5-1. The function $\hat{U}$ depends on the values of the other constants in the definition of the region G_κ. If $\kappa \leqslant \kappa_0$, then $G_\kappa \subseteq G_{\kappa_0}$, provided $x_0 \leqslant 1$ which will be assumed from now on.

Theorem 7.5-1 implies that $U(x,\sigma)x^\sigma$ is bounded in G_κ for $\sigma > \sigma_0$, $\kappa \leqslant \kappa_0$. The functions K and k in formula (3.3-9) are now asymptotic to zero in the sense of Theorem 7.5-1, with the same restraint index κ. More specifically, for all $N \geqslant 0$ and for certain constants τ_0, the functions

$K(x,\epsilon)x^{\tau+N/\kappa}\epsilon^{-N}$ and $k(x,\epsilon)x^{\tau+N/\kappa}\epsilon^{-N}$ are bounded in G_κ, if $\tau > \tau_0$, $\kappa \leqslant \kappa_0$. (Remember that $m = 1$, now.) The number τ_0 may be different from σ_0, but K and k have the same asymptotic orders of magnitude as $x \to 0$, $\epsilon \to 0$ in G_κ. The asymptotic sizes of $\hat{U}$ and K can also be expressed by the inequalities

$$|\hat{U}(x,\epsilon)| \leqslant \text{const.}|x|^{-\sigma} \qquad \text{in } G_{\kappa_0},$$

$$|K(x,\epsilon)| \leqslant \text{const.}|x|^{-\tau-N/\kappa_0}\epsilon^N \qquad \text{in } G_{\kappa_0}. \tag{7.6-2}$$

Here $|\cdots|$ denotes a norm for matrices with entries in $\mathbb{C}$. The second constant in (7.6-2) depends on N.

For $(x,\epsilon) \in G_\kappa$ one has $(\epsilon/c)^\kappa \leqslant |x| \leqslant x_1$, i.e., $\epsilon \leqslant |x|^{1/\kappa}c$, and (7.6-2) implies, therefore, that

$$|K(x,\epsilon)| \leqslant \text{const.}|x|^{-\tau-1/\kappa+N(1/\kappa-1/\kappa_0)}\epsilon, \tag{7.6-3}$$

for $(x,\epsilon) \in G_\kappa$. The factor ϵ has been split off in this formula to explain the application of the Contraction Mapping Theorem, below. In order to deal with an integral equation with bounded coefficients the unknown function z in (3.3-8) is replaced by

$$z^* := zx^\sigma.$$

If $\kappa < \kappa_0$, the integer N can be chosen so large that the exponent in (7.6-3) is positive. Now write the integral equation (3.3-8) in the symbolic form

$$z^* = Tz^*. \tag{7.6-4}$$

Let G_κ^* be a subset of G_κ obtained from G_κ in (7.5-6) by replacing the sector S of (7.5-4) by a compact subset S^* depending on ϵ and assume that S^* is accessible, in the sense of Definition 3.3-2 with respect to q_k, for all ϵ in G^* and for some $\kappa < \kappa_0$. It follows from inequalities (7.6-2) and (7.6-3) that for each sufficiently small positive ϵ the transformation T maps vector functions of x that are bounded and holomorphic in S^* into functions with the same property. These functions form a Banach space. If v and w are two such vectors, the same estimate as in Section 3.3 leads to the inequality

$$\|Tw - Tv\| \leqslant \text{const. } \epsilon\|w - v\|.$$

Therefore, T is a contraction mapping for all sufficiently small $\epsilon > 0$, and the integral equation (7.6-4) has then a unique solution z^* bounded and holomorphic in G_κ^*.

It must still be shown that this solution is asymptotic to zero. The integral equation for z^* now becomes an expression for z^* of the form

$$z^*(x,\epsilon) = B(x,\epsilon)\int_{\Gamma_x} b_1(x,t,\epsilon)t^{-\tau}(t^{-1/\kappa_0}\epsilon)^N dt$$

$$+ \int_{\Gamma_x} b_2(x,t,\epsilon)t^{-\tau}(t^{-1/\kappa_0}\epsilon)^N dt, \tag{7.6-5}$$

with B bounded in G_κ^* and b_1, b_2 bounded for $(x,\epsilon) \in G_\kappa^*$ and $t \in \Gamma_x$, i.e., on all paths $\gamma_{jx}, j = 1, 2, \ldots, n$. These paths are uniformly bounded in length, therefore (7.6-5) implies

$$|z^*(x,\epsilon)| \leqslant \text{const.} \sum_{j=1}^{n} \left| \int_{\gamma_{jx}} |t|^{-\tau-N/\kappa_0} |dt| \right| \epsilon^N. \tag{7.6-6}$$

To appraise these integrals observe that $|t| \geqslant (\epsilon/c)^\kappa$ on γ_{jx}. Therefore,

$$|z^*(x,\epsilon)| \leqslant \text{const.} \, \epsilon^{-\kappa\tau+(1-\kappa/\kappa_0)N}$$

for $x \in G_\kappa$, and for $\kappa < \kappa_0$ the solution z is, thus, asymptotic to zero in the strong sense that

$$|z(x,\epsilon)| \leqslant k_N \epsilon^N \qquad \text{for all} \quad N \geqslant 0, \quad (x,\epsilon) \in G_\kappa, \quad \kappa < \kappa_0. \tag{7.6-7}$$

The constants k_N depend on κ.

This essentially completes the proof of Theorem 7.6-1 below. The final steps are explained after the formal statement of the theorem, which will be preceded by a brief review of the hypotheses:

Let the matrices A_r, $r = 0, 1, \ldots,$ be holomorphic functions of x in $|x| \leqslant x_1$ and let the matrix A have the asymptotic expansion

$$A(x,\epsilon) \sim \sum_{r=0}^{\infty} A_r(x)\epsilon^r, \tag{7.6-8}$$

valid as $\epsilon \to 0$ in

$$E = \{\epsilon \in C/0 < |\epsilon|, \delta_1 \leqslant \arg\epsilon \leqslant \delta_2\}, \tag{7.6-9}$$

uniformly for $|x| \leqslant x_1$. Consider a Basic Formal Matrix Solution

$$\left(\sum_{r=0}^{\infty} Y_r(x)\epsilon^{r/m} \right) e^{Q(x,\epsilon)} \tag{7.6-10}$$

(see formula (2.8-12)) of the formal differential equation

$$\epsilon^h \frac{dy}{dx} = \left(\sum_{r=0}^{\infty} A_r(x)\epsilon^r \right) y \tag{7.6-11}$$

(see (2.8-11)). Then, for some numbers σ_0 and $\kappa_0 > 0$, each function

$$Y_r(x)x^{\sigma+r/\kappa} \tag{7.6-12}$$

is bounded in $|x| \leqslant x_1$, whenever $\sigma > \sigma_0$, $\kappa \leqslant \kappa_0$ (see Theorem 7.4-1).

The matrix Q is diagonal,

$$Q = \text{diag}(q_1, q_2, \ldots, q_n).$$

The columns of the matrix (7.6-10),

$$\left(\sum_{r=0}^{\infty} y_{rk}(x)\epsilon^{r/m} \right) e^{q_k(x,\epsilon)}, \qquad k = 1, 2, \ldots, n, \tag{7.6-13}$$

are independent formal vector solutions of (7.6-11).

Theorem 7.6-1. *Let κ_0 be the restraint index of the series $\sum_{r=0}^{\infty} Y_r(x)\epsilon^{r/m}$ in Theorem 2.8-1 and let $\kappa < \kappa_0$. Let G_κ be a set of pairs (x,ϵ) as defined in*

(7.5-6) (*but with ϵ replaced by $\epsilon^{1/m}$*). *Assume that in subset $G_\kappa^* \subset G_\kappa$ the set*

$$D^*(\epsilon') = \{ x/(x,\epsilon) \in G_\kappa^*, \epsilon = \epsilon' \}$$

in $\mathbb{C}_L$ is accessible for q_k whenever $\epsilon' \in E$, [cf. (7.6-9)]. Then the differential equation

$$\epsilon^h \frac{dy}{dx} = A(x,\epsilon)y$$

[$A(x,\epsilon)$ as in (7.6-8)] possesses a vector solution of the form

$$y = \hat{y}_k(x,\epsilon) = \hat{y}_k(x,\epsilon)e^{q_k(x,\epsilon)} \tag{7.6-14}$$

in which $\hat{y}_k(x,\epsilon)$ has an asymptotic power series expansion

$$\hat{y}_k(x,\epsilon) \sim \sum_{r=0}^{\infty} y_{rk}(x)\epsilon^{r/m}$$

in the sense that, for $N = 0, 1, 2, \ldots,$

$$\left| \hat{y}_k(x,\epsilon) - \sum_{r=0}^{N} y_{rk}\epsilon^{r/m} \right| \leqslant C_N x^{-\sigma}(x^{-1/\kappa_0}\epsilon^{1/m})^{N+1} \tag{7.6-15}$$

for $(x,\epsilon) \in G_\kappa^*$, σ a real constant, c_N a positive constant.

PROOF. By analogy with Section 3.2, the $\hat{y}_k$ of (7.6-14) is the difference between z as in (7.6-7) and a function $\hat{u}_k$ which is constructed by the Borel–Ritt theorem so as to have the power series in (7.6-13) as its asymptotic expansion. However, this asymptotic expansion is now taken to be valid in the sense of Theorem 7.5-1. Therefore, for $(x,\epsilon) \in G_\kappa^*$,

$$\left| \hat{y}(x_k,\epsilon) - \sum_{r=0}^{N} y_{rk}(x)\epsilon^{r/m} \right| = \left| z(x,\epsilon) + \hat{u}_k(x,\epsilon) - \sum_{r=0}^{N} y_{rk}(x)\epsilon^{r/m} \right|$$

$$\leqslant k_{N+1}|\epsilon|^{N+1} + c_N|x|^{-\sigma}|x^{-1/\kappa}\epsilon|^{N+1}.$$

Here, (7.5-7) and (7.6-7) have been used. The last inequality implies (7.6-15). Thus, Theorem 7.6-1 is proved.

7.7. Asymptotic Solutions in Expanding Regions: A Local Theorem

To apply Theorem 7.6-1 one must find accessible domains. First, an extended form of Theorem 3.4-2 will be proved. As in Chapter III let q_k be the exponent in (3.2-14) and define $q_{jk} := q_j - q_k$. From the construction of the formal solution in Chapter II it is clear that, while q_{jk} may contain logarithmic terms, its derivative, if $x = 0$ is a turning point, has the form

$$\frac{dq_{jk}(x,\epsilon)}{dx} = x^{-\sigma_{jk}}\epsilon^{-h_{jk}/m} \sum_{s=0}^{s_0} g_{jk,s}(x)\epsilon^{s/m}. \tag{7.7-1}$$

Here, σ_{jk} is a rational number, not necessarily positive, h_{jk} and m are positive integers, s_0 is a non-negative integer, and the function $g_{ik,s}$ are, for $|x| \leqslant x_0$, holomorphic in some fractional power of x. Also, $g_{jk,0}(0) \neq 0$. Therefore dq_{jk}/dx has the form

$$\frac{dq_{jk}}{dx} = x^{-\sigma_{jk}} \epsilon^{-h_{jk}/m} \left(g_{jk,0}(0) + o(1) + O(\epsilon^{1/m}) \right) \qquad (7.7\text{-}2)$$

as $x \to 0$, $\epsilon \to 0$, and it can be assumed that x_0 and ϵ_0 are so small that $dq_{jk}/dx \neq 0$ for $0 < |x| \leqslant x_0$, $0 < \epsilon \leqslant \epsilon_0$ and $1 \leqslant j \leqslant n$, $j \neq k$. Then all points $|x| \leqslant x_0$ are there asymptotically simple for q_k, according to Definition 3.4-1.

Now choose a fixed value of ϵ and a point $x = x_1$ in $(\epsilon/c)^{\kappa} \leqslant |x| \leqslant x_0$. To construct an accessible domain which contains x_1, consider the domain

$$S = \left\{ x \Big/ \left(\frac{\epsilon}{c} \right)^{\kappa} \leqslant |x| \leqslant x_0, |\arg x - \arg x_1| \leqslant \delta \right\}, \qquad (7.7\text{-}3)$$

and subject it to the mapping $\xi = q_{jk}(x, \epsilon)$, which depends, of course, on j, k, and ϵ. If $\sigma_{jk} \neq 1$, then, for small x_0 and ϵ, the image $\xi(S)$ differs as little as desired from a domain bounded by two concentric circular arcs around the origin and two radial segments, and if δ in (7.7-3) is sufficiently small a point can be found on the boundary of $\xi(S)$ from which every point in that domain can be reached by a path along which $\text{Re}\,\xi$ decreases. The preimage of this path in the x-plane is then progressive in the sense of Definition 3.3-1. The angle δ may be chosen independent of ϵ.

If $\sigma_{jk} = 1$, the mapping $\xi = q_{jk}(x, \epsilon)$ produces a domain $\xi(S)$ close to a rectangle and the same reasoning can be applied. Hence, if δ is sufficiently small, the domain S is accessible and Theorem 7.6-1 applies. This result is stated below in more precise language.

Theorem 7.7-1. *Let (x_1, ϵ') be a point of the set G_κ defined in (7.5-6). Then this point belongs to a set of the form*

$$G_\kappa^* = \left\{ (x, \epsilon) \Big/ \left(\frac{\epsilon}{c} \right)^{\kappa} < |x| < x_0, |\arg x - \arg x_1| < \delta, 0 < \epsilon < \epsilon_0 \right\}$$

in which the asymptotic relation (7.6-15) is true, provided x_0, ϵ_0, and δ are sufficiently small.

In many applications the possibly narrow unknown sector of width 2δ in this theorem can be replaced by a known larger sector. Two such instances have already been stated and used before, namely, in Theorems 6.2-1 and 6.3-1. The construction of the accessible domains which are the justification for those theorems is straightforward and can be omitted.

CHAPTER VIII
Connection Problems

8.1. Introduction

The results of Chapter VII extend the regions in which the solutions of the differential equations are asymptotically known, but they do not solve the connection problems, central or lateral, as defined in Section 7.1. Only for relatively simple equations such as those analyzed in Chapter VI are the theorems of Chapter VII of decisive usefulness, by leading to a reduction of the given equation to well known special equations in full neighborhoods of the turning point. In general, less elegant methods such as the stretchings and matchings to be investigated in this chapter have to be turned to. The existing theory is still quite incomplete and rather involved. Most of these investigations are best studied in the original papers such as Sibuya [77], [78]; Iwano and Sibuya [29]; Iwano [26], [28]; Olver [62], [63], [65]; Nakano [49]; and Nishimoto [53]–[58]. In this account only differential equations of the form

$$\epsilon^h \frac{dy}{dx} = \begin{pmatrix} 0 & 1 \\ \phi(x,\epsilon) & 0 \end{pmatrix} y \qquad (8.1\text{-}1)$$

with a turning point at $x = 0$ will be discussed. The known results for systems of higher order are less satisfying and more complicated.

Let

$$\phi(x,\epsilon) = \sum_{r=0}^{\infty} \phi_r(x)\epsilon^r, \qquad |x| \leqslant x_0, \quad |\epsilon| \leqslant \epsilon_0 \qquad (8.1\text{-}2)$$

with

$$\phi_r(x) = \sum_{s=0}^{\infty} \phi_{rs} x^s, \qquad |x| \leqslant x_0, \quad r = 0, 1, \ldots . \qquad (8.1\text{-}3)$$

No serious new difficulties arise if the series in (8.1-2) is only an asymptotic representation of $\phi(x, \epsilon)$. For simplicity's sake convergence will be assumed here.

The differential equation (8.1-1) has a turning point at $x = 0$ if $\phi_{00} = 0$. If

$$\phi_{00} = \phi_{01} = \cdots \phi_{0,m-1} = 0, \qquad \phi_{0m} \neq 0, \tag{8.1-4}$$

the turning point is said to be of order m.

For a preliminary illustration of the difficulties to be encountered, consider the very simple equation

$$\epsilon \frac{dy}{dx} = \begin{pmatrix} 0 & 1 \\ x^m & 0 \end{pmatrix} y. \tag{8.1-5}$$

From Chapter VII it is known that the asymptotic expansions for the solutions derived in Chapters II and III are valid, in weakened form, in certain regions that depend on ϵ so as to approach the origin at the rate $O(\epsilon^\kappa)$ where κ may be any positive number less than the "restraint index" κ_0. In Sections 8.2 and 8.3 a method for calculating κ_0 will be explained. For the special equation (8.1-5) it turns out to be $\kappa_0 = 2/(m + 2)$. The solution derived in Chapters II and III are commonly called *outer* solutions. It is a natural idea to subject the differential equation to the stretching transformation

$$x = t\epsilon^\kappa \tag{8.1-6}$$

with a suitable exponent κ and to apply the method of Chapter II and III to the stretched equation in the hope that the new solution will then be asymptotically known in regions that overlap with the region in which the outer solution is known, and that a finite number of successive stretchings will lead to an asymptotic knowledge of the solutions at the turning point.

The stretched equation resulting from (8.1-5) is

$$\epsilon^{1-\kappa} \frac{dy}{dt} = \begin{pmatrix} 0 & 1 \\ t^m \epsilon^{\kappa m} & 0 \end{pmatrix} y. \tag{8.1-7}$$

If the program of Chapter II is followed, the next step toward the solution of (8.1-7) is the shearing transformation

$$y = \begin{pmatrix} 1 & 1 \\ 0 & \epsilon^{\kappa m/2} \end{pmatrix} v.$$

It produces the equation

$$\epsilon^{1-\kappa/\kappa_0} \frac{dv}{dt} = \begin{pmatrix} 0 & 1 \\ t^m & 0 \end{pmatrix} v, \tag{8.1-8}$$

which differs from (8.1-5) only by the size of the exponent of ϵ. It is clear that for $\kappa < \kappa_0$ no finite number of repetitions of the same procedure will ever lead to a knowledge of the solution at $x = 0$.

If $\kappa = \kappa_0$, (8.1-8) becomes

$$\frac{dv}{dt} = \begin{pmatrix} 0 & 1 \\ t^m & 0 \end{pmatrix} v. \tag{8.1-9}$$

This differential equation can be solved by series in ascending powers of t that converge in every finite disk in the t-plane. In terms of x and ϵ this is then a convergent series in powers of some fractional power of ϵ. However, no matter how large that disk in the t-plane, its image in the x-plane will not contain points which, for small ϵ, have distance of $O(\epsilon^\kappa)$, $\kappa > \kappa_0$ of $x = 0$. To obtain an overlap that will permit matching solutions of the differential equation (8.1-9) must be determined that are known, both for finite t *and also near $t = \infty$*. This is a general situation. *Central connection problems in the theory with a small parameter can only be solved with the help of connection formulas from the asymptotic theory for some related differential equation, as the independent variable tends to infinity.* Such matching techniques will be taken up later.

8.2. Stretching and Parameter Shearing

The combined effect of the stretching

$$x = t\epsilon^\kappa, \qquad \kappa > 0 \tag{8.2-1}$$

and the shearing

$$y = \begin{pmatrix} 1 & 0 \\ 0 & \epsilon^\alpha \end{pmatrix} v, \qquad \alpha > 0 \tag{8.2-2}$$

on the differential equation (8.1-1) is, in the general case, the new equation

$$\epsilon^{h-\kappa-\alpha} \frac{dv}{dt} = \begin{bmatrix} 0 & 1 \\ \displaystyle\sum_{r,s=0}^{\infty} \phi_{rs} \epsilon^{r+\kappa s - 2\alpha} t^s & 0 \end{bmatrix} v. \tag{8.2-3}$$

In the task of finding choices for κ and α for which (8.2-3) is especially simple, a geometric interpretation in an auxiliary (X, Y)-plane will be helpful. In that plane consider the set $\{P_{rs}\}$ of those points with non-negative integral coordinates $X = r$, $Y = s$ for which ϕ_{rs} in (8.2-3) is not zero. Note that $(0, m)$ is the lowest such point on the Y-axis, because of (8.1-4). It is undesirable to have negative powers of ϵ in the right hand member of (8.2-3). This means that κ and α *must* be chosen so that no point of $\{P_{rs}\}$ lies below the line l with equation

$$X + \kappa Y - 2\alpha = 0. \tag{8.2-4}$$

There are, of course, infinitely many such lines. In searching for the "best line," two cases must be distinguished. The first, simpler, case occurs if among those lines these is one for which, in addition, $h - \kappa - \alpha \leqslant 0$, for then (8.2-3) does no longer depend in a singular manner on ϵ, at $\epsilon = 0$. Writing the last inequality in the form $2h + \kappa(-2) - 2\alpha \leqslant 0$ one sees that it is satisfied whenever the line (8.2-4) is below—or at least not above—the point $(X, Y) = (2h, -2)$.

In the second case, i.e., when the singularity in ϵ cannot be so easily eliminated, it is best to choose the line (8.2-4) so that it contains the point $(0, m)$ and at least one other point of the set $\{P_{rs}\}$. (The condition that no point of $\{P_{rs}\}$ is below the line must, of course, be maintained.) For the corresponding terms of the summation in (8.2-3) the exponent s is less than m and the resulting new differential equation *has a turning point of order less than m at the origin.*

It has thus been shown that the differential equation (8.1-1) can always be significantly simplified by an appropriate choice of κ and α in the stretching and shearing transformations (8.2-1), (8.2-2). Either the new differential equation is no longer singular in ϵ (or in some rational power of ϵ) at $\epsilon = 0$, or it has a turning point of lower order at the origin than (8.1-1) had.

In the first case the differential equation can be solved in a t-neighborhood of the origin by means of series in powers of ϵ. In the second case an outer solution of the new equation can be found by the method of Chapters II and III. The restraint index of the equation determines how close to the origin this new asymptotic solution is valid. This process ends after at most m iterations. It produces a set of annuli in each of which asymptotic solutions can be found. For the sake of simpler reference the first case may be called "fully reducible" by stretching and shearing. The second case is then "partially reducible."

If it were true that these annuli overlap, a full neighborhood of the turning point would be covered by such open annuli, which would greatly facilitate the explicit calculation of connection formulas. It suffices to analyze the first of those transformations, the stretching and shearing (8.2-1), (8.2-2).

The relative size of the number κ in (8.2-1), as it results from the description given, and of the restraint index κ_0 of the original equation (8.1-1) is what matters. If $\kappa < \kappa_0$, then the outer solution of the original equation is valid in regions of the x-plane where $t = x\epsilon^{-\kappa}$ is small enough to use the outer solution of the new differential equation.

Unfortunately, this is not the case. In the next section it will be proved that $\kappa = \kappa_0$. Therefore, there is a gap between the zones with $|x| \geqslant M_1 \epsilon^{\kappa_1}$ ($\kappa_1 < \kappa_0$, M_1 a constant) where the outer solution of the original equation has a known asymptotic expansion and zones where $|t| \leqslant m_1$, i.e., $|x| \leqslant m_1 \epsilon^{\kappa} = m_1 \epsilon^{\kappa_0}$, in which the stretched equation can be asymptotically solved. As was mentioned before, explicit matching requires the asymptotic solution of the stretched equation in regions of the t-plane that extend to infinity as $\epsilon \to 0$. Such a procedure will be illustrated in Sections 8.4 and 8.5. At this point it will be helpful to illustrate the method described here by an example somewhat more typical than (8.1-5). Consider the equation

$$\epsilon \frac{dy}{dx} = \begin{pmatrix} 0 & 1 \\ x^6 + \epsilon x & 0 \end{pmatrix} y. \tag{8.2-5}$$

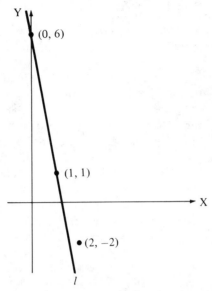

Figure 8.1. Stretching diagram for equation (8.2-9).

Figure 8.1 shows the (X, Y)-plane with the two points of the set $\{P_{rs}\}$ and the point $(X, Y) = (2h, -2) = (2, -2)$ as well as the line l. From (8.2-4) one sees here that

$$\kappa = \tfrac{1}{5}, \qquad \alpha = \tfrac{3}{5}.$$

The transformations (8.2-1), (8.2-2) take equation (8.2-5) into

$$\mu \frac{dv}{dt} = \begin{pmatrix} 0 & 1 \\ t + t^6 & 0 \end{pmatrix} v \tag{8.2-6}$$

with $\mu = \epsilon^{1/5}$.

For the new equation the diagram in the (X, Y)-plane has the appearance shown in Fig. 8.2. For the new line l one has

$$\kappa = \tfrac{2}{3}, \qquad \alpha = \tfrac{1}{3}.$$

The corresponding transformation,

$$t = s\mu^{2/3}, \qquad v = \begin{pmatrix} 1 & 0 \\ 0 & \mu^{1/3} \end{pmatrix} w$$

takes (8.2-6) into

$$\frac{dw}{ds} = \begin{pmatrix} 0 & 1 \\ s + \mu^{10/3}s^6 & 0 \end{pmatrix} w.$$

The method described in this section resembles in many—but not in all

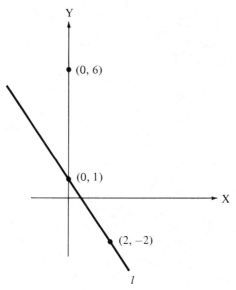

Figure 8.2. Second stretching for equation (8.2-9).

—respects the procedure of Iwano and Sibuya in [26]–[29]. Those authors make use of transformations of the form

$$y = \begin{pmatrix} 1 & 0 \\ 0 & x^{\beta} \end{pmatrix} z \qquad (8.2\text{-}7)$$

in combination with the changes of variables (8.2-1), (8.2-2). Their analysis applies to scalar differential equations of any order and to some more general systems. For equations of order two, "shearing" transformations of the independent variable such as (8.2-7) are not needed and, as far as I can see, do not simplify matters. I do not know if this is true also for equations of higher order.

Before leaving the subject of this section a feature of equation (8.2-6) should be mentioned which illustrates a common phenomenon and pushes still farther away the aim of a complete asymptotic solution of turning point problems, even for equations of order two. As was pointed out before, equation (8.2-6) must be solved asymptotically in large regions of the t-plane if a connection with the outer solutions of equation (8.2-5) is desired. Now, equation (8.2-6) has turning points in the t-plane besides $t = 0$, namely, the five zeros of the equation $t^5 + 1 = 0$. Such points may be called *secondary turning points*. If such secondary turning points are analyzed one often finds that there exist additional turning points, infinitesimally close to the secondary ones. I do not even know whether the successive turning points that appear in this manner are finite in number. This complicated subject will be left aside in the present account.

8.3. Calculation of the Restraint Index

As in the special cases analyzed in Chapter VI, the restraint index of equation (8.1-1) can be calculated by studying the orders of the singularities at $x = 0$ in the process of constructing formal solutions described in Chapter II. The first of those reduction steps was the diagonalization of the leading coefficient of the matrix in (8.1-1). It can be carried out by the transformation

$$y = \begin{pmatrix} 1 & 1 \\ \phi_0^{1/2} & -\phi_0^{1/2} \end{pmatrix} v.$$

A short calculation leads to

$$\epsilon^h v' = \left\{ \phi_0^{1/2} \begin{pmatrix} 1 & 0 \\ 0 & -1 \end{pmatrix} - \frac{\phi_0'}{4\phi_0} \begin{pmatrix} 1 & -1 \\ 1 & 1 \end{pmatrix} \epsilon^h \right.$$
$$\left. + \frac{1}{2} \phi_0^{-1/2} \sum_{r=1}^{\infty} \phi_r \begin{pmatrix} 1 & 1 \\ -1 & -1 \end{pmatrix} \epsilon^r \right\} v. \qquad (8.3\text{-}1)$$

For notational abbreviation let this be written as

$$\epsilon^h v' = \left(\sum_{r=0}^{\infty} A_r \epsilon^r \right) v.$$

Proceeding as in the proof of Theorem 6.1-1, one finds that the formal transformation

$$v = \left\{ I + \sum_{r=1}^{\infty} \begin{pmatrix} 0 & p_r^{12} \\ p_r^{21} & 0 \end{pmatrix} \epsilon^r \right\} z,$$

which takes (8.3-1) into the diagonalized form

$$\epsilon^h z' = \left\{ A_0 + \sum_{r=1}^{\infty} \begin{pmatrix} b_r^{11} & 0 \\ 0 & b_r^{22} \end{pmatrix} \epsilon^r \right\} z,$$

satisfies the relations

$$b_r^{11} = h_r^{11}, \qquad b_r^{22} = h_r^{22}$$
$$p_r^{12} = -\tfrac{1}{2}\phi_0^{-1/2} h_r^{12}, \qquad p_r^{21} = -\tfrac{1}{2}\phi_0^{-1/2} h_r^{21}, \qquad r \geqslant 1 \qquad (8.3\text{-}2)$$
$$p_0^{jk} = \delta_{jk}, \qquad j,k = 1,2; \qquad b_0^{11} = \phi^{1/2}, \qquad b_0^{22} = -\phi^{1/2},$$

where

$$\begin{pmatrix} h_r^{11} & h_r^{12} \\ h_r^{21} & h_r^{22} \end{pmatrix} = H_r := -\sum_{s=1}^{r-1} (P_s B_{r-s} - A_{r-s} P_s) + A_r - P_{r-h}', \qquad r > 0,$$

$$(8.3\text{-}3)$$

with

$$B_r = \begin{pmatrix} b_r^{11} & 0 \\ 0 & b_r^{22} \end{pmatrix}, \qquad P_r = \begin{pmatrix} 0 & p_r^{21} \\ p_r^{21} & 0 \end{pmatrix}, \qquad r > 0,$$

$$B_0 = A_0, \qquad P_0 = I.$$

(8.3-4)

In (8.3-3) it is understood that $P_s = 0$ for $s < 0$, by definition, and the summation in the formula is absent for $r = 1$. In particular,

$$H_1 = A_1.$$

The calculation of the restraint index requires the calculation of the rate of growth of the B_r and P_r, as $x \to 0$. These functions are meromorphic in $x^{1/2}$; therefore there are numbers $k_r, r = 1, 2, \ldots$, integral multiples of $\frac{1}{2}$, such that

$$H_r(x) = \tilde{H}_r(x) x^{-k_r}$$

(8.3-5)

with $\tilde{H}_r(x)$ bounded and not zero at $x = 0$. For the calculation of the k_r one must introduce the numbers β_r defined by the relation

$$\phi_r(x) = \tilde{\phi}_r(x) x^{\beta_r}$$

with $\tilde{\phi}_r(x)$ bounded and not zero at $x = 0$, and, likewise,

$$A_r(x) = \tilde{A}_r(x) x^{-\alpha_r}.$$

Observe that for each $r \geqslant 0$, the point (r, β_r) is the lowest point of the set $\{P_{rs}\}$ on the line $X = r$. If $\phi_r \equiv 0$, define $\beta_r = \infty$. The β_r are non-negative integers, and by (8.3-1),

$$\alpha_0 = -\tfrac{1}{2}m, \qquad \alpha_r = \tfrac{1}{2}m - \beta_r, \qquad r \neq h,$$

$$\alpha_h = \max(1, \tfrac{1}{2}m - \beta_h).$$

(8.3-6)

By (8.3-2), and (8.3-5) the growth rates of B_r and P_r are k_r and $k_r + \tfrac{1}{2}m$, respectively. If the diagonalized differential equation is then formally solved by the method of Chapter II, it is seen that the restraint index of the differential equation is

$$\kappa_0 = \inf_{r \to \infty} \frac{r}{k_r},$$

(8.3-7)

according to Definition 7.4-2. The k_r can be successively calculated from the formulas

$$k_r = \begin{cases} \alpha_1 & \text{for} \quad r = 1; \\ \max_{s=1,2,\ldots,r-1} \left[k_s + k_{r-s} + \tfrac{1}{2}m, k_s + \tfrac{1}{2}m + \alpha_{r-s}, \alpha_r \right], & r = 2, \ldots, h; \\ \max_{s=1,2,\ldots,r-1} \left[k_s + k_{r-s} + \tfrac{1}{2}m, k_s + \tfrac{1}{2}m + \alpha_{r-s}, \alpha_r, k_{r-h} + \tfrac{1}{2}m + 1 \right], \\ \qquad\qquad\qquad\qquad\qquad\qquad\qquad\qquad\qquad r > h. \end{cases}$$

They follow from (8.3-3). For $h = 1$, the second of the three lines in the right-hand member is absent.

A first observation is that the term $k_r + \frac{1}{2}m + \alpha_{r-s}$ in the recursion formula for k_r can be omitted, since $k_s + k_{r-s} + \frac{1}{2}m - (k_s + \frac{1}{2}m + \alpha_{r-s}) = k_{r-s} - \alpha_{r-s} \geqslant 0$. Thus the formula for k_r simplifies into

$$
k_r = \begin{cases}
\alpha_1 & \text{for } r = 1; \\
\displaystyle\max_{s=1,2,\ldots,s-1} \left[k_s + k_{r-s} + \tfrac{1}{2}m, \alpha_r \right], & r = 2, 3, \ldots, h; \\
\displaystyle\max_{s=1,2,\ldots,s-1} \left[k_s + k_{r-s} + \tfrac{1}{2}m, \alpha_r, k_{r-h} + \tfrac{1}{2}(m+2) \right], & r > h.
\end{cases}
$$

$$(8.3\text{-}8)$$

The number κ_0 defined by (8.3-7) and (8.3-3) must now be compared with the number κ in the stretching transformation (8.2-1) when it is determined by the construction in Section 8.2. Remember that $-1/\kappa$ is the slope of the line l in (8.2-4). In the fully reducible case the line passes through $(0, m)$ and $(2h, -2)$, and then κ turns out to have the value

$$
\kappa = \frac{2h}{m+2}. \tag{8.3-9}
$$

In the other case the line l is steeper. Let (r_0, s_0) be a point of $\{P_{rs}\}$ on the line l. Then

$$
\kappa = \frac{r_0}{m - s_0} \tag{8.3-10}
$$

and

$$
\frac{2h}{m+2} > \frac{r_0}{m - s_0}.
$$

Theorem 8.3-1. *If κ is the constant in (8.2-1) found by the rule described in Section 8.2, and if κ_0 is the restraint index of equation (8.1-1), then*

$$
\kappa = \kappa_0.
$$

PROOF. The arguments are somewhat more simply explained if k_r in (8.3-8) is replaced by

$$
p_r := k_r + \tfrac{1}{2}m. \tag{8.3-11}
$$

Then (8.3-8) reads, if (8.3-6) is also made use of,

$$
p_r = \begin{cases}
\alpha_1 + \tfrac{1}{2}m; \\
\displaystyle\max_{s=1,2,\ldots,r-1} \left\{ p_s + p_{r-s}, m - \beta_r \right\}, & r = 2, 3, \ldots, h; \\
\displaystyle\max_{s=1,2,\ldots,r-1} \left\{ p_s + p_{r-s}, m - \beta_r, p_{r-h} + \tfrac{1}{2}(m+2) \right\}, & r > h.
\end{cases}
$$

$$(8.3\text{-}12)$$

The two formulas (8.3-9), (8.3-10) can be combined into one as follows. In both cases the line l contains the point $(0, m)$. If it does not contain the point $(2h, -2)$ it contains one or more of the points (r, β_r), with β_r defined after (8.3-5), namely, the ones for which the slope

$$-\frac{m - \beta_r}{r}, \qquad r = 0, 1, \ldots,$$

is a minimum. Formula (8.3-10) applies when this minimum is less than $-(m + 2)/2h$, the slope of l in the fully reducible case. Hence, not all points (r, β_r) need to be considered: The minimum of $-(m - \beta_r)/r$ is assumed for some r in the interval $0 < r \le 2mh/(m + 2)$ and for $\beta_r < m$. Since $-1/\kappa$ is the slope of that line, one has in all cases, for $\beta_r < m$,

$$\kappa = \min\left[\frac{2h}{m + 2}, \frac{r}{m - \beta_r} \right], \qquad 0 < r < \frac{m}{m + 2} 2h. \tag{8.3-13}$$

Formula (8.3-7) is equivalent to

$$\kappa_0 = \inf_{r > 0} \frac{r}{p_r}. \tag{8.3-14}$$

To prove the theorem in the more interesting partially reducible case it amply suffices to show that

$$p_r = \frac{r}{\kappa} \qquad \text{for all} \quad r \ge 1. \tag{8.3-15}$$

By (8.3-6) and (8.3-12),

$$p_1 = m - \beta_1.$$

Therefore, (8.3-13) implies the truth of (8.3-15) for $r = 1$. Assume it to be true up to $p_{\rho-1}$, $\rho > 1$. Then (8.3-12) implies

$$p_\rho = \begin{cases} \displaystyle\max_{s = 1,2,\ldots,\rho-1}\left\{ \frac{\rho}{\kappa}, m - \beta_r \right\}, & \rho = 2, 3, \ldots, h. \\[2ex] \displaystyle\max_{s = 1,\ldots,\rho-1}\left\{ \frac{\rho}{\kappa}, m - \beta_s, \frac{\rho - h}{\kappa} + \frac{m + 2}{2} \right\}, & \rho > h. \end{cases} \tag{8.3-16}$$

But (8.3-13) implies that $\kappa \le s/(m - \beta_s)$ for all $\beta_s < m$, i.e., $s/\kappa \ge m - \beta_s$, and therefore $\rho/\kappa \ge m - \beta_s$ for all $s \le \rho$. Finally, the difference of the first and third term in braces in the right-hand term of (8.3-16) is

$$h/\kappa - (m + 2)/2,$$

which is non-negative by (8.3-13). Thus, (8.3-16) means that $p_\rho = \rho/\kappa$, and Theorem 8.3-1 for the partially reducible case is proved by induction.

If $\kappa = 2h/(m + 2)$ in (8.3-13), the argument begins with the relation $p_1 < 1/\kappa$, and then the induction proceeds as before, completing the proof of the theorem.

8.4. Inner and Outer Solutions for a Particular nth-Order System

For systems of higher order than two the difficulties of the method described in this chapter, so far, are so great that they have been completely overcome in a few cases only. One of those is the Somerfeld–Orr equation mentioned in the historical introduction. The techniques employed there are specifically adapted to that problem and require a book by themselves. Such a book has been published [9], and that topic will be omitted from this account.

Instead, this section and the next will sketch an application of the stretching and matching method to a relatively simple equation of order n with a first-order turning point at $x = 0$. The details of this rather long analysis can be found in the papers [88], [92], and [94]. Other turning point problems for differential equations of order higher than two have been studied by Iwano [26], [27], Iwano and Sibuya [29], and Nishimoto [54]–[56].

The differential equation to be studied is

$$\epsilon \frac{dy}{dx} = A(x)y, \qquad (8.4\text{-}1)$$

with

$$A(x) = \begin{bmatrix} 0 & 1 & 0 & \cdots & 0 & 0 \\ 0 & 0 & 1 & \cdots & 0 & 0 \\ \cdot & \cdot & \cdot & \cdots & \cdot & \cdot \\ 0 & 0 & 0 & \cdots & 0 & 1 \\ x & a_{n-1}(x) & a_{n-2}(x) & \cdots & a_2(x) & 0 \end{bmatrix}. \qquad (8.4\text{-}2)$$

If

$$a_j(0) = 0, \qquad j = 2, 3, \ldots, n-1, \qquad (8.4\text{-}3)$$

then the differential equation has a turning point at $x = 0$. In Theorem 5.2-1 it was stated without proof that a certain more general class of differential equations is formally equivalent to (8.4-1) with (8.4-2), (8.4-3) in a full neighborhood of the turning point at $x = 0$. In Ref. 92 I have proved, as part of a more general result, that the formal equivalence implies holomorphic equivalence in certain closed sectors of the x-plane with vertex at $x = 0$. Therefore, the illustration of the matching procedure to be given is of greater generality than equations (8.4-1), (8.4-2), and (8.4-3) indicate.

The first part of the process of asymptotic solution of equation (8.4-1) is the formal calculation of the series solution by the method of Chapter II. This is now a routine matter, because the eigenvalues $\lambda_1(x), \ldots, \lambda_n(x)$ of $A(x)$ are distinct in a punctured disk $0 < |x| \leqslant x_0$ and are all equal to zero at $x = 0$. They have the form $\lambda_j(x) = x^{1/n}(1 + O(x))$ with the n determinations of $x^{1/n}$. Set

$$T(x) = \begin{bmatrix} 1 & 1 & \cdots & 1 \\ \lambda_1(x) & \lambda_2(x) & \cdots & \lambda_n(x) \\ \lambda_1^2(x) & \lambda_2^2(x) & \cdots & \lambda_n^2(x) \\ & \cdots & \\ \lambda_1^{n-1}(x) & \lambda_2^{n-1}(x) & \cdots & \lambda_n^{n-1}(x) \end{bmatrix},$$

then the change of variable

$$y = T(x)v$$

produces a new differential equation, whose leading coefficient matrix is

$$\Lambda(x) = T^{-1}AT = \operatorname{diag}(\lambda_1, \lambda_2, \ldots, \lambda_n).$$

The details of the complete formal diagonalization of the differential equation can be found in [94]. That paper also shows, at the same time, that the restraint index for the equation is

$$\kappa_0 = \frac{n}{n+1}.$$

The concepts of the theory behind these matters should be clear by now, and a reference to the paper [94] for further details must suffice.

The formal matrix solution of the differential equations (8.4-1)–(8.4-3) has, by Theorem 2.6-1, the form

$$Y(x, \epsilon) = \hat{Y}(x, \epsilon) \exp\left\{ \frac{1}{\epsilon} \int_0^x \Lambda(t)\, dt \right\}. \tag{8.4-4}$$

The coefficients in the formal series

$$\hat{Y}(x, \epsilon) = \sum_{r=0}^{\infty} Y_r(x)\epsilon^r$$

have the property that for all r

$$\lim_{x \to 0} Y_r(x) x^{r/\kappa} = 0$$

whenever $\kappa < \kappa_0 = n/(n + 1)$, but that this limit relation is false for infinitely many r if $\kappa > n/(n + 1)$.

In [94] can be found more precise information on the nature of the function $Y_r(x)$. They can be expanded in convergent series near $x = 0$. These series contain fractional powers of x and, in general, also of $\log x$.

Theorems 7.6-1 and 7.7-1 guarantee the existence of a matrix solution $Y(x, \epsilon)$ of the differential equation which is asymptotically represented by the expression $Y(x, \epsilon)$ of (8.4-4) in certain, possibly narrow, sectors of the x-plane. For the relatively simple differential equation of this section wider regions of accessibility can be constructed, as shown in [92], and the regions of asymptotic validity of the outer formal solutions have the form

$$G = \left\{ (x, \epsilon) / M\epsilon^\kappa < |x| < m, \vartheta_1 < \arg x < \vartheta_1 + \frac{n}{n+1} \pi, 0 < \epsilon < \epsilon_0 \right\}.$$

$$(8.4\text{-}5)$$

Here the constants κ, M, m, ϵ_0 must satisfy certain inequalities, in particular, $\kappa < \kappa_0 = n/(n+1)$, and m may have to be small. The angle ϑ_1 is arbitrary. The asymptotic representation is valid in the extended sense explained in Chapter VII.

By trial and error, or by a reasoning analogous to what was done for $n = 2$ in the preceding section, one finds that the appropriate stretching and shearing transformations in this example are

$$x = t\epsilon^{n/(n+1)}, \qquad y = \Omega(\epsilon^{1/(n+1)})v, \tag{8.4-6}$$

where

$$\Omega(s) := \operatorname{diag}(1, s, \ldots, s^{n-1}).$$

They take (8.4-1) formally into

$$\frac{dv}{dt} = H(t, \epsilon)v \tag{8.4-7}$$

with

$$H(t, \epsilon) = \sum_{r=0}^{\infty} H_r(t)\epsilon^{r/(n+1)}. \tag{8.4-8}$$

A short calculation, omitted here, shows that

$$H_0(t) = \begin{bmatrix} 0 & 1 & 0 & 0 & \cdots & 0 \\ 0 & 0 & 1 & 0 & \cdots & 0 \\ & & \cdots & & & \\ 0 & 0 & 0 & 0 & \cdots & 1 \\ t & 0 & 0 & 0 & \cdots & 0 \end{bmatrix} \tag{8.4-9}$$

and that $H_r(t)$ is a polynomial in t whose degree does not exceed the number $1 + r/n$.

If the coefficients a_j of equation (8.4-2) are holomorphic at $x = 0$, it is easy to show that the power series in (8.4-8) has a positive radius of convergence, and represents the function H in a sufficiently small disk of the t-plane. It follows then from standard theorems for linear differential equations that (8.4-7) has fundamental solutions which are holomorphic in $\epsilon^{1/(n+1)}$ at $\epsilon = 0$ for all t in that disk. This result falls, however, far short of what is needed for the central connection problem. One wishes to find solutions of (8.4-7) that are known, not only in a fixed bounded disk of the t-plane, but in a domain that contains the origin and expands so rapidly, as $\epsilon \to 0$, that its image in the x-plane has for all small ϵ, a nonempty intersection with a region where the outer solution is asymptotically known. The latter regions approach the origin at best at a rate $O(\epsilon^\kappa)$ with $\kappa < n/(n+1)$. Therefore, the inner solutions must be known in regions that contain points whose approach to the origin, when mapped onto the x-plane by (8.4-6), is not faster, and preferably slower, than $O(\epsilon^\kappa)$. The images in the t-plane of such points have distance $O(\epsilon^{\kappa-n/(n+1)})$ from the origin. Since $\kappa > 0$ may have any value less than $n(n + 1)$, it suffices to find a solution for equation (8.4-7) that is asymptotically known in some region of the t-plane which expands at the rate $O(\epsilon^{-\alpha})$ for some positive α. The solution found in [94] does have this property with $\alpha = n/(n + 1)(n + 2)$.

To simplify the notation, set

$$\mu := \epsilon^{1/(n+1)}.$$

It is plausible to attempt the solution of equation (8.4-7) by replacing v with the matrix series

$$\sum_{r=0}^{\infty} V_r(t)\mu^r \tag{8.4-10}$$

in the equation, and then to identify successively the powers of μ in the two members. The recursive sequence of equations so obtained is

$$\frac{dV_0}{dt} = H_0(t)V_0,$$

$$\frac{dV_r}{dt} = H_0(t)V_r + \sum_{s=0}^{r} H_s(t)V_{r-s}(t), \qquad r > 0. \tag{8.4-11}$$

If the series in (8.4-10) is to represent a solution of (8.4-7), both for finite t and for values of t that are allowed to grow, as ϵ tends to zero, at such a rate that the corresponding value of x remains in a region of validity of the outer solution, the solutions of the differential equations (8.4-11) have to be chosen, successively, in a particular way. The necessity of invoking results from the parameterless asymptotic theory has been emphasized in earlier sections. Fortunately, equations (8.4-11) for $r = 0$ belong to the small class of differential equations whose asymptotic theory as $t \to \infty$ is completely known. One reference in which the equation for V_0 appears as a special case is the paper [82] by Turrittin. The properties of the nonhomogeneous equations (8.4-11) for $r > 0$ can then be derived with the help of the variation of parameters formula.

Once these solutions V_r have been suitably calculated one can prove that (8.4-10) represents a solution of the stretched equation asymptotically in the desired region. The details can be found in the sources quoted before. Here it must suffice to state the result as a theorem.

Theorem 8.4-1. *Define Q by $Q(t) := [n/(n + 1)]\Omega(\omega)t^{(n+1)/n}$, $\omega = e^{2\pi i/n}$ (Ω was defined after formula (8.4-6)). Equation (8.4-7) possesses a fundamental matrix solution V of the form*

$$V(t,\epsilon) = \Omega(t^{1/n})t^{(1-n)/2n}\hat{V}(t,\epsilon)e^{(1/\epsilon)Q(t)}, \tag{8.4-12}$$

where the matrix $\hat{V}(t,\epsilon)$ *has the following properties:*

(1) $\hat{V}(t,\epsilon) = \tilde{V}(t,\epsilon) t^{D(\epsilon)}$. $\qquad\qquad\qquad\qquad\qquad\qquad$ (8.4-13)

The matrix $D(\epsilon)$ is diagonal and has an expansion

$$D(\epsilon) \sim \sum_{r=1}^{\infty} D_r \mu^r, \qquad \mu := \epsilon^{1/(n+1)}, \quad \mu \to 0+; \qquad (8.4\text{-}14)$$

(2) $\tilde{V}$ *is bounded for*

$$|\mu t^{(n+2)/n}| < \eta_0, \qquad 0 < \epsilon < \epsilon_0, \quad |t| \geq t_0 > 0, \quad t \in \Sigma \qquad (8.4\text{-}15)$$

where Σ *is the sector*

$$\Sigma = \begin{cases} \left\{ t \Big/ |\arg t| \leq \dfrac{n\pi}{2(n+1)} \right\}, & \text{for odd } n \\[4mm] \left\{ t \Big/ -\dfrac{\pi}{2} \leq \arg \leq \dfrac{(n-1)\pi}{2(n+1)} \right\}, & \text{for even } n. \end{cases} \qquad (8.4\text{-}16)$$

The constants η_0, t_0 *can be chosen arbitrarily, but the solution depends on that choice.*

(3) $\tilde{V}$ *has, for* $t \in \Sigma$, $t \geq t_0$ *an asymptotic expansion*

$$\tilde{V}(t,\epsilon) \sim \sum_{r=0}^{\infty} \tilde{V}_r(t) \left(\mu t^{(n+2)/n} \right)^r \qquad \text{as} \quad \mu t^{(n+2)/2} \to 0. \qquad (8.4\text{-}17)$$

(4) *Each matrix* $\tilde{V}_r(t)$ *has an asymptotic expansion*

$$\tilde{V}_r(t) \sim \sum_{s=0}^{\infty} V_{rs} t^{-s/n} \qquad \text{as} \quad t \to \infty \quad \text{in } \Sigma. \qquad (8.4\text{-}18)$$

REMARKS.

(1) In analogy with situations met before, the relation (8.4-17) means that

$$\left| \tilde{V}(t,\epsilon) - \sum_{r=0}^{N} V_r(t) \left(\mu t^{(n+2)/n} \right)^r \right| \leq C_N |\mu t^{(n+2)/n}|^{N+1}. \qquad (8.4\text{-}19)$$

For all $N \geq 0$ and for (t,ϵ) satisfying the inequalities (8.4-15). The constants C_N depend on the domain defined by (8.4-15).

(2) Note that Theorem 8.4-1 does not contain information on the solution V in the disk $|t| < t_0$. This question will be taken up later.

(3) It is true that the sequence of differential equations (8.4-11) can be satisfied by many different functions V_r, but only one choice yields a solution of equation (8.4-6) with the properties of Theorem 8.4-1. A proof of this fact for $n = 2$ is contained in [93], Lemma 3.5. The Remark 2 on page 71 of [94] is not correct.

By utilizing the symmetry properties of equation (8.4-11) for $r = 0$, theorems analogous to Theorem 8.4-1 can be stated for each of a number of sectors different from Σ, which together cover all directions in the t-plane.

8.5. Calculation of a Central Connection Matrix

The outer solutions and the inner solutions of the differential equation (8.4-1) are obtainable from each other by right-hand multiplication with a matrix independent of x but possibly depending on ϵ. In this section this matrix will be asymptotically calculated for small ϵ, and, as in the preceding section, many not quite trivial calculations will be omitted with a reference to the paper [94].

Consider an outer solution $Y(x, \epsilon)$ that has the formal expression (8.4-4) as its asymptotic expansion in a region of the form (8.4-5) with

$$\frac{n}{n+2} < \kappa < \frac{n}{n+1},$$

and an inner solution

$$U(x, \epsilon) = \Omega(\mu)V(t, \epsilon)$$

with

$$\mu = \epsilon^{1/(n+1)}, \qquad x = t\mu^n,$$

and with V being a solution of the form described in Theorem 8.4-1. Let ϑ_1 in (8.4-5) be chosen so that G overlaps with the (x, ϵ) region in which Theorem 8.4-1 supplies the asymptotic expansion of V.

The solutions Y and U are related by an identity of the form

$$Y(x, \epsilon) = U(x, \epsilon)\Gamma(\epsilon), \tag{8.5-1}$$

i.e.,

$$\Gamma(\epsilon) = U^{-1}(x, \epsilon)Y(x, \epsilon). \tag{8.5-2}$$

The obvious approach of replacing x by some fixed point in the region where both the inner and the outer solution have known asymptotic expansion is unsatisfactory. It is better to search for an expression for $\Gamma(\epsilon)$ from which the spurious appearance of the number x has been eliminated. This requires an asymptotic analysis of the right member of (8.5-2) in that common region of validity.

The region of the (x, ϵ)-space in which U has a known expansion was defined in formula (8.4-15). The choice of the constants η_0, t_0, ϵ_0 is arbitrary (except that all points obeying the inequalities (8.4-15) must belong to values of x at which the coefficients a_j in (8.4-2) are holomorphic). Choose η_0, t_0, ϵ_0 and call the corresponding region S_i, the "inner domain." Similarly, G in (8.4-5) defines the "exterior domain" and $S_m = S_i \cap G$ may be called the "intermediate domain." It can be verified that for suitable choices of the constants occurring in the definitions of G and S_i, the domain S_m is not empty for any small ϵ.

A first simplification of $\Gamma(\epsilon)$ in (8.5-2) is a consequence of the fact that for small x the exponential functions $\exp\{(1/\epsilon)\int_0^x \Lambda(t)\,dt\}$ in (8.4-4), $\exp\{(1/\epsilon)Q(t)\}$ in (8.4-12) are almost equal. This fact entails—as can be shown by a short calculation—that the matrix $\hat{\Gamma}(x,\epsilon)$ defined by the relation

$$\Gamma(\epsilon) = e^{-Q(t)}\hat{\Gamma}(x,\epsilon)e^{Q(t)} \qquad (8.5\text{-}3)$$

is bounded in S_m.

The principal step in the simplification of the connecting formula (8.5-1) is the recognition of the fact that, in a certain sense, only the diagonal entries of $\Gamma(\epsilon)$ matter. To show this, denote the entries of Γ and $\hat{\Gamma}$ in (8.5-3) by γ_{jk} and $\hat{\gamma}_{jk}$ ($j,k = 1,2,\ldots,n$). Formula (8.5-3) and the definition of $Q(t)$ in Theorem 8.4-1 imply that

$$\gamma_{kk}(\epsilon) = \hat{\gamma}_{kk}(\epsilon)$$

and

$$\gamma_{jk}(\epsilon) = \hat{\gamma}_{jk}(x,\epsilon)\exp\left\{\frac{n}{n+1}(\omega^{j-1} - \omega^{k-1})t^{(n+1)/n}\right\}. \qquad (8.5\text{-}4)$$

From this relation one derives readily the theorem below.

Theorem 8.5-1. *The off-diagonal entries of the connection matrix $\Gamma(\epsilon)$ are asymptotic to zero, as $\epsilon \to 0 +$.*

PROOF. Let ϑ_{jk} be an angle for which

$$\text{Re}(\omega^{j-1} - \omega^{k-1})\exp i\,\frac{n+1}{n}\,\vartheta_{jk} = 0.$$

A simple calculation proves that none of these directions coincides with those of the boundary rays of the sector Σ defined in (8.4-16). (The distinction between even and odd n has to be taken into account in this connection.) Therefore, Σ contains for each pair (j,k), $j \neq k$, directions in which the exponent in (8.5-4) has negative real part. Let $\hat{\vartheta}_{jk}$ be such a direction and take t on that ray;

$$t = t_{jk}(\epsilon) = |t_{jk}(\epsilon)|e^{i\hat{\vartheta}_{jk}},$$

with $|t_{jk}(\epsilon)|$ depending on ϵ in such a way that the inequalities (8.4-15) are satisfied and that at the same time

$$x = x_{jk}(\epsilon) = t_{jk}(\epsilon)\mu^n.$$

is in G. Then this point belongs to the intermediate domain S_m. In terms of the variable x the width of the intermediate region is given by

$$M\epsilon^\kappa < |x| < \eta_0^{n/(n+2)}\epsilon^{[n/(n+1)][2/(n+2)]}.$$

Hence, there are many ways of choosing $x_{jk}(\epsilon)$ in that region so that $t_{jk}(\epsilon)$ tends to infinity, as $\epsilon \to 0$. For instance, one may set $|x| = \epsilon^{\kappa_1}$ with $\kappa_1 > 0$ being some number in the interval $\kappa > \kappa_1 > 2n/(n+1)(n+2)$. Then $|t_{jk}(\epsilon)| = O(\epsilon^{\kappa_1 - n/(n+1)})$. The exponent is negative, because $\kappa_1 < \kappa < n/(n+1)$. Inserting this value for t in formula (8.5-4) and remembering that $\hat{\gamma}_{jk}(x,\epsilon)$ is bounded uniformly for $0 < \epsilon < \epsilon_0$, $x \in S_m$, one concludes that Theorem 8.5-1 is true.

Theorem 8.5-1 reduces the task of calculating the outer solution in the inner domain to the calculation of the n products $u_{kk}^{-1}(x, \epsilon) y_{kk}(x, \epsilon)$ at one point, instead of having to calculate the whole matrix $U^{-1}Y$. This is much easier, but it still requires extensive computations, which can be found in [94]. The essence of those arguments is again to take advantage of the fact that $u_{kk}^{-1}(x, \epsilon) y_{kk}(x, \epsilon) = \gamma_{kk}(\epsilon)$, and to replace x by a variable point in S_m which depends on ϵ. The point chosen in [94] is such that $|x| = \epsilon^{\alpha}$ with $n/(n+2) < \alpha < n/(n+1)$. The result of these calculations is stated below.

Theorem 8.5-2. *The diagonal entries $\gamma_{kk}(\epsilon)$ of the matrix $\Gamma(\epsilon)$ defined in (8.5-2) have an asymptotic expansion of the form*

$$\gamma_{kk}(\epsilon) \sim \epsilon^{c_k(\epsilon)} \sum_{r=0}^{\infty} \gamma_{kk,r} \epsilon^{r/(n+1)}, \qquad \epsilon \to 0+,$$

in which

$$c_k(\epsilon) \sim \sum_{r=0}^{\infty} c_{kr} \epsilon^{r}.$$

The numbers $\gamma_{kk,r}$ and c_{kr} can be explicitly calculated.

It is, of course, possible to expand $\epsilon^{c_k(\epsilon)} = \exp\{\log(\epsilon) c_k(\epsilon)\}$ into an asymptotic series involving powers of ϵ and multiply that series with $\sum_{r=0}^{\infty} \gamma_{kk,r} \epsilon^{r/(n+1)}$, but the form given in Theorem 8.5-2 shows the structure of the result more clearly.

The explicit calculation of more than the first term of the series appearing in Theorem 8.5-2 is very cumbersome. I believe it would be a worthwhile endeavor for specialists to program the algebraic manipulations for an electronic computer.

Knowing the asymptotic expansion of the connection matrix $\Gamma(\epsilon)$ one can now calculate the asymptotic expansion of the outer solution in the inner region S_i. However, this still does not solve the connection problem for the equations (8.4-1)–(8.4-3) completely, as has been pointed out already, since the continuation of the solution into the disk $|t| \leqslant t_0$, i.e., $|x| \leqslant t_0 \epsilon^{1/(n+1)}$, has not yet been calculated.

To simplify the necessary calculations assume that $t_0 < 1$ in (8.4-15). Then the solution V of equation (8.4-7) in Theorem 8.4-1 has the value

$$V(1, \epsilon) = \tilde{V}(1, \epsilon) e^{[n/(n+1)] \Omega(\omega)} \tag{8.5-5}$$

at $t = 1$, and

$$\tilde{V}(1, \epsilon) \sim \sum_{r=0}^{\infty} \tilde{V}_r(1) \mu^r \qquad \text{as} \quad \mu \to 0+. \tag{8.5-6}$$

On the other hand, the formal solution obtained for equation (8.4-7) by solving successively the equations (8.4-11) so as to satisfy the initial conditions

$$V_0(1) = I, \qquad V_r(1) = 0, \qquad r = 1, 2, \ldots,$$

produces a formal solution of (8.4-7),

$$W(t,\epsilon) = \sum_{r=0}^{\infty} W_r(t)\epsilon^{r/(n+1)}, \qquad (8.5\text{-}7)$$

in which the W_r are entire functions with $W(1) = I$, $W_r(1) = 0$, $r > 0$. The series in (8.5-7) converges in any prescribed disk $|t - 1| \leqslant t_1$, if $|\epsilon| \leqslant \epsilon_1$ and ϵ_1 is chosen small enough. Hence, the analytic continuation of the solution V into the neighborhood of the origin is given by

$$V(t,\epsilon) = W(t,\epsilon)\tilde{V}(1,\epsilon)e^{[n/(n+1)]\Omega(\omega)}.$$

8.6. Connection Formulas Calculated Through Uniform Simplification

For systems of two equations the uniform reduction method of Chapters V and VI solves, in principle, the connection problems at turning points of order one and two. The numerical calculation of even the second term of the series in ϵ for the connection coefficients is, however, still an unpleasant, if straightforward, task. For the sake of concreteness these calculations will be briefly recapitulated for the differential equation

$$\epsilon \frac{dy}{dx} = \begin{pmatrix} 0 & 1 \\ a(x) & y \end{pmatrix} y, \qquad (8.6\text{-}1)$$

where

$$a(x_1) = 0, \qquad a'(x_1) \neq 0 \qquad (8.6\text{-}2)$$

and then applied to a concrete example. This material is mostly taken from [93].

The transformation of Lemma 5.1-2 is, in slightly changed notation,

$$t = \left[\frac{3}{2} \int_{x_1}^{x} a^{1/2} \, ds \right]^{2/3} \qquad (8.6\text{-}3)$$

$$y = \begin{bmatrix} 1 & 0 \\ 0 & \omega \end{bmatrix} w, \qquad \omega = \frac{dt}{dx}. \qquad (8.6\text{-}4)$$

It takes the differential equation (8.6-1) into

$$\epsilon \frac{dw}{dt} = \begin{bmatrix} 0 & 1 \\ t & -\epsilon \frac{d^2 t}{dx^2} \left(\frac{dx}{dt} \right)^2 \end{bmatrix} w. \qquad (8.6\text{-}5)$$

By Theorem 5.2-1 this equation can be formally simplified into

$$\epsilon \frac{dz}{dt} = \begin{pmatrix} 0 & 1 \\ t & 0 \end{pmatrix} z \qquad (8.6\text{-}6)$$

through a transformation

$$w = \mathbf{P}z, \tag{8.6-7}$$

where

$$\mathbf{P}(t, \epsilon) = \sum_{r=0}^{\infty} P_r(t)\epsilon^r \tag{8.6-8}$$

and all P_r are holomorphic at $t = 0$. Instead of carrying out the steps described in Section 5.2 it is simpler to calculate the P_r directly, as was done in [93]. If

$$P_r =: \begin{bmatrix} p_{11}^r & p_{12}^r \\ p_{21}^r & p_{22}^r \end{bmatrix}$$

and if

$$q(t) := \left(\frac{dx}{dt} \right)^{1/2}, \tag{8.6-9}$$

insertion of (8.6-7) into (8.6-6) shows that the transformed differential equation has the simple form (8.6-6) if the p_{jk}^r (here r is not an exponent) are solutions of the differential equations

$$p_{21}^r - t p_{12}^r = \dot{p}_{11}^{r-1}, \qquad p_{22}^r - p_{11}^r = \dot{p}_{12}^{r-1}, \qquad r = 0, 1, \ldots . \tag{8.6-10}$$

Here, $p_{jk}^{-1} \equiv 0$, by definition. The dot indicates differentiation with respect to t. If p_{21}^r and p_{12}^r are eliminated from these equations, p_{11}^r, p_{12}^r can be calculated successively by the variation of parameters formula, and p_{21}^r, p_{22}^r are then determined from (8.6-10). One finds the general solution

$$p_{11}^r(t) = c_{1r}q(t) - \frac{1}{2}q(t)\int_{\alpha_{1r}}^t q(s)\frac{d}{ds}\left[\dot{p}^{r-1}(s)q^{-2}(s) \right]ds,$$

$$r = 0, 1, \ldots ,$$

$$p_{12}^r(t) = c_{2r}t^{-1/2}q(t) - \int_{\alpha_{2r}}^t s^{-1/2}q(s)\frac{d}{ds}\left[\dot{p}_{11}^{r-1}(s)q^{-2}(s) \right]ds,$$

$$\tag{8.6-11}$$

in which the constants c_{jr}, α_{jr} must be chosen so that all p_{jk}^r are holomorphic at $t = 0$. It can be verified that $\alpha_{1r} = \alpha_{2r} = 0$, $c_{10} = 1$, $c_{20} = 0$, and $c_{1r} = c_{2r} = 0$ for $r > 0$ are possible, but not unique, choices.

Set, for abbreviation,

$$h(t) = \frac{1}{2}\int_0^t s^{-1/2}q(s)\frac{d}{ds}\left[\dot{q}(s)q^{-2}(s) \right]d\tau. \tag{8.6-12}$$

Then the beginning of the series in (8.6-8) turns out to be

$$\mathbf{P}(t, \epsilon) = q(t)I + \begin{bmatrix} 0 & -t^{1/2}q(t)h(t) \\ -t^{1/2}q(t)h(t) & 0 \end{bmatrix}\epsilon + \cdots . \tag{8.6-13}$$

To the formal transformation (8.6-7), there corresponds, according to

Theorem 6.5-1, an analytic transformation of (8.6-5) into (8.6-6),

$$w = P(t, \epsilon)z \tag{8.6-14}$$

with

$$P(t, \epsilon) \sim \mathbf{P}(t, \epsilon) \qquad \text{as} \quad \epsilon \to 0+ \quad \text{for} \quad |t| \leqslant t_0. \tag{8.6-15}$$

The simplified differential equation (8.6-6) can, of course, be explicitly solved in terms of Airy functions. One fundamental matrix solution is

$$Z(t, \epsilon) = 2\sqrt{\pi} \begin{bmatrix} e^{-\pi i/6} \epsilon^{-1/6} \mathrm{Ai}(e^{-2i/3} \epsilon^{-2/3} t) & e^{-\pi i/3} \epsilon^{-1/6} \mathrm{Ai}(e^{-4\pi i/3} \epsilon^{-2/3} t) \\ e^{-7\pi i} \epsilon^{1/6} \mathrm{Ai}'(e^{-2\pi i/3} \epsilon^{-2/3} t) & e^{\pi i/3} \epsilon^{1/6} \mathrm{Ai}'(e^{-4\pi i/3} \epsilon^{-2/3} t) \end{bmatrix}. \tag{8.6-16}$$

Here $\mathrm{Ai}'(\tau) := d\,\mathrm{Ai}(\tau)/d\tau$. If $\epsilon^{-2/3} t$ is not too large this matrix can be conveniently calculated from the convergent power series for Airy's function. For $\epsilon^{-2/3} t$ large, $Z(t, \epsilon)$ has the asymptotic form

$$Z(t, \epsilon) = \begin{bmatrix} t^{-1/4} & 0 \\ 0 & t^{1/4} \end{bmatrix} \left\{ \begin{pmatrix} 1 & 1 \\ 1 & -1 \end{pmatrix} - \frac{1}{48} \begin{pmatrix} -5 & 5 \\ 7 & 7 \end{pmatrix} g^{-3/2} \epsilon + O(t^{-3} \epsilon^2) \right\}$$

$$\times \begin{bmatrix} \exp\left[\frac{2}{3\epsilon} t^{3/2} \right] & 0 \\ 0 & \exp\left[-\frac{2}{3\epsilon} t^{3/2} \right] \end{bmatrix}, \tag{8.6-17}$$

which follows from (8.6-16) and from the asymptotic expansion for Airy's function. This representation is valid in

$$\pi/3 < \arg t < 5\pi/3. \tag{8.6-18}$$

Fundamental matrix solutions having asymptotic expansions analogous to (8.6-17) in the other two pairs of the three sectors bounded by the separation lines can easily be written down. The one in (8.6-16) has been chosen, because it contains a recessive and a dominant vector solution in a domain symmetric to the real axis.

The fundamental matrix solution

$$W = PZ$$

of equation (8.6-5) with P as in (8.6-14), (8.6-15) and Z as in (8.6-16) is now asymptotically known, as $\epsilon \to 0+$, in all of $|t| \leqslant t_0$. The central and lateral connection formulas are immediate consequences of the corresponding formulas for the solution of (8.6-6) by Airy functions. By means of the transformations (8.6-3) and (8.6-4) one finally returns to connection formulas for (8.6-1).

To be still more concrete, take the example of the scalar differential equation

$$\epsilon^2 \frac{d^2 u}{dx^2} - (x^3 - 1)u = 0 \tag{8.6-19}$$

and consider the task of calculating two independent solutions with known asymptotic behavior, as $\epsilon \to 0+$, in a neighborhood of the turning point at $x = 1$. With

$$y = \begin{pmatrix} u \\ \epsilon \dfrac{du}{dx} \end{pmatrix}$$

the corresponding system is

$$\epsilon y' = \begin{pmatrix} 0 & 1 \\ x^3 - 1 & 0 \end{pmatrix} y.$$

In (8.6-3) one must take $x_1 = 1$. Of the three distinct functions of x defined by (8.6-3) the one which maps $x > 1$ onto a curve in the sector (8.6-18) has the advantage that the asymptotic expansion (8.6-17) is valid for $x > 1$. As

$$a(x) = 3(x - 1)\left[1 + (x - 1) + \tfrac{1}{3}(x - 1)^2\right],$$

one has

$$a(x)^{1/2} = \sqrt{3}\,(x - 1)^{1/2}\left[1 + \tfrac{1}{2}(x - 1) + \tfrac{5}{24}(x - 1)^2 + O\big((x - 1)^3\big)\right]$$

and

$$\int_1^x a^{1/2}(s)\,ds = \frac{2}{\sqrt{3}}\,(x - 1)^{3/2}\left[1 + \tfrac{3}{10}(x - 1) + \tfrac{15}{168}(x - 1)^2 + O\big((x - 1)^3\big)\right].$$

This function is two-valued. For $x > 1$ the two branches are real, one positive, one negative. The function t of x defined in (8.6-3) has therefore three values for each x, one of which is real for $x > 1$, the other two are obtained by multiplying it, respectively, with $e^{2\pi i/3}$ or $e^{-2\pi i/3}$. Both of the latter map the ray $x > 1$ into rays in the sector (8.6-18). If the ray $\arg t = \tfrac{2}{3}$ is chosen, one has

$$t(x) = e^{2\pi i/3}\left(\frac{9}{2}\right)^3 \frac{6}{5}\left[1 + \frac{1}{5}(x - 1) + \frac{23}{2625}(x - 1)^2 + O\big((x - 1)^3\big)\right]$$

(8.6-20)

and furthermore,

$$\frac{dt}{dx} = e^{2\pi i/3}\left(\frac{9}{2}\right)^3 \frac{6}{5}\left[1 + \frac{2}{5}(x - 1) + \frac{23}{875}(x - 1)^2 + O\big((t - 1)^3\big)\right].$$

Then

$$\frac{dx}{dt} = e^{-2\pi i/3}\frac{20}{9^3 \cdot 3}\left[1 - \frac{2}{5}(x - 1) + \frac{477}{875}(x - 1)^2 + O\big((x - 1)^3\big)\right],$$

$$q = (\dot{x})^{1/2} = e^{2\pi i/3}\sqrt{\frac{20}{9^3 \cdot 3}}\left[1 - \frac{1}{5}(x - 1) + \frac{2083}{3500}(x - 1)^2 + O\big((x - 1)^3\big)\right],$$

$$\dot{q} = \frac{dq}{dx} \cdot \frac{dx}{dt} = -\frac{1}{5}\left(\frac{20}{9^3 \cdot 3}\right)^{3/2}\left[1 - \frac{2223}{3500}(x - 1) + O\big((x - 1)^2\big)\right].$$

Assume that one wishes to calculate, both for $x > 1$ and at $x = 1$, the approximate values for small $\epsilon > 0$, of a solution of (8.6-19) that is recessive for $x > 1$. From (8.6-14) and (8.6-13) one sees that, with $W = \{w_{jk}\}$, $Z = \{z_{jk}\}$, one has

$$w_{11} = qz_{11} + (q - t^{1/2}qh\epsilon)z_{21} + O(\epsilon^2 z_{11}) + O(\epsilon^2 z_{21})$$

and, by (8.6-16),

$$w_{11} = q2\sqrt{\pi}\, e^{-\pi i/6}\epsilon^{-1/6}\mathrm{Ai}/e^{-2\pi i/3}\epsilon^{-2/3}t)$$
$$+ (q - t^{1/2}qh\epsilon)2\sqrt{\pi}\, e^{-7\pi i/6}\epsilon^{1/6}2\sqrt{\pi}\,\mathrm{Ai}'(e^{2\pi i/3}\epsilon^{-2/3}t)$$
$$+ O(\epsilon^2 z_{11}) + O(\epsilon^2 z_{21}). \tag{8.6-21}$$

For t in the sector (8.6-18) and different from zero, w_{11} can now be approximately found from (8.6-17):

$$u(x,\epsilon) = w_{11}(t,\epsilon)$$
$$= q(t)\left[(t^{-1/4} + t^{1/4}) + \tfrac{5}{48}t^{-7} + \tfrac{7}{48}t^{-5/4} - t^{-3/4}h(t)) + O(\epsilon^2)\right]$$
$$\times \exp\{\tfrac{2}{3}t^{3/2}\}. \tag{8.6-22}$$

If t, q, and h are replaced by their respective expressions in terms of x, by means of (8.6-3), (8.6-9), and (8.6-12), $u(x,\epsilon) = w_{11}(t,\epsilon)$ is the desired asymptotic approximation. For $x > 1$, the argument of t is $2\pi/3$, by (8.6-20), so that (8.6-22) is indeed recessive, there.

To find the value of the same solution at the turning point $x = 1$, i.e., at $t = 0$ one returns to (8.6-21). Thus,

$$u(0,\epsilon) = 2\sqrt{\pi}\, q(0)e^{-\pi i/6}\epsilon^{-1/6}\left(\mathrm{Ai}(0) - \mathrm{Ai}'(0)\epsilon^{1/3} + O(\epsilon^2)\right)$$

$$= 2\sqrt{\pi}\, i\epsilon^{-1/6}\left[\sqrt{\frac{20}{9^3 \cdot 3}}\,\frac{1}{3^{2/3}\Gamma(\tfrac{2}{3})} - \frac{1}{3^{1/3}\Gamma(\tfrac{1}{3})}\epsilon^{1/3} + O(\epsilon^2)\right].$$

The method described in this section is applicable whenever two conditions are satisfied: The given differential equation must be holomorphically similar, *in a full neighborhood of the turning point*, to a decisively simpler differential equation *and* the connection problems for this simplified problem must have a known solution. This program has been carried out for a certain class of second-order equations. It is not surprising that these arguments are long and intricate. References to the pertinent more specialized literature must suffice here.

Sibuya has extended the uniform simplification method of Chapters V and VI to a general type of turning point problem of any order ([77] and [78]). To do this he made use of the formal theory of Hanson and Russell [19], [20] described here in Section 5.2, which reduces the differential equation to one with polynomial coefficients. He developed a global asymptotic theory for the simplified differential equation by a method first

used in the paper [25] by Hsieh and Sibuya and then considerably ampli-
fied by Sibuya in his book [78]. This asymptotic theory of differential
equations with polynomial coefficients includes the differential equation of
Airy as a very special case. With the help of the global theory of the
simplified equation the formal reduction to such a simplified equation can
be developed into a uniformly valid analytic reduction, more or less as was
done here in Chapter VI for turning points of order one. The task of
transforming the known connection formulas for the simplified equation
back into such formulas for the original equation is then a routine matter.

This theory applies to bounded neighborhoods of a turning point only. In
Chapters IX and X some extensions to unbounded regions will be studied.

CHAPTER IX
Fedoryuk's Global Theory of Second-Order Equations

9.1. Global Formal Solutions of $\epsilon^2 u'' = a(x)u$

In Section 2.3 a general method for finding formal solutions was illustrated by applying it to the simple equation

$$\epsilon y' = \begin{pmatrix} 0 & 1 \\ a(x) & 0 \end{pmatrix} y, \tag{9.1-1}$$

which is equivalent to the scalar equation $\epsilon^2 u'' - a(x)u = 0$. The result was the Liouville–Green approximation (2.3-24), which is the matrix form of

$$\mathbf{u}(x,\epsilon) = \left[(a(x))^{-1/4} + \ \dots \ \right] \exp\left\{ \pm \frac{1}{\epsilon} \int (a(x))^{1/2} dx \right\}. \tag{9.1-2}$$

The dots here indicate a formal series in powers of ϵ. In [11] and in several additional papers Fedoryuk, with some contributions by Evgrafov, has developed a global theory of the equation (9.1-1) in the special case that a is a polynomial or an entire function with some very restrictive properties. If

$$a(x) = x^m a^*(x), \tag{9.1-3}$$

m a non-negative integer and $a^*(x)$ a polynomial in x^{-1} with nonzero constant term, the coefficients in the formal solutions of (9.1-1) have a simple, useful structure at $x = \infty$. The presentation in this chapter is based on [11] with some nontrivial modifications.

Theorem 9.1-1. *If (9.1-3) is true, then the differential equation (9.1-1) has a*

Basic Formal Matrix Solution

$$Y(x, \epsilon) = \left(\sum_{r=0}^{\infty} Y_r(x) \epsilon^r \right) e^{Q(x, \epsilon)} \tag{9.1-4}$$

for which

$$Y_r(x) = O(x^{m/4 - r(m+2)/2}) \qquad as \quad x \to \infty,$$

$$Q(x, \epsilon) = \frac{1}{\epsilon} \begin{bmatrix} \int^x a^{1/2}(t) \, dt & 0 \\ 0 & -\int^x a^{1/2}(t) \, dt \end{bmatrix}.$$

More generally,

$$\frac{d^j Y_r}{dx^j} = O(x^{m/4 - r(m+2)/2 - j}), \qquad j = 0, 1, \ldots .$$

Also,

$$Y_0(x) = \begin{bmatrix} (a(x))^{-1/4} & 0 \\ 0 & (a(x))^{1/4} \end{bmatrix} \begin{bmatrix} 1 & 1 \\ 1 & -1 \end{bmatrix}.$$

PROOF. The transformation $y = T(x)V$ of (2.3-1) which diagonalizes the leading coefficient matrix of the differential equation has here the order of magnitude $T(x) = O(x^{m/2})$. Assume that this transformation has already been performed and that the notation is changed back to that of (2.1-2), as in the Proof of Theorem 2.3-1, then, in the notation of that proof, as $x \to \infty$,

$$P_0 = I, \qquad P_1(x) = O(x^{-1 - m/2}),$$

$$J(x) = B_0(x) = A_0(x) = O(x^{m/2}), \qquad A_r(x) = 0 \qquad for \quad r > 1$$

$$H_1(x) = B_1(x) = O(x^{-1}).$$

I claim that, for all $r \geq 0$,

$$H_r(x) = O(x^{m/2 - r(m+2)/2}) \tag{9.1-5}$$

$$P_r(x) = O(x^{-r(m+2)/2}) \qquad as \quad x \to \infty \tag{9.1-6}$$

$$B_r(x) = O(x^{m/2 - r(m+2)/2}). \tag{9.1-7}$$

These relations have just been verified for $r = 0, 1$. Assume them to be true for $r \leq k - 1, k > 1$. Then, by formula (2.3-16) it follows that (9.1-5) is true for $r = k$. The formulas (9.1-6), (9.1-7) are then obtained from (2.3-15). A formal solution of the formally diagonalized differential equation (2.3-22) can be calculated as follows:

$$\mathbf{W}(x,\epsilon) = \exp\left\{ \sum_{r=0}^{\infty} \left(\int B_r(x)\,dx \right) \epsilon^r \right\}$$

$$= \exp\left(\frac{1}{\epsilon} \int B_0(x)\,dx \right) \exp\left(\int B_1(x)\,dx \right) \exp\left\{ \sum_{r=2}^{\infty} \int (B_r(x)\,dx)\epsilon^{r-1} \right\}$$

$$= \begin{bmatrix} \exp\left(\dfrac{1}{\epsilon} \int \sqrt{a(x)}\ dx \right) & 0 \\ 0 & \exp\left(-\dfrac{1}{\epsilon} \int \sqrt{a(x)}\ dx \right) \end{bmatrix} (a(x))^{-1/4}$$

$$\times \sum_{s=0}^{\infty} \frac{1}{s!} \left(\sum_{r=2}^{\infty} \int_{\infty}^{x} B_r(t)\,dt\ \epsilon^{r+1} \right)^s.$$

Since

$$\int_{\infty}^{x} B_r(t)\,dt = O(x - (r-1)m + 2/2) \qquad \text{for} \quad r > 1,$$

the reordering of the double series above shows that, for an appropriate choice of the constants of integration,

$$\mathbf{W}(x,\epsilon) = \left(\sum_{r=0}^{\infty} W_r(x)\epsilon^r \right) e^{Q(x,\epsilon)} \tag{9.1-8}$$

with

$$W_r(x) = O(x^{-m/4 - r(m+2)/2}).$$

Now, formally one has

$$\mathbf{Y} = T(x) \left(\sum_{r=0}^{\infty} P_r \epsilon^r \right) \mathbf{W}.$$

The orders of magnitude of the P_r are given by (9.1-6), and $T(x) = O(x^{m/2})$. Expansion of the product of series completes the proof of Theorem 9.1-1 for $j = 0$. The validity for $j = 1, 2, \ldots$ is proved in the same way. The explicit formula for $Y_0(x)$ was given in formula (2.3-24).

The fact, just proved that the terms of the series in (9.1-4) shrink, at $x = \infty$, increasingly fast as r increases, enables Fedoryuk to construct solutions that have the expression (9.1-4) as their asymptotic solutions, for $\epsilon \to 0$, in unbounded regions of the x-plane. The description of these regions requires some geometric preliminaries.

9.2. Separation Curves for $\epsilon^2 u'' = a(x)u$

It is clear from the results of Chapters II and III that the functions $\text{Re}[\int a^{1/2}(x)\,dx]$ will play an important role in the global asymptotic study of the differential equation (9.1-1). The definition of these functions depends on the determination of the square root and on the path of integration, but the family of all the curves along which $\operatorname{Re}\int a^{1/2}(x)\,dx$ is constant is the same no matter how these choices are made.

The expression $a^{1/2}(x)$ has two opposite values at all points of $\mathbb{C}$, except at the zeros $c_1, c_2, \ldots, c_m$ of a. It defines a single-valued function on its two-sheeted Riemann surface $\mathbb{C}_a$. In a simply connected region D of $\mathbb{C}$ which contains none of the points $c_j, j = 1, \ldots, m$, the expression

$$\operatorname{Re}\int_{x_0}^{x} a^{1/2}(t)\,dt$$

along a path in D represents two harmonic functions, depending on the determination of the square root. Let

$$F(x) := \operatorname{Re}\int_{x_0}^{x} a^{1/2}(t)\,dt, \qquad x_0, x \in D, \tag{9.2-1}$$

denote one of these two functions. If

$$x = \mu + i\nu, \qquad \mu, \nu \text{ real}, \quad x \in D,$$

the derivatives $F_\mu := \partial F/\partial\mu$, $F_\nu := \partial F/\partial\nu$ define a vector field (F_μ, F_ν) orthogonal to the curves $F(x) = \text{const.}$ Therefore, $(-F_\nu, F_\mu)$ is a vector field tangential to these curves. Since $\int_{x_0}^{x} a^{1/2}\,dt$ is an analytic function,

$$F_\nu = -\frac{\partial}{\partial\mu}\operatorname{Im}\int_{x}^{x} a^{1/2}\,dt.$$

Thus the curves $F = \text{const.}$ are the integral curves of the differential equations

$$\frac{d\mu}{ds} = \frac{\partial}{\partial\mu}\left(\operatorname{Im}\int_{x_0}^{x} a^{1/2}\,dt\right), \qquad \frac{\partial\nu}{\partial s} = \frac{\partial}{\partial\mu}\left(\operatorname{Re}\int_{x_0}^{x} a^{1/2}\,dt\right).$$

Now,

$$\frac{\partial}{\partial\mu}\left(\operatorname{Re}\int_{x_0}^{x} a^{1/2}\,dt\right) + i\frac{\partial}{\partial\mu}\left(\operatorname{Im}\int_{x_0}^{x} a^{1/2}\,dt\right)$$

$$= \frac{\partial}{\partial\mu}\left(\int_{x_0}^{x} a^{1/2}\,dt\right) = \frac{\partial}{\partial x}\left(\int_{x_0}^{x} a^{1/2}\,dt\right) = \operatorname{Re}(a^{1/2}) + i\operatorname{Im}(a^{1/2}),$$

so that the differential equations can be written, more simply, as

$$\frac{d\mu}{ds} = \operatorname{Im}(a^{1/2}(x)), \qquad \frac{d\nu}{ds} = \operatorname{Re}(a^{1/2}(x)). \tag{9.2-2}$$

By the uniqueness theorem for ordinary differential equations, there passes exactly one curve of the family $F = \text{const.}$ through each point of D. In a region with a zero c_j of a, the right side members of (9.2-2) have, in general, a branch point and then F cannot be defined as a single-valued function there. Even if $a^{1/2}$ is holomorphic at $x = c_j$, that point is a critical point of the differential equation (9.2-2), and it may be a common end point of more than one arc of the set where $F = $ const. The situation is described in the lemma, below.

Lemma 9.2-1. *Let* $c_1 = \mu_1 + iv_1$ *be a point where* a *has a zero of order* q. *Then* c_1 *is a common point of* $q + 2$ *distinct arcs on which* $\mathrm{Re} \int_{x_0}^x a^{1/2} dt$ $= $ const. *Neighboring such arcs meet at the angle* $2\pi/(q + 2)$.

PROOF. As the family of curves $\mathrm{Re} \int_{x_0}^x a^{1/2} dt = $ const. does not depend on x_0, on the choice of path of integration or the determination of the square root (only the *value* of the constant may change), no generality is lost by taking $x_0 = c_1$. Let

$$a(x) = (x - c_1)^q a^*(x), \qquad a^*(c_1) \neq 0$$

with a^* holomorphic at $x = c_1$. Then

$$\int_{c_1}^x a^{1/2} dt = (x - c_1)^{(q+2)/2} b(x), \qquad b(c_1) \neq 0,$$

and b is holomorphic at $x = c_1$. For the curves through $x = c_1$ on which $\mathrm{Re} \int_{c_1}^x a^{1/2} dt = $ const., the constant must be zero, and the equation of these curves becomes

$$0 = \mathrm{Re}\{(x - c_1)^{(q+2)/2} b(x)\},$$

or

$$0 = \mathrm{Re}\{(x - c_1)^{(q+2)/2} [b(c_1) + O(x - c_1)]\}, \qquad (9.2\text{-}3)$$

as $x - c_1 \to 0$. Set $x - c_1 = re^{i\theta}$, $b(c_1) = \rho e^{i\alpha}$. Then (9.2-3) can be written

$$r^{(q+2)/2} \left[\rho \cos\left(\frac{q+2}{2} \theta + \alpha \right) + rO(1) \right] = 0,$$

as $r \to 0$, or

$$\cos\left(\frac{q+2}{2} \theta + \alpha \right) + rO(1) = 0, \qquad r \to 0.$$

The derivative of $rO(1)$ with respect to θ vanishes at $r = 0$. By the implicit function theorem the last equation is, therefore, satisfied for $q + 2$ values of θ that tend, respectively, to

$$\theta_j = -\frac{2\alpha}{q+2} + \frac{2j+1}{q+2} \pi, \qquad j = 0, 1, \ldots, q + 1, \qquad (9.2\text{-}4)$$

as $r \to 0$. This completes the proof of the lemma.

Those of the arcs $\mathrm{Re} \int_{x_0}^x a^{1/2} dt = $ const. that pass through a zero of a are called Stokes arcs by Fedoryuk. Here, they will be called "separation arcs," a term more in line with the terminology previously employed.

Theorem 9.2-1.. *If a is a polynominal, no set of separation arcs for the differential equation* (9.1-1) *bounds a bounded region in* $\mathbb{C}$.

PROOF. Since the separation arcs satisfy the system of differential equations (9.2-2) they cannot end, or intersect, except at zeros of a. For a proof of the lemma by contradiction, assume that D is a bounded region whose boundary ∂D consists of separation arcs. Let $\beta \geqslant 0$ be the number of zeros of a in D and $\gamma \geqslant 0$ the number of such zeros on ∂D. If $\beta = 0$ then formula (9.2-1) defines a function F that is harmonic in D and continuous in $\overline{D}$. As F must be constant on ∂D, it must be constant identically. This is not the case, and a contradiction has been reached when $\beta = 0$. If $\beta > 0$, consider the graph formed by all separation arcs in $\overline{D}$. If this graph contains a closed contour other than ∂D, the reasoning can be repeated with that smaller region. After a finite number of steps a region is reached that either contains no zeros of a or no closed contours other than its boundary. The first case has been shown before to be impossible. The second case cannot occur, because then every zero of a in D can be connected with the boundary of D by at least three chains of separation arcs. They, together with ∂D divide $\overline{D}$ into smaller regions bounded by separation arcs, contrary to the assumption.

A knowledge of the graph formed by the separation arcs is important for a global asymptotic analysis of differential equations of type (9.1-1). The differential equation (9.2-2) is useful for this task. Figures 9.1 to 9.6 illustrate this matter for a few simple polynomials. They are based on formula (9.2-4), its analog at $x = \infty$ and on the differential equations (9.2-2).

For a global asymptotic theory of the differential equation (9.1-1) the quantity

$$\xi := \int_{x_0}^x a^{1/2} dt \qquad (9.2\text{-}5)$$

is essential, as was pointed out before. If x_0 is one of the zeros of a, then in any region bounded by separation arcs that has no zeros of a in its interior ξ is defined as a holomorphic function of x once the sign of the square root is chosen, provided the paths of integration are limited to that region. Of particular interest in Fedoryuk's theory are those regions in the x-plane whose image in the ξ-plane covers the whole plane in a univalent way except for the image of the boundary. Such regions he calls "canonical." The images of the separation arcs are finite or infinite segments parallel to the imaginary ξ-axis. In Figs. 9.7 and 9.8, two examples of ξ-images of canonical domains are illustrated. Separation curves, turning points and regions in the x-plane and their images in the ξ-plane are denoted by the same letters in those figures.

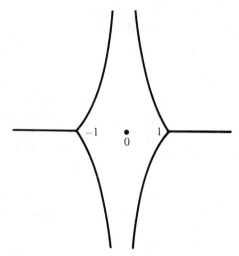

Figure 9.1. Separation curves and Stokes regions for $a(x) = 1 - x^2$.

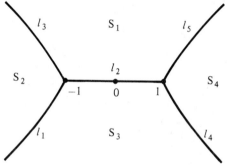

Figure 9.2. Separation curves and Stokes regions for $a(x) = x^2 - 1$.

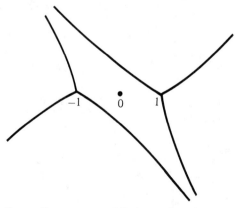

Figure 9.3. Separation curves and Stokes regions for $a(x) = i(x^2 - 1)$.

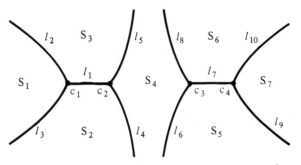

Figure 9.4. Separation curves and Stokes regions for $a(x) = \prod\limits_{k=1}^{4} (x - c_k)$,

$$c_1 < c_2 < \cdots < c_4.$$

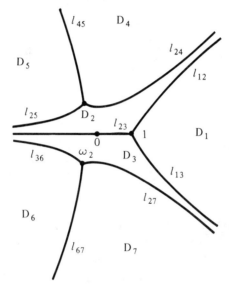

Figure 9.5. Separation curves and Stokes regions for $a(x) = x^3 - 1$.

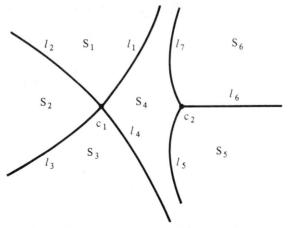

Figure 9.6. Separation curves and Stokes regions for
$a(x) = (x - c_1)^2(c_2 - x),\ c_1 < c_2$.

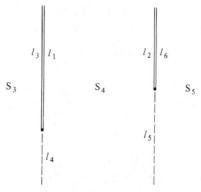

Figure 9.7. Image in the ξ-plane of the canonical domain $s_3 \cup l_4 \cup s_4 \cup l_5$ for $a(x) = (x - c_1)^2(c_2 - x)$.

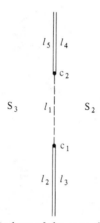

Figure 9.8. Image in the ξ-plane of the canonical domain $s_3 \cup l_1 \cup s_2$ for

$$a(x) = \prod_{k=1}^{4} (x - c_k).$$

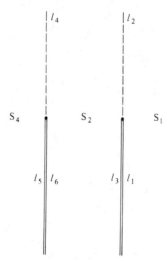

Figure 9.9. Image in the ξ-plane of the canonical domain $S_1 \cup l_2 \cup S_2 \cup S_4 \cup S_4$ for $a(x) = 1 - x^2$.

9.3. A Global Asymptotic Existence Theorem for $\epsilon^2 u'' = a(x)u$

Again, the differential equation to be studied is (9.1-1) with a being a polynomial of degree m. Let D be a canonical domain and denote by $\xi(D)$ its image in the ξ-plane. The subset of D consisting of points at distance greater than some number $\delta > 0$ from the boundary of D will be called D_δ, and δ must be so small that D_δ is simply connected. Recall that by Theorem 9.1-1 the coefficients Y_r in the formal solution

$$\mathbf{Y}(x,\epsilon) = \hat{\mathbf{Y}}(x,\epsilon) e^{Q(x,\epsilon)} = \sum_{r=0}^{\infty} Y_r(x)\epsilon^r \exp\left\{\frac{1}{\epsilon}\begin{pmatrix} \xi(x) & 0 \\ 0 & -\xi(x) \end{pmatrix}\right\} \quad (9.3\text{-}1)$$

have the property

$$|Y_r^{(j)}(x)| \leq c_r(1 + |x|^{m/4 - r(m+2)/2 - j}), \qquad j \geq 0, \quad r \geq 0. \quad (9.3\text{-}2)$$

The constants c_r depend on δ, but not on x or r.

The existence proof to be given here resembles the one in Chapter III, but it differs sufficiently to justify a separate account. For one, D is now an unbounded region and, in addition, there is now the aim to exploit the inequalities (9.3-2) for a *doubly asymptotic* approximation, i.e., the remainder should be small both for small ϵ *and* for large x. The Borel–Ritt theorem does not lend itself easily to this situation. Therefore, the partial sums

$$\hat{Y}^{(N)} := \sum_{r=0}^{N} Y_r \epsilon^r \quad (9.3\text{-}3)$$

will here take the place of a function that has $\hat{\mathbf{Y}}$ as its complete asymptotic expansion, like the ones available in Chapter III.

By its construction, $\hat{\mathbf{Y}}$ satisfies formally the differential equation

$$\epsilon \hat{\mathbf{Y}}' + \epsilon \hat{\mathbf{Y}} Q' - A\hat{\mathbf{Y}} = 0. \quad (9.3\text{-}4)$$

Since

$$Q'(x,\epsilon) = \epsilon^{-1} a^{1/2}(x)\begin{pmatrix} 1 & 0 \\ 0 & -1 \end{pmatrix},$$

$$A(x,\epsilon) = A_0(x) = \begin{pmatrix} 0 & 1 \\ a(x) & 0 \end{pmatrix},$$

and, therefore, $Y_r(\epsilon Q') - AY_r = Y'_{r-1}, r = 1, 2, \ldots,$ for the present simple differential equation, one finds that to (9.3-4) there corresponds the relation

$$\epsilon(\hat{Y}^{(N)})' + \hat{Y}^{(N)}(\epsilon Q') - A\hat{Y}^{(N)} = \epsilon^N Y'_N. \quad (9.3\text{-}5)$$

Let

$$y_j(x,\epsilon) = \hat{y}_j(x,\epsilon)\exp\{(-1)^j \epsilon^{-1}\xi(x)\} = \hat{y}_j(x,\epsilon) e^{q_j(x,\epsilon)}, \qquad j = 1, 2, \quad (9.3\text{-}6)$$

be the two columns of the matrix $\mathbf{Y}(x, \epsilon)$. Then (9.3-4) is equivalent to the two formal vectorial differential equations

$$\epsilon \hat{\mathbf{y}}_j' = (A - \epsilon q_j' I) \hat{\mathbf{y}}_j, \qquad j = 1, 2, \tag{9.3-7}$$

while $\hat{\mathbf{y}}_j^{(N)}$ satisfies, by (9.3-6), the differential equation

$$\epsilon (\hat{\mathbf{y}}_j^{(N)})' = (A - \epsilon q_j' I) \hat{\mathbf{y}}_j^{(N)} + \epsilon^N \mathbf{y}_{Nj}', \tag{9.3-8}$$

where $\mathbf{y}_{Nj}$ is the jth column of Y_N, $j = 1, 2$.

The existence proof consists in showing that (9.3-7) can be satisfied as an analytic—not only formal—differential equation by a vector $\hat{\mathbf{y}}_j$ which differs by terms of order $O(\epsilon^N)$ from $\hat{\mathbf{y}}_j^{(N)}$. Equivalently, it must be shown that the differential equations

$$\epsilon u_j' + \epsilon q' u_j - A u_j = \epsilon^N \mathbf{y}_{Nj}', \qquad j = 1, 2, \tag{9.3-9}$$

have "small" solutions of order $O(\epsilon^N)$.

In analogy with the procedure in Chapter III, a first step toward the asymptotic solution of (9.3-9) is to introduce the matrix

$$
\begin{aligned}
A^{(N)} &:= \epsilon (Y^{(N)})' (Y^{(N)})^{-1} \\
&= \epsilon [(\hat{Y}^{(N)})' + \hat{Y}^{(N)} Q'] (\hat{Y}^{(N)})^{-1} \\
&= A + \epsilon^N Y_N' (\hat{Y}^{(N)})^{-1} \tag{9.3-10}
\end{aligned}
$$

where

$$Y^{(N)} := \hat{Y}^{(N)} e^Q.$$

Here, (9.3-5) has been used. Then the differential equation (9.3-9) can also be written

$$\epsilon u_j' = (A^{(N)} - \epsilon q_j') u_j - \epsilon^N [Y_N' (\hat{Y}^{(N)})^{-1} u_j - \mathbf{y}_{Nj}']. \tag{9.3-11}$$

The advantage of (9.3-11) over (9.3-8) is that the homogeneous equation

$$\epsilon z' = (A^{(N)} - \epsilon q_j' I) z$$

has the known fundamental matrix solution

$$Z = Y^{(N)} e^{-q_j} = \hat{Y}^{(N)} e^{Q - q_j'}.$$

which will now be used to convert (9.3-11) into an equivalent integral equation by the variation of parameter formula. If $A^{(N)}$ were replaced by A, the kernel of the integral equation would not be known, at this stage.

The integral equations in question have the form

$$u_j(x, \epsilon) = \hat{Y}^{(N)}(x, \epsilon) \int_{\gamma_j(x)} K_j(x, t, \epsilon) \epsilon^{N-1} [F(t, \epsilon) u_j(t, \epsilon) + g_j(t, \epsilon)] \, dt,$$

$$j = 1, 2, \tag{9.3-12}$$

in which the following abbreviations have been introduced:

$$K_1(x,t,\epsilon) = \exp\left\{\frac{1}{\epsilon}\begin{pmatrix} 0 & 0 \\ 0 & \frac{2}{3}(\xi(t) - \xi(x)) \end{pmatrix}\right\},$$

$$K_2(x,t,\epsilon) = \exp\left\{\frac{1}{\epsilon}\begin{pmatrix} \frac{2}{3}(\xi(x) - \xi(t)) & 0 \\ 0 & 0 \end{pmatrix}\right\},$$

(9.3-13)

$$F(t,\epsilon) = \left(\hat{Y}^{(N)}(t,\epsilon)\right)^{-1}\hat{Y}'_N(t)\left(\hat{Y}^{(N)}(t,\epsilon)\right)^{-1},$$

$$g_j(t,\epsilon) = \left(\hat{Y}^{(N)}(t,\epsilon)\right)^{-1}y'_{Nj},$$

(9.3-14)

and $\gamma_j(x)$ is a path of integration in the t-plane that will be described presently.

I claim that the domain D_δ is "accessible" for $j = 1$ as well as for $j = 2$, in the sense of that term in Chapter III. To show this, consider the subset $\xi^*(D)$ of $\xi(D)$ consisting of those of its points that have distance at least δ^* from the boundary and choose $\delta^* > 0$ so small that the preimage of $\xi^*(D)$ contains D_δ. Every point of $\xi^*(D)$ can be reached from the point at infinity on the positive real axis by a path along which the real part decreases. Similarly, all those points can be reached from the point at infinity on the negative real axis by paths along which that real part increases. The preimages of these paths in the t-plane are progressive; the former one for $j = 1$, the second for $j = 2$. This follows from (9.3-13). They will be the paths $\gamma_j(x)$ in (9.3-12).

The arguments from here on resemble so much the ones in the proof of Theorem 3.3-1 that a sketch of the necessary modifications will suffice. The functions K_j in (9.3-13) satisfy the inequality

$$|K_j(x,t,\epsilon)| \leqslant 1 \qquad \text{for} \quad t \in \gamma_j(x), \quad x \in D_\delta, \quad 0 < \epsilon \leqslant \epsilon_0$$

where $|\cdots|$ indicates the matrix norm defined by

$$|M| = \max_{\mu = 1,2} \sum_{\nu=1}^{2} |m_{\mu\nu}|.$$

Also, the formulas (9.3-13) and (9.3-14) and Theorem 9.1-1 imply

$$|F(t,\epsilon)| \leqslant c(|t| + 1)(|t| + 1)^{3m/4 - N(m+2)/2 - 1},$$

$$|g(t,\epsilon)| \leqslant c(|t| + 1)(|t| + 1)^{m/2 - N(m+2)/2 - 1}$$

(9.3-15)

for $t \in D_\delta$, $0 < \epsilon \leqslant \epsilon_0$. The constant c may depend on N. Therefore, the integral in (9.3-12) exists and defines a bounded holomorphic function in D_δ whenever u_j is there bounded and holomorphic, at least if N is taken positive. The existence of a unique holomorphic and bounded solution of the integral equation (9.3-12) can now be derived from a contraction mapping theorem or, directly, by Picard iteration. Some details are omitted.

Insertion of that solution into (9.3-12) then shows that this proves a preliminary result, which will be stated as a lemma.

Lemma 9.3-1. *For every $N \geqslant 1$ the differential equation* (9.3-9) *possesses a solution* $u_j = w_j^{(N)}(x, \epsilon)$ *of the form*

$$w_j^{(N)} = \epsilon^{N-1} \eta_j^{(N)}(x, \epsilon),$$

with $\eta_j^{(N)}(x, \epsilon)$ *uniformly bounded for* $x \in D_\delta$, $0 < \epsilon \leqslant \epsilon_0$.

By the preceding arguments this lemma immediately implies the next one.

Lemma 9.3-2. *The differential equation* (9.1-1) *possesses, for* $N = 0, 1, \ldots$ *vector solutions* $y = y_j^{(N)}$ *of the form*

$$y_j^{(N)}(x, \epsilon) = \left[\sum_{r=0}^{N} y_{rj}(x)\epsilon^r + \epsilon^{N+1} \chi_j^{(N)}(x, \epsilon) \right] \exp\left\{ (-1)^j \frac{1}{\epsilon} \xi(x) \right\}, \qquad j = 1, 2$$

with $\chi^{(N)}$ *bounded for* $x \in D_\delta$, $0 < \epsilon \leqslant \epsilon_0$.

PROOF. By Lemma 9.3-1 and the preceding explanations,

$$\left[\sum_{r=0}^{M} y_{rj}(x)\epsilon^r + \epsilon^{M-1} \eta_j(x, \epsilon) \right] \exp\left\{ (-1)^j \frac{1}{\epsilon} \xi(x, \epsilon) \right\}$$

is for $M = 2, 3, \ldots$, a solution of (9.1-1). Take $M = N + 2$, then

$$\left[\sum_{r=0}^{N} y_{rj}(x)\epsilon^r + \epsilon^{N+1}\left(\eta_j^{(N+2)}(x, \epsilon) + y_{N+1}(x) + y_{N+2}(x)\epsilon \right) \right]$$

$$\times \exp\left\{ (-1)^j \frac{1}{\epsilon} \xi(x) \right\}$$

is the same solution. The right-hand factor of ϵ^{N+1} is $\chi^{(N)}$.

Lemma 9.3-2 falls short of the aim of this section in two respects: The solution appears to depend on N and it has not yet been shown to have the doubly asymptotic property, as $x \to \infty$, mentioned before. Therefore, the function $\chi_j^{(N)}$ in Lemma 9.3-2 must be appraised for large x. To that end consider a circle $|\xi| = R$ with R so large that the circle crosses all images of unbounded separation arcs on the boundary of the canonical domain D. These crossing points divide the circle $|\xi| = R$ into arcs. One of these arcs intersects the positive ξ-axis, and one intersects the negative ξ-axis. If the boundary of $\xi(D)$ lies in one half-plane these two arcs are one and the same, otherwise they are distinct. Each of these two (or one) arcs, together with the ξ-images of the separation arcs through their endpoints, bound a subregion of $\xi(D)$ whose preimages in D may be called D^1 and D^2. Analogously one defines regions D_δ^1, D_δ^2. These regions depend on R, as well. When all parts of the boundary of $\xi(D)$ lie in a half-plane $\text{Im}\,\xi$

$\geqslant$ const. or Im $\xi \leqslant$ const., one has $D_\delta^1 = D_\delta^2$. The canonical region D will then be called *consistent*. Otherwise, it is called *inconsistent*. If R is sufficiently large and x is in one of these subregions D_δ^j, the paths of integration $\gamma_j(x)$ can be changed without changing the value of the integral so that on these paths $|x| < |t|$, in addition to Re t being monotone.

Lemma 9.3-3. *For $x \in D_\delta^j$, $0 < \epsilon \leqslant \epsilon_0$, $j = 1, 2$, the function $\chi_j^{(N)}$ of Lemma 9.3-2 satisfies the inequality*

$$|\chi_j^{(N)}(x,\epsilon)| \leqslant c_N |x|^{m/4 - N(m+2)/2}$$

(c_N a constant).

PROOF. Multiply and divide the integral in (9.3-12) by

$$|x|^{3m/4 - N(m+2)/2 + 1}.$$

Thanks to the inequalities (9.3-15) one finds that for $N \geqslant 1$

$$|\eta_j^{(N)}(x,\epsilon)| \leqslant c_N^* |x|^{3m/4 - N(m+2)/2 + 1}, \qquad x \in D_\delta^j.$$

Then proceed as in the proof of Lemma 9.3-2 and refer to Theorem 9.1-1. The expression $|x| + 1$ has been replaced by $|x|$ because this does not change the orders of magnitudes, as $x \to \infty$.

The quickest, though not the most elementary, way to prove that the solutions $y_j^{(N)}$ in Lemma 9.3-2 are the same for all N is to appeal to the asymptotic theory of analytic linear differential equations without a parameter near $x = \infty$. For any fixed $\epsilon \neq 0$ the system (9.1-1) has an irregular singularity at infinity. The asymptotic form of the solution, as $x \to \infty$, has been even more thoroughly studied than the asymptotic dependence on a parameter. The theorem below, a proof of which can, e.g., be found in [61] Ch. IX, §11, can be used in the present context.

Theorem 9.3-1. *For every fixed $\epsilon > 0$ the scalar differential equation $\epsilon^2 u'' = a(x)u$ has two solutions $u = w_j$, $j = 1, 2$, of the form*

$$w_j(x,\epsilon) = [a(x)]^{-1/4} \exp\left\{(-1)^j \frac{1}{\epsilon} \xi(x)\right\} [1 + \zeta_j(x,\epsilon)],$$

where

$$\lim_{x \to \infty} \zeta_j(x,\epsilon) = 0,$$

if x tends to infinity in a direction in which $\mathrm{Re}((-1)^j \xi(x)) \to -\infty$.

Every solution of $\epsilon^2 u'' = a(x)u$ is a linear combination *with constant coefficients* of w_1 and w_2 of Theorem 9.3-1. Except for a scalar nonzero factor that may depend on ϵ, w_j is the only solution that tends to zero as $x \to \infty$ in directions for which $\mathrm{Re}((-1)^j \xi(x)) \to -\infty$. Now, the first component $y_{j1}^{(N)}$ of the vector $y_j^{(N)}$ in Lemma 9.3-2 *is* a solution of $\epsilon^2 u'' = a(x)u$. Hence, for all M, N with $0 \leqslant M < N$,

$$y_{j1}^{(N)}(x,\epsilon) - y_{j1}^{(M)}(x,\epsilon) = k_j^{MN}(\epsilon) w_j(x,\epsilon), \qquad k_j^{MN}(\epsilon) \text{ independent of } x.$$

However, by Lemma 9.3-3 and Theorem 9.1-1,

$$\lim_{\substack{x \to \infty \\ x \in D_\delta^j}} \left[y_{j1}^{(N)}(x,\epsilon) - y_{j1}^{(M)}(x,\epsilon) \right] \exp\left\{ (-1)^{j+1} \frac{1}{\epsilon} \xi(x) \right\} x^{m/4} = 0$$

(at least for $0 < M < N$), while Theorem 9.3-1 implies

$$\lim_{\substack{x \to \infty \\ x \in D_\delta^j}} w_j(x,\epsilon) \exp\left\{ (-1)^{j+1} \frac{1}{\epsilon} \xi(x) \right\} x^{m/4} = \left[a^*(\infty) \right]^{-1/4} \neq 0$$

so that $k_j^{MN}(\epsilon)$ must be zero for all ϵ. The theorem below summarizes the results of this section.

Theorem 9.3-2. *Let $a(x)$ be a polynomial of degree m, let D be a canonical domain for the differential equation*

$$\epsilon y' = \begin{pmatrix} 0 & 1 \\ a(x) & 0 \end{pmatrix} y,$$

and let D_δ^1 and D_δ^2 be the subregions of D defined above. Then the formal matrix solution (9.1-4) represents asymptotically, as $\epsilon \to 0$, a fundamental solution $\hat{Y}(x,\epsilon) e^{Q(x,\epsilon)}$ of the differential equation, uniformly in D_δ. The jth ($j = 1,2$) column of that matrix solution is also asymptotic for large x, in the sense that the jth column of

$$\hat{Y}(x,\epsilon) - \sum_{r=0}^{N} Y_r(x) \epsilon^r$$

is small of order $O(|x|^{m/4 - (N+1)(m+2)/2})$, uniformly for $0 < \epsilon \leqslant \epsilon_0$, as $x \to \infty$ in D_δ^j. If the canonical domain D is consistent, then $D_\delta^1 = D_\delta^2$. (D_δ was defined at the beginning of Section 9.3. D_δ^1, D_δ^2 were introduced after Lemma 9.3-2.)

To illustrate the meaning of Theorem 9.3-2 take $a(x) = 1 - x^2$ and denote the separation arcs and the regions they bound as in Fig. 9.3. One canonical domain is $D = S_1 \cup l_2 \cup S_2 \cup l_4 \cup S_4$. Define the branch of the function

$$\xi(x) = \int_1^x (1 - t^2)^{1/2} dt$$

in D by the condition that $\xi(S_1)$ is a *right* half-plane. In the picture of $\xi(D)$ in Fig. 9.10 the images of the S_j and l_j with the ξ-plane are again denoted by S_j and l_j. It is seen that D is a consistent domain. Accordingly, $D_\delta^1 = D_\delta^2$. In Fig. 9.10, D is the unshaded part of the x-plane. The subregion of D_δ above the semicircle with radius R drawn in Fig. 9.10 is D_δ^1 (and D_δ^2). In this case the formal expression $\hat{Y} e^Q$, (9.1-4), describes asymptotically a fundamental matrix solution $\hat{Y} e^Q$ not only for the passage to the limit as $\epsilon \to 0 +$, but also for the passage to the limit as $x \to \infty$ in the upper half-plane.

By contrast, define D by

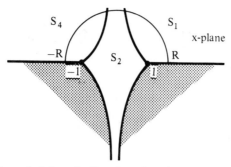

Figure 9.10. Canonical domain $S_1 \cup l_2 \cup S_2 \cup l_4 \cup S_4$ for $a(x) = 1 - x^2$.

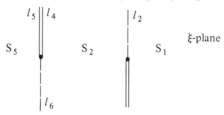

Figure 9.11. Image in the ξ-plane of the canonical domain
$S_1 \cup l_2 \cup S_2 \cup l_6 \cup S_5$ for $a(x) = 1 - x^2$.

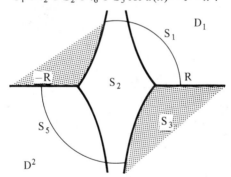

Figure 9.12. Canonical domain $S_1 \cup l_2 \cup S_2 \cup l_6 \cup S_5$ for $a(x) = 1 - x^2$.

$$D = S_1 \cup l_2 \cup S_2 \cup l_6 \cup S_5.$$

Now, $\xi(D)$ has the boundaries shown in Fig. 9.11. Thus, this D is inconsistent. It is the unshaded part of the x-plane in Fig. 9.12. The region D_δ^1 lies in the unbounded part of $S_1 \cup S_2$ in the lower half-plane. Now, the first column of $\hat{Y}e^Q$ represents a vector solution asymptotically, as $x \to \infty$ inside S_1, while the second column of $\hat{Y}e^Q$ represents a vector solution as $x \to \infty$ inside S_5. Therefore, Ye^Q is *not* an asymptotic expression for a fundamental matrix solution, as $x \to \infty$ anywhere in the x-plane. It *is* an asymptotic expression as $\epsilon \to 0 +$, uniformly in all of D_δ. Obviously, consistent fundamental domains are preferable for asymptotic analysis in unbounded regions.

CHAPTER X
Doubly Asymptotic Expansions

10.1. Introduction

In the preceding chapter the concept of "doubly asymptotic expansions" was introduced through an important special case. Whenever in the applications of the asymptotic theory one has to characterize functions of x and ϵ by their behavior near $x = \infty$ in some unbounded region of the x-plane, the knowledge of doubly asymptotic expansions is extremely helpful. Unfortunately, all known results of some generality in this direction require that the coefficients of the differential equation to be solved be polynomials in x. The matter has been explored in a number of papers by Leung, listed in the bibliography. The presentation here is based on Leung's work.

Definition 10.1-1. Let $D \subset \mathbb{C}$ be an unbounded region and let E be the sectorial domain

$$E = \{ \epsilon / 0 < |\epsilon| \leqslant \epsilon_0, |\arg \epsilon| \leqslant \delta \}.$$

A scalar function f of the two variable x, ϵ holomorphic for $x \in D$, $\epsilon \in E$ is said to have a doubly asymptotic power series representation in $D \times E$, as $\epsilon \to 0$ in E if it can be represented by a series of the form $\sum_{r=0}^{\infty} f_r(x) \epsilon^r$ in the sense that, for $N = 0, 1, \ldots,$

$$\left| f(x, \epsilon) - \sum_{r=0}^{N} f_r(x) \epsilon^r \right| \leqslant K_N |x|^{-\mu_N} |\epsilon|^{N+1}$$

for all $x \in D$, $\epsilon \in E$. The K_N, μ_N are certain constants, and $\lim_{N \to \infty} \mu_N = +\infty$.

The definition can obviously be extended to vector or matrix-valued functions.

The theory will be developed for certain systems of the form

$$\epsilon y' = A(x)y, \qquad \epsilon > 0, \qquad (10.1\text{-}1)$$

with

$$A(x) = \begin{bmatrix} 0 & 1 & 0 & \cdots & 0 \\ 0 & 0 & 1 & \cdots & 0 \\ & & \cdots & \cdots & \\ 0 & 0 & 0 & \cdots & 1 \\ a_n(x) & a_{n-1}(a) & & \cdots & a_1(x) \end{bmatrix}. \qquad (10.1\text{-}2)$$

The functions a_k are assumed to be polynomials of degrees m_k so that

$$a_k(x) = x^{m_k} p_k(x^{-1}), \qquad p_k(0) \neq 0, \quad k = 1, \ldots, n, \qquad (10.1\text{-}3)$$

except when $a_k \equiv 0$, in which case $p_k \equiv 0$, by definition, and m_k may be defined as zero. The p_k are polynomials in x^{-1} of degree not exceeding m_k.

Even with these severe restrictions the nature of the solutions near $x = \infty$ can be too complicated for a systematic analysis.

In the second-order equations dealt with in Chapters V and VI, there was only one coefficient, the one corresponding to a_n in (10.1-2). In [42] Leung requires that for large x the coefficient a_n, $n \geq 2$, be dominant in the sense that the first approximation to the eigenvalues of $A(x)$, as $x \to \infty$, should be $(a_n)^{1/n}$ or, more precisely, that these eigenvalues have the form

$$x^{m_n/n} \left[(p_n(0))^{1/n} + o(1) \right], \qquad (10.1\text{-}4)$$

as $x \to \infty$. The expression (10.1-4) represents n distinct holomorphic functions near (but not "at") $x = \infty$. These functions have convergent expansions in fractional powers of x for large x, by the theory of algebraic functions.

Lemma 10.1-1. *If the exponent m_n in (10.1-3) is positive, then the expansions of the eigenvalues λ_j of the matrix $A(x)$ in (10.1-2) for large x have the form*

$$\lambda_j(x) = x^{m_n/n} \sum_{s=0}^{\infty} \lambda_{js} x^{-s/m}, \qquad j = 1, 2, \ldots, n, \qquad (10.1\text{-}5)$$

for $|x| \geq x_0$ (x_0 a constant), with $\lambda_{j0}^n = p_n(0)$, if and only if the other exponents m_k satisfy the inequalities

$$m_k < m_n \frac{k}{n}, \qquad k = 1, \ldots, n-1. \qquad (10.1\text{-}6)$$

PROOF. One way of proving the lemma consists in the construction of the Newton–Puiseux polygon for the polynomial [23]

$$\det(\lambda I - A(x)) = \lambda^n - a_1(x)\lambda^{n-1} \ldots - a_n(x). \qquad (10.1\text{-}7)$$

Equivalently, one may insert the series in (10.1-5) for λ into (10.1-7), collect like powers of x and take into account the inequalities (10.1-6). If, and only if, these inequalities are true, the leading term in the resulting expansion is

$$x^{m_n}(\lambda_{j0}^n - p_n(0)).$$

which shows that one must set $\lambda_{j0} = p_n^{1/n}(0)$. The conditions for the λ_{js}, $s > 0$, are recursive in character and can be satisfied successively for $s = 1, 2, \ldots$. The easy details of this calculation are omitted.

In a later paper [44], Leung has replaced conditions (10.1-6) by less stringent hypotheses. The theorems (and the proofs) of the asymptotic expansion theorems for the solutions of the differential equation then become much more complicated.

10.2. Formal Solutions for Large Values of the Independent Variable

The differential equation (10.1-1) with (10.1-2), (10.1-3), and (10.1-4) being true admits formal series solutions which can be calculated by the method of Chapter II. If, in this process the nature of the coefficients at $x = \infty$ is studied, just as their nature at $x = 0$ was studied in Chapter VII, it turns out that there is an analog of the restraint index in this situation. In fact, the series in powers of ϵ appearing in the formal solution has coefficients whose order of magnitude, for $x \to \infty$ tends linearly to $-\infty$ as the exponent r of ϵ grows to $+\infty$. Hence, rather than slowing down the precision of the asymptotic approximation, the dependence on x can be expected to *improve* the usefulness of the asymptotic series, as x gets larger. That analog of the restraint index might again be called by the same name, because, if

$$f_r(x) = O(x^{-r/\kappa}), \qquad r = 1, 2, \ldots,$$

then the shrinking of the functions, as $x \to \infty$, is slower when κ is larger.

Thanks to the simplifying hypothesis (10.1-6) the restraint index can be calculated explicitly. The first step is the formal diagonalization of $A_0(x)$ by a transformation

$$y = T(x)v. \tag{10.2-1}$$

Since the eigenvalues $\lambda_j(x)$ of $A(x)$ are distinct in $x \geqslant x_0$ for large x_0, the Vandermonde matrix of the λ_j is a possible choice for T, i.e., one can set

$$T = \{t_{jk}\} = \{\lambda_j^{k-1}\}. \tag{10.2-2}$$

Each column of this matrix is an eigenvector of $A_0(x)$. The transformed differential equation is

$$\epsilon \frac{dv}{dx} = \left(\Lambda(x) - \epsilon T^{-1} \frac{dT}{dx} \right) v, \tag{10.2-3}$$

where

$$\Lambda = \mathrm{diag}(\lambda_1, \lambda_2, \ldots, \lambda_n). \tag{10.2-4}$$

Because of (10.1-5), the matrix T in (10.2-2) has a convergent series in descending powers of $x^{1/n}$ of the form

$$T(x) = \text{diag}\{1, px^\rho, (px^\rho)^2, \ldots, (px^\rho)^{n-1}\}[K + O(x^{-1/n})], \quad (10.2-5)$$

where the matrix K and the numbers ρ and p are defined by

$$K = \{k_{jk}\} = \{e^{(2\pi i/n)(j-1)(k-1)}\},$$

$$\rho = m_n/n, \qquad p = (p_n(0))^{1/n}.$$

A short calculation leads to

$$T^{-1}(x)T'(x) = x^{-1}K^{-1}\begin{bmatrix} 0 & 0 & 0 & \cdots & 0 \\ 0 & \rho & 0 & \cdots & 0 \\ 0 & 0 & 2\rho & \cdots & 0 \\ & & \cdot & & \\ & & & \cdot & \\ 0 & 0 & \cdots & & (n-1)\rho \end{bmatrix}K + O(x^{-1-1/n}).$$

Therefore, equation (10.2-3) can be written as

$$\epsilon \frac{dv}{dx} = x^\rho(B_0(x) + \epsilon x^{-(\rho+1)}B_1)v, \quad (10.2-6)$$

where B_0 and B_1 are holomorphic functions of $x^{-1/n}$ at $x = \infty$. Moreover, $B_0(x)$ is diagonal, and its diagonal entries at $x = \infty$ are the n distinct nth roots of $p_n(0)$ [see (10.1-3)].

The successive diagonalization of the coefficients in (10.2-6) can be carried out as in Chapters II and VII, but in view of the particularly simple dependence on ϵ in (10.2-6) it is easier to follow a variant of the formal diagonalization procedure which is due to Turrittin in [83] and was also used by Leung in the present context. This method will now be described. Anticipating that, as in (10.2-6), the magnitudes of the coefficients will decrease by the factor $x^{-(\rho+1)}$ from term to term, one performs successively transformations of the form

$$v = (I + \mu R_1(x))v_1$$

$$\vdots \qquad\qquad\qquad (10.2-7)$$

$$v_{r+1} = (I + \mu^r R_r(x))v_r$$

in this method, where

$$\mu = x^{-(\rho+1)}\epsilon, \quad (10.2-8)$$

and the matrices R_r must be successively calculated so as to be holomorphic in $x^{-1/n}$ and to diagonalize the rth coefficient in the differential equation. For $r = 1$ the transformed differential equation becomes

$$\epsilon \frac{dv_1}{dx} = x^\rho\big[(I - \mu R_1 + \cdots)(B_0 + \mu B_1)(I + \mu R_1),$$

$$- \epsilon x^{-\rho}(I - \mu R_1 + \cdots)\big(\mu R_1' - (\rho+1)x^{-1}\mu R_1\big)\big]v_1, \quad (10.2-9)$$

where the dots indicate a geometric series of μ. The coefficient matrix in the right member can be formally arranged in powers of μ, if the fact that at infinity $O(R_1'(x))$ is $O(R_1(x)x^{-1})$ is used. The coefficient of μ in this series is $x^p(-R_1B_0 + B_1 + B_0R_1)$. The condition that this be a diagonal matrix can be satisfied by choosing R_1 so that

$$R_1B_0 - B_0R_1 = B_1^0 \tag{10.2-10}$$

if B_1^0 denotes the matrix obtained from B_1 by replacing its diagonal entries with zeros. Thanks to the fact that $B_0(x)$ is diagonal with distinct diagonal entries for large x, equation (10.2-10) yields unique functions for the off-diagonal entries of R_1. Its diagonal entries are arbitrary and may be taken as zero. The matrix R_1 is then holomorphic in $x^{-1/n}$ at $x = \infty$, and so is the diagonal matrix $-R_1B_0 + B_1 + B_0R_1$.

The matrices R_r, $r = 2, 3, \ldots$, can be calculated successively by the same argument. The details of the induction argument are simple and need no elaboration here. The formal product $T\prod_{s=1}^{\infty}(I + \mu^s R_s(x))$ can be expanded into a series

$$\mathbf{P}(x, \epsilon) = \sum_{r=0}^{\infty} P_r(x)\epsilon^r$$

and the formal diagonalization theorem below has been proved.

Theorem 10.2-1. *If the coefficient matrix $A(x)$ of the differential equation*

$$\epsilon y' = A(x)y$$

is a polynomial with the properties (10.1-2), (10.1-3), (10.1-6), then the differential equation is formally equivalent, for sufficiently large x, to a formal differential equation with diagonal coefficient matrix. The transformation

$$y = \left(\sum_{r=0}^{\infty} P_r(x)\epsilon^r\right)\mathbf{w} \tag{10.2-11}$$

and the transformed differential equation

$$\epsilon\mathbf{w}' = \left(\sum_{r=0}^{\infty} D_r(x)\epsilon^r\right)\mathbf{w} \tag{10.2-12}$$

have the following properties:

(i) $P_0(x) = T(x)$ $[T$ *is defined in* (10.2-2)$]$;

$P_r(x) = T(x)\hat{P}_r(x)x^{-(\rho+1)r}$, $\rho = m_n/n$;

The matrices $\hat{P}_r(x)$ are holomorphic functions of $x^{-1/n}$ in $|x| \geq x_0$, for some $x_0 > 0$;

$$D_0(x) = x^\rho B_0(x) = \Lambda(x),$$

$[\Lambda$ *as in* (10.2-4), (10.2-5)$]$,

(ii) $D_r(x) = x^{\rho-(\rho+1)r}\tilde{D}_r(x)$, \tag{10.2-13}

and the $\hat{D}_r(x)$ are holomorphic functions of $x^{-1/n}$ in $|x| \geq x_0$.

As was done several times before, the formally diagonalized differential equation (10.2-12) can be formally integrated. Left-hand multiplication with the series in (10.2-11) then yields a formal solution of the original differential equation (10.1-1). The new feature here is that the calculation shows the orders of magnitude of the coefficients in the resulting series, as stated in the corollary below.

Corollary to Theorem 10.2-1. *The differential equation* (10.1-1) *with the properties required in Theorem* 10.2-1 *possesses a formal matrix solution*

$$\left(\sum_{r=0}^{\infty} Y_r(x)\epsilon^r \right)\exp\left\{ \frac{1}{\epsilon} \int^x \Lambda(t)\, dt \right\} \tag{10.2-14}$$

in which the matrices Y_r *have series representations of the form*

$$Y_0(x) = T(x)\exp\left\{ \int_1^x \hat{D}_1(t)t^{-1}\, dt \right\}, \tag{10.2-15}$$

$$Y_r(x) = T(x)\hat{Y}_r(x), \qquad r \geqslant 1 \tag{10.2-16}$$

where $\hat{Y}_r(x)$ *is a series in descending powers of* $x^{-1/n}$ *beginning with the power* $x^{-(\rho+1)r}$ *and convergent for* $|x| \geqslant x_0$.

The verification of the corollary offers no difficulties.

10.3. Asymptotic Solutions for Large Values of the Independent Variable

Corresponding to any point in $|x| \geqslant x_0$ there exist solutions of the differential equation (10.1-1) that are asymptotically represented by the expression in formula (10.2-14) in a neighborhood of that point. The aim of this section is higher: namely, to prove the existence of solutions which have that asymptotic form in unbounded domains and even are there doubly asymptotic. While the proof does not require strikingly new ideas, the necessary modifications of the existence proofs in Chapters III and VII are substantial enough to justify an exposition in some detail.

To simplify the notation it will be assumed that in (10.1-3)

$$p_n(0) = 1. \tag{10.3-1}$$

This does not affect the generality, since the shearing transformation

$$y = \text{diag}\left(1, (p_n(0))^{1/n}, \ldots (p_n(0))^{(n-1)/n}\right) y^*$$

together with the substitution

$$\epsilon = \epsilon^*(p_n(0))^{1/n}$$

produces this property in the differential equation (10.1-1).

To begin the existence proof, the formal series $\sum_{r=0}^{\infty} Y_r(x)\epsilon^r$ in the corollary to Theorem 10.2-1 must be shown to be the asymptotic expansion —in the doubly asymptotic sense—of some function $U(x,\epsilon)$. This requires a slight modification of the generalized Borel–Ritt theorem 7.5-1. It will suffice to state the result in the present terminology and to omit the proof.

If $Y_r, r = 1, 2, \ldots,$ is defined by

$$\hat{Y}_r(x) = \tilde{Y}_r(x) x^{-(\rho+1)r}, \qquad r = 1, 2, \ldots t$$

then, by the corollary above, $\tilde{Y}_r$ has a representation as an ascending series in non-negative powers of $x^{-1/n}$, convergent for $|x| \geqslant x_0$. The function Y_0 in (10.2-15) is somewhat different. It may involve logarithms. If $|x| \geqslant x_0$ the functions $\tilde{Y}_r$ are holomorphic on the n-sheeted Riemann surface with a branch point of order n at $x = \infty$. The function $Y_0(x)$ is singlevalued, for $|x| \geqslant x_0$ on the logarithmic Riemann surface $\mathbb{C}_L$ over $\mathbb{C}$.

Lemma 10.3-1 (Borel–Ritt Theorem on Doubly Asymptotic Series). *Assume that the matrices $Y_r, r = 1, 2, \ldots$ are holomorphic functions of $x^{-1/n}$ for $|x| \geqslant x_0 > 0$. Let*

$$S = \{x \in \mathbb{C} \,|\, |x| \geqslant x_0, \vartheta_1 < \arg x < \vartheta_2\}, \qquad 0 < \epsilon \leqslant \epsilon_0, \quad \delta > 0$$

with $\vartheta_1, \vartheta_2, \epsilon_0, \delta$ arbitrary. Then there exists, for any given $c > 0$, a matrix-valued function U of x and ϵ holomorphic in both variables in the set

$$G = \{(x,\epsilon) \,|\, x \in S, 0 < \epsilon \leqslant \epsilon_0, |x^{-(\rho+1)}\epsilon| \leqslant c\}$$

so that U has in G the asymptotic expansion

$$U(x,\epsilon) \sim \sum_{r=1}^{\infty} Y_r(x)(x^{-(\rho+1)}\epsilon)^r$$

in the sense that for $N = 1, 2, \ldots$

$$\left| U(x,\epsilon) - \sum_{r=1}^{N} Y_r(x)(x^{-(\rho+1)}\epsilon)^r \right|$$

$$\leqslant c_N |x^{-(\rho+1)}\epsilon|^{N+1}, \qquad (x,\epsilon) \in G, \qquad 0 < \epsilon \leqslant \epsilon_0, \quad (10.3\text{-}2)$$

with certain c_N. The function U depends on G.

By virtue of Lemma 10.3-1 and the Corollary to Theorem 10.2-1 the formal expression (10.2-14) represents asymptotically the function

$$\hat{U}(x,\epsilon) \exp\left\{ \frac{1}{\epsilon} \int^x \Lambda(t)\, dt \right\}, \tag{10.3-3}$$

where

$$\hat{U}(x,\epsilon) = Y_0(x) + T(x)U(x,\epsilon). \tag{10.3-4}$$

The procedure for constructing true solutions that are asymptotically equal to the function in (10.3-3) is an adaption of the technique introduced in Chapter III. Section 3.2 carries over with almost no changes, and so does Section 3.3—except for minor differences of notation—through Definition 3.2-2. The analog of Theorem 3.3-1, however, is not immediately apparent, because the region involved is unbounded.

Therefore, the remaining problem is to show that the integral equation (3.3-8) possesses solutions in certain unbounded domains, and that these solutions are asymptotic to zero in the doubly asymptotic sense, i.e., that in these regions

$$|z(x,\epsilon)| \leqslant C_N |(x^{-(\rho+1)}\epsilon)|^N, \qquad N = 0, 1, \ldots,$$

with certain constants C_N.

The details of the argument to be given resemble those employed by the author in the asymptotic theory without a parameter, as x tends to infinity [90]. The integral equation corresponding to (3.3-8) can be written

$$z(x,\epsilon) = U(x,\epsilon) \int_{\Gamma_x} \exp\left\{ \frac{1}{\epsilon} \int_t^x (\Lambda(s) - I\lambda_k(s)) \, ds \right\}$$
$$\times \left[K(t,\epsilon) z(t,\epsilon) + g(t,\epsilon) \right] dt \qquad (10.3\text{-}5)$$

with

$$K(t,\epsilon) \sim 0, \qquad g(t,\epsilon) \sim 0 \qquad \text{for} \quad (x,\epsilon) \in G. \qquad (10.3\text{-}6)$$

[G was defined in Lemma 10.3-1. The relations (10.3-6) are to be understood in the doubly asymptotic sense.] The dependence of z on the subscript k is not set in evidence in (10.3-5). As in Chapter III, the symbol Γ_x stands for a set of n paths of integration γ_{jx}, $j = 1, 2, \ldots, n$, in the t-plane, each of which starts at some fixed point, not necessarily the same for all γ_{jx}, and ends at $t = x$. The paths must be chosen so that they are "progressive" in the sense of Definition 3.3-1, i.e., the integrals

$$\int_t^x (\lambda_j(s) - \lambda_k(s)) \, ds \qquad (10.3\text{-}7)$$

must have nonpositive real parts for $t \in \gamma_j$. However, in contrast to the context in Chapter III the possibility of γ_{jx} starting at $t = \infty$ is not ruled out.

The progressive paths γ_{jx} must lie in what was called an "accessible" region in Definition 3.3-2. The construction of such regions, which will now be described, follows closely the one in [90], section 14.3.

Define q_j by

$$q_j(x) := \int^x \lambda_j(s) \, ds. \qquad (10.3\text{-}8)$$

The indefinite integral in (10.3-8) may be made unique by the stipulation that the series for $q_j(x)$ obtained by termwise integration of the series in (10.1-5) should not contain a constant term. Set

$$q_{jk} := q_j - q_k, \qquad j = 1, 2, \ldots, n.$$

The curves

$$\operatorname{Re} q_{jk}(x) = 0$$

are, by definition, the "separation curves" for $x = \infty$. One has, using (10.3-1),

$$q_{jk}(x) = \frac{1}{\rho + 1} x^{\rho+1} (e^{2\pi i j/n} - e^{2\pi i k/n})[1 + O(x^{-1/n})] \qquad \text{as} \quad x \to \infty.$$

$$(10.3-9)$$

Therefore, the asymptotic directions of the separation curves at infinity are

$$(\rho + 1)^{-1}[(l + \tfrac{1}{2})\pi - \arg(e^{2\pi i j/n} - e^{2\pi i k/n})],$$

$$l = 0, \pm 1, \pm 2, \ldots, \quad j, k = 1, 2, \ldots, n, \quad j \neq k.$$

After simplification these angles are seen to be

$$\frac{(j+k) + nl}{m_n + n}\pi, \qquad j, k = 1, 2, \ldots n, \quad j \neq k, \quad l = 0, \pm 1, \ldots . \qquad (10.3\text{-}10)$$

They are best interpreted on C_L, but they are single valued on the Riemann domain of p sheets with branch point at $x = \infty$, if pn is the least common multiple of n and $2(n + m_n)$. They are uniquely defined on such a domain, if one of the p directions that project onto the positive x-axis is arbitrarily assigned the angle $\arg x = 0$. Two consecutive directions (10.3-10) form the angle $\pi/(m_n + n)$. On such a Riemann domain there are $2p$ arcs on which $\operatorname{Re} q_{jk}(x) = 0$ for each fixed pair (j, k). Their projections onto $\mathbb{C}$ have two distinct directions that are opposite to each other. On the other hand, every multiple of $\pi/(n + m_n)$ is the direction of several separation arcs at infinity.

Similarly to what was done in Section 7.3, the next step is the transformation

$$\xi = x^{\rho+1}. \qquad (10.3\text{-}11)$$

If one stipulates that $\arg \xi = (\rho + 1)\arg x$, then (10.3-11) defines a holomorphic one-to-one mapping of C_L onto itself. The images of the separation arcs have the asymptotic directions $(j + k)\pi/n + l\pi, l = \pm 1, \pm 2, \ldots$.

Now assume, *temporarily*, that only the leading term in the series for $\lambda_j(x)$ in (10.1-5) is present. Then the separation arcs are strictly the straight rays from the origin with the direction s (10.3-10), and their images are the rays

$$\arg \xi = (j + k)\pi/n \qquad (\operatorname{mod} \pi). \qquad (10.3\text{-}12)$$

Let S be an open sector of C_L with vertex at the origin and with central angle $\pi(\rho + 1)$ and such that its bounding rays are not in the set (10.3-12).

Then, S contains, for each pair (j, k) with $j \neq k$ exactly one separation ray. The image $\xi(S)$ of that sector is a half plane containing one ray $\arg \xi = (j + k)\pi/n$ for each pair (j, k) $(j \neq k)$. The sector S therefore contains, for each pair (j, k), directions along which $\operatorname{Re} q_{jk}(x)$ increases and rays along which it decreases. This property is shared by all straight lines parallel to these rays. Hence, there can be found, corresponding to every such sector $\xi(S)$ and to every pair (j,k), $(j \neq k)$, a direction in $\xi(S)$ such that along a ray from infinity to any given point $\xi \in \xi(S)$ the real part of $\xi(e^{2\pi ij/n} - e^{2\pi ik/n})$ $(j \neq k)$ *decreases*. This means that the real part of $q_{jk}(x)$ in (10.3-8) decreases along the preimages of these rays, provided $O(x^{-1/n}) \equiv 0$ in that formula. This makes it plausible that, even without this condition, these preimages, when taken as the paths γ_{jx} in (10.3-5) for given $k \neq j$, will yield an exponential factor with negative real part for every component of the vector in the integrand if $t \in \gamma_{jx}$ $x \in S$, except, of course, when $t = x$ or $j = k$, provided x is limited to the part of S for which $|x| \geq x_0$ with x_0 sufficiently large. That this actually is so was proved in [42]. The reader is referred to that paper. Let $S_0 = \{x/x \in S, |x| \geq x_0\}$. For $j = k$, the exponential factor in (10.3-5) is absent, and one may choose as path of integration γ_{kx} for that component of the vector in the right member of (10.3-5) any path in the t-plane which goes from $t = \infty$ to $t = x$ and which lies in G (recall Theorem 10.3-1) so that the relations (10.3-6) are valid.

For any vector-valued function z holomorphic and bounded for $x \in S_0$, $0 < \epsilon \leq \epsilon_0$, the integral in the right member of (10.3-5) exists because of (10.3-6). To derive the existence of a unique bounded solution of that integral equation by a contraction mapping theorem is now a simple technical matter, thanks to the relations (10.3-6). It sufficiently resembles the arguments in Sections 3.3 and 3.4 to justify omission of the details. They can be found in Leung's paper [41]. The reasoning also resembles that given in [90] for the parameterless case. That argument does not yet show directly that the solution is asymptotically zero in the sense of Definition 10.1-1. That property now follows immediately by insertion of the solution z into (10.3-5). In fact, $\hat{U}$ is uniformly bounded for $x \in S_0$, $0 < \epsilon \leq \epsilon_0$, as is the exponential factor in the integral, because of the choice of the paths γ_{jx}. Therefore,

$$|z(x,\epsilon)| \leq \text{const.} \sum_{j=1}^{n} \int_{\gamma_{jk}} |(K(t,\epsilon)z(t,\epsilon) + g(t,\epsilon))_j| \, dt$$

in S_0, where the subscript j indicates the jth component of the vector in parenthesis. The constant depends on S_0 and ϵ_0, but not on x or ϵ. Because of (10.3-6) the last inequality implies, with

$$\|z\| = \sup_{\substack{x \in S_0 \\ 0 < \epsilon \leq \epsilon_0}} |z(x,\epsilon)|,$$

that

$$|z(x,\epsilon)| \leq k_N |x|^{-(\rho+1)} |\epsilon|^{N+1} (\|z\| + 1)$$

for all $N \geq 0$, uniformly for $x \in S_0$, $0 < \epsilon \leq \epsilon_0$, with some constants k_N. Since $\|z\|$ is finite, it has been proved that $z \sim 0$ in the desired sense.

Now return to Section 3.2 and to the derivation of the integral equation (3.3-8) for z, there. It was proved there that the existence of a solution of (3.3-8) asymptotic to zero in the sense of the ordinary Borel–Ritt theorem (Theorem 3.2-1) in some region implies the existence of a corresponding vector solution of the original differential equation having the formal series solution as its asymptotic representation. That argument holds literally in the present situation with the representation being true in the doubly asymptotic sense. Here is the statement of the ensuing theorem.

Theorem 10.3-1.   *If the coefficient matrix $A(x)$ of the differential equation*

$$\epsilon y' = A(x) y \qquad (10.3\text{-}13)$$

is a polynomial with the properties (10.1-2), (10.1-3), and (10.1-6), then the formal matrix solution (10.2-14) is an asymptotic representation of a true solution, in the doubly asymptotic sense, in any region

$$S_0 = \left\{ x / |x| > x_0, \vartheta_0 < \arg x < \vartheta_0 + \frac{\pi}{m_n + n} \right\},$$

provided x_0 is sufficiently large and ϑ_0 is not the direction at $x = \infty$ of a separation arc. The doubly asymptotic property means that if

$$y_k(x, \epsilon) = \sum_{r=0}^{\infty} y_{rk}(x) \epsilon^r \exp\left\{ \frac{1}{\epsilon} \lambda_k(t) \, dt \right\} \qquad (k = 1, 2, \ldots, n)$$

is a column of the formal matrix solution (10.2-14) then there exists a true vector solution $y_k(x, \epsilon) = \hat{y}_k(x, \epsilon) e^{(1/\epsilon) \int^x \lambda_k(t) \, dt}$ for which

$$\left| \hat{y}_k(x, \epsilon) - \sum_{r=0}^{N} y_{rk}(x) \epsilon^r \right| \leq C_{Nk} |x^{-n/n + m_n} \epsilon|^{N+1}$$

with constants C_{Nk}, for $N = 0, 1, 2, \ldots$, and for all $x \in S_0$, $0 < \epsilon \leq \epsilon_0$. The solution depends on S_0.

10.4. Some Properties of Doubly Asymptotic Solutions

The connection problems in the asymptotic theory of differential equations when the independent variable tends to a point where the coefficients have a pole—usually the point at infinity—have been thoroughly explored. Many of these results carry over to the passage to the limit, as $x \to \infty$, when doubly asymptotic expansions are available. This will be the case whenever it is only the order of magnitude of the terms in the asymptotic expansion that is relevant, without requiring that they are strictly powers of x, as in the standard parameterless theory.

Even in the theory without a parameter most solutions are not uniquely characterized by their asymptotic expansions, a fact which often makes meaningless the question as to the asymptotic nature of a solution in sectors outside the one where the solution is given through its asymptotic expansion.

The simplest situation in the parameterless theory arises when the solution under study is asymptotically known, as $x \to \infty$, in sectors where it is recessive, i.e., where all other linearly independent solutions are larger, in the sense of order of magnitude, as $x \to \infty$. It is almost obvious then that such a solution is uniquely determined by its asymptotic expansion (see e.g., [90] section 15).

In the asymptotic theory with respect to a parameter, as developed in Chapters II and III, a vector solution of the form $y_1(x,\epsilon) = \hat{y}_1(x,\epsilon) e^{q_1(x,\epsilon)}$ which is recessive in a region of the x-plane in the sense that in that region all independent other solutions are "larger" in order of magnitude, as $\epsilon \to 0$, is still far from unique, because if

$$y_2(x,\epsilon) = \hat{y}_2(x,\epsilon) e^{q_2(x,\epsilon)}$$

and

$$\operatorname{Re} q_2(x,\epsilon) > \operatorname{Re} q_1(x,\epsilon),$$

a solution

$$y_1(x,\epsilon) + e^{q_1(x_2,\epsilon) - q_2(x_2,\epsilon)} y_2(x,\epsilon)$$

will be asymptotically indistinguishable from y_1, as $\epsilon \to \infty$, in some region depending on the choice of the point x_2. This is the underlying reason why almost all investigation of connection problems for asymptotic expansions with respect to a parameter have to rely, at some point, on results from the parameterless theory or even on particular properties of special functions.

For the doubly asymptotic expansion of the form described in Theorem 10.3-1 more precise statements can be made.

To get some insight into such problems, consider four consecutive separation arcs C_0, C_1, C_2, C_3 near $x = \infty$ and denote by $\Sigma_1, \Sigma_2, \Sigma_3$ the sectorial regions bounded in succession by these arcs and by arcs of the circle $|x| = x_0$. Corresponding to each of these separation arcs C_ν, $\nu = 0, 1, 2, 3$, one can construct, in many ways, a region of the type called S_0 in Theorem 10.3-1. Let S_0^ν, $\nu = 0, 1, 2, 3$, be four such regions such that $C_\nu \subset S_0^\nu$. Define q_j by

$$q_j(x) = \frac{1}{\epsilon} \int^x \lambda_j(t)\,dt, \qquad j = 1, 2, \ldots, n,$$

with the constant of integration as in (10.3-8), and set

$$q_{jk} = q_j - q_k .$$

Let those functions be labeled so that

$$q_{jk}(x,\epsilon) < 0 \qquad \text{for} \quad j < k, \quad x \in \Sigma_2 .$$

Now assume that y is a vector solution of the differential equation (10.3-13) of Theorem 10.3-1 which has the form

$$y(x,\epsilon) = \hat{y}(x,\epsilon) e^{q_1(x,\epsilon)},$$

where

$$\hat{y}(x,\epsilon) \sim \sum_{r=0}^{\infty} y_{r1}(x)\epsilon^r,$$

in the doubly asymptotic sense, for x on some ray $\arg x = \vartheta$ in Σ_2. The regions S_0^1 and S_0^2 can be chosen so that the ray $\arg x = \vartheta$ is in $S_0^1 \cap S_0^2$. If $y_j^2, j = 1, 2, \ldots, n$, is a fundamental system of vector solutions of (10.3-13) with the properties described in Theorem 10.3-1, there must exist an identity

$$y(x,\epsilon) = \sum_{j=1}^{n} c_j(\epsilon) y_j^2(x,\epsilon) \tag{10.4-1}$$

with functions c_j of ϵ alone. On the ray $\arg x = \vartheta$ that identity can be written in the form

$$\hat{y}(x,\epsilon) - c_1(\epsilon)\hat{y}_1^2(x,\epsilon) = \sum_{j=2}^{n} c_j(\epsilon)\hat{y}_j^2(x,\epsilon)e^{q_{j1}(x,\epsilon)}, \tag{10.4-2}$$

where $\hat{y}$ and $\hat{y}_j^2$ have on $\arg x = \vartheta$ doubly asymptotic series in powers of ϵ with coefficients that have the magnitude of positive or negative powers of x. Since $\operatorname{Re} q_{j1}(x,\epsilon) > 0$ on that ray, for $j > 1$, the $c_j(\epsilon)$ must be identically zero, for $j > 1$, if (10.4-2) is to be true, and then it follows that $c_1(\epsilon) \equiv 1$, by virtue of Theorem 10.3-1.

It has thus been proved that $y = y_1^2$. The same argument shows that $y = y_1^1$, so that y possesses the doubly asymptotic expansion

$$y(x,\epsilon) \sim \left(\sum_{r=0}^{\infty} y_{r1}(x)\epsilon^r \right) e^{q_1(x,\epsilon)} \tag{10.4-3}$$

in all of $S_0^1 \cup S_0^2$. This is true even if y is no longer recessive in the parts of $S_0^1 \cup S_0^2$ that lie in Σ_1 or Σ_3. Finally, consider another ray, $\arg x = \vartheta^*$ in Σ_1. If ϑ^* is close enough to the direction of C_1 at infinity, a region S_1^{1*} with the properties of S_0 in Theorem 10.3-1 can be found whose boundary is as close to C_0 as one pleases. A repetition of the previous argument, with ϑ^*, S_0^{1*} replacing ϑ, S_0^1 proves that the asymptotic representation (10.4-3) extends to S_0^{1*} as well. Proceeding symmetrically in Σ_3 the asymptotic relation (10.4-3) is proved to be valid inside the whole set $\Sigma_1 \cup \overline{\Sigma}_2 \cup \Sigma_3$. Here is a complete statement of this result.

Theorem 10.4-1. *If a vector solution y of the differential equation* (10.3-13) *has a doubly asymptotic expansion*

$$y(x,\epsilon) \sim \left(\sum_{r=0}^{\infty} y_{r1}(x)\epsilon^r \right) e^{q_1(x,\epsilon)} \tag{10.4-4}$$

as $\epsilon \to 0+$ and $x \to \infty$ on the ray $\arg x = \vartheta$ and if the solution is recessive in the sense that

$$\operatorname{Re} q_1(x,\epsilon) < \operatorname{Re} q_j(x,\epsilon), \qquad j = 2, 3, \ldots, n,$$

for $\arg x = \vartheta$, $|x| \geqslant x_0$, *then that doubly asymptotic expansion is valid, for* $|x| \geqslant x_0$, *in any sector that contains in its interior the two separation arcs nearest to* $\arg x = \vartheta$ *but no other separation arcs.*

In different, less precise language: Near $x = \infty$, the separation arcs divide the neighborhood of ∞ on C_L into curvilinear sectors. A solution that is recessive in part of one of these sectors and has there a doubly asymptotic expansion maintains this expansion in a region consisting of three adjacent such sectors.

Even a recessive doubly asymptotic solution is not quite uniquely determined by its asymptotic expansion: If

$$y(x,\epsilon) = \hat{y}(x,\epsilon)e^{q_1(x,\epsilon)}, \tag{10.4-5}$$

with $\operatorname{Re} q_1(x,\epsilon)(x,\epsilon) < \operatorname{Re} q_j(x,\epsilon)$, for $j > 1$, when x is in some sector Σ and if $\hat{y}$ has the doubly asymptotic expansion

$$\hat{y}(x,\epsilon) \sim \sum_{r=0}^{\infty} y_r(x)\epsilon^r, \qquad \epsilon \to 0+, \quad x \to \infty, \tag{10.4-6}$$

for $x \in \Sigma$, then

$$y^*(x,\epsilon) = \hat{y}(x,\epsilon)(1 + e^{-\kappa/\epsilon})e^{q_1(x,\epsilon)}, \qquad \kappa > 0,$$

has the same asymptotic expansion as $y(x,\epsilon)$ in Σ and, in fact, everywhere in $|x| > x_0$. However, the freedom of choice for $\hat{y}(x,\epsilon)$ in (10.4-5) is very limited, as is shown by the simple theorem below.

Theorem 10.4-2. *If*

$$y(x,\epsilon) = \hat{y}(x,\epsilon)e^{q_1(x,\epsilon)},$$

$$y^*(x,\epsilon) = \hat{y}^*(x,\epsilon)e^{q_1(x,\epsilon)}$$

are two doubly asymptotic solutions of the differential equation (10.3-1) *which are recessive in the same sector* Σ *and if*

$$\begin{aligned} \hat{y}(x,\epsilon) &\sim \sum_{r=0}^{\infty} y_r(x)\epsilon^r, \\ \hat{y}^*(x,\epsilon) &\sim \sum_{r=0}^{\infty} y_r^*(x)\epsilon^r, \end{aligned} \tag{10.4-7}$$

for $\epsilon \to 0+$, $x \in \Sigma$, $x \to \infty$, *then there is a constant* c_0 *independent of* ϵ *such that*

$$y_r^*(x) = c_0 y_r(x).$$

PROOF.   There must exist an identity

$$\hat{y}^*(x,\epsilon)e^{q_1(x,\epsilon)} = c(\epsilon)\hat{y}(x,\epsilon)e^{q_1(x,\epsilon)} + \sum_{j=1}^{n} c_j(\epsilon)\hat{y}_j(x,\epsilon)e^{q_j(x,\epsilon)}$$

in which $\operatorname{Re} q_j(x,\epsilon) > \operatorname{Re} q_1(x,\epsilon)$, $j = 2, 3, \ldots, n$, and the $y_j(x,\epsilon)$ have doubly asymptotic expansions in, at least, a subsector of Σ. The functions c, c_j are scalars. As in the proof of Theorem 10.4-1 one concludes that $c_j \equiv 0$ for $j = 2, 3, \ldots, n$. In the remaining vectorial identity,

$$\hat{y}^*(x,\epsilon) = c(\epsilon)\hat{y}(x,\epsilon),\qquad(10.4\text{-}8)$$

observe that at least one component of the vector y_0 in (10.4-7) is not identically zero, so that $c(\epsilon)$ must have an asymptotic series in powers of ϵ, say,

$$c(\epsilon) \sim \sum_{r=0}^{\infty} c_r \epsilon^r.$$

Insertion of that expansion and of (10.4-7) into (10.4-8) leads to the recursive relations

$$y_0^*(x) = c_0 y_0(x),$$

$$y_r^*(x) = c_0 y_r(x) + \sum_{\substack{\alpha+\beta=r \\ \beta < r}} c_\alpha y_\beta(x),\qquad r > 0.$$

For $r = 1$ this means that

$$y_1^*(x) = c_0 y_1(x) + c_1 y_0(x).$$

Now y_1^* and y_1 are smaller at infinity, in the sense of order or magnitude, than y_0. Hence $c_1 \equiv 0$. A simple induction shows that $c_r \equiv 0$ for $r = 1$, $2, \ldots$. This completes the proof.

10.5. Central Connection Problems in Unbounded Regions

The uniform reduction to Airy's equation of systems of the form

$$\epsilon \frac{dy}{dx} = \begin{pmatrix} 0 & 1 \\ a(x) & 0 \end{pmatrix} y\qquad(10.5\text{-}1)$$

in a neighborhood of a turning point of order one, i.e., of a point x_1 where

$$a(x_1) = 0,\qquad a'(x_1) \neq 0\qquad(10.5\text{-}2)$$

solves the connection problems at that point by means of the well-known connection formulas for Airy's function. Section 8.6 contains the details of these calculations. That theory is strictly local: it applies to a neighborhood of x_1 only.

On the other hand, Fedoryuk and Evgrafov's work described in Chapter IX and the theory of the present chapter solve the differential equation (10.5-1) asymptotically near $x = \infty$ when a is a polynomial. These solutions are essentially uniquely determined, except for a constant scalar factor, by their doubly asymptotic property in sectors where they are recessive. That was proved in Theorem 10.4-2. Unfortunately, the solutions introduced in Chapter VIII through the uniform reduction method are, in general, not the analytic continuations of the doubly asymptotic solution of the present chapter. Therefore, there remains a nontrivial central connection problem whose solution will be briefly sketched in this section. Details can be found in [93].

Consider, therefore, the case that a in (10.5-1) is a polynomial of degree m, and let D be a canonical domain containing the simple turning point x_1 on its boundary. For simplicity, set $x_1 = 0$.

The reader is referred to Section 8.6 for the details of the formal transformation of the differential equation into Airy's equation. The main point was that the coefficients $P_r(t)$ in (8.6-8) had to be successively calculated from the formulas (8.6-11) so as to be holomorphic at $t = 0$. The doubly asymptotic solutions introduced in the present chapter were calculated by a different method. They could, however, have been also determined by a procedure analogous to that of Chapter VIII: calculate the coefficients $P_r(t)$ in (8.6-8) from their recursion formula in such a way that they satisfy inequalities of the form

$$|P_r(t)| \leqslant c_r|t|^{-r\beta}, \qquad |t| \geqslant t_1, \tag{10.5-3}$$

with certain constants c_r and β. These inequalities define the P_r uniquely up to a common scalar constant factor independent of ϵ. They are the solutions of (8.6-11) obtained by taking $P_0(t) = q(t)I$, as in Section 8.6 and then choosing the constants in (8.6-11) equal to $c_{jk} = 0$, $\alpha_{jk} = \infty$, $j, k = 1, 2$. Let the formal series so obtained be denoted by

$$\mathbf{P}^F(t, \epsilon) = \sum_{r=0}^{\infty} P_r^F(t)\epsilon^r \tag{10.5-4}$$

(F for Fedoryuk).

It turns out that $\mathbf{P}^F$ differs from the series $\mathbf{P}$ in (8.6-13) in the second term already. One finds

$$\mathbf{P}^F(t, \epsilon) = q(t)I$$
$$+ \begin{bmatrix} 0 & t^{-1/2}q(t)(h(\infty) - h(t)) \\ t^{1/2}q(t)(h(\infty) - h(t)) + \dot{q}(t) & 0 \end{bmatrix}\epsilon$$
$$+ \cdots \tag{10.5-5}$$

[$h(t)$ was defined in (8.6-12)].

To the formal series $\mathbf{P}^F$ there corresponds a matrix P^F which has it as its asymptotic expansion in the t-plane image of the subset D_δ of the fundamental domain whose distance from the boundary ∂D is $\delta > 0$, provided D is a consistent domain. Then the transformation

$$W = P^F(t, \epsilon)Z \tag{10.5-6}$$

takes equation (8.6-5) into Airy's equation (8.6-6).

For the sake of clarity let the matrix $P(t, \epsilon)$ in (8.6-15) be denoted now by $P^L(t, \epsilon)$ (L for Langer). Then the transformation

$$W = P^L(t, \epsilon)Z \qquad (10.5\text{-}7)$$

also transforms (8.6-5) into (8.6-6). If Z in (10.5-5) and (10.5-6) is replaced by the particular fundamental matrix solution (8.6-16) of (8.6-6), the matrices

$$W^F := P^F Z, \qquad W^L := P^L Z$$

are two fundamental solutions of equation (8.6-5). The connection matrix $C(\epsilon)$ defined by $W^L = W^F C$, i.e.,

$$C(\epsilon) := \left(W^F(t, \epsilon) \right)^{-1} W^L(t, \epsilon) = Z^{-1}(t, \epsilon) \left(P^F(t, \epsilon) \right)^{-1} P^L(t, \epsilon) Z(t, \epsilon)$$

$$(10.5\text{-}8)$$

can now be calculated by taking for t some point in a region where both P^L and P^F have known asymptotic expansions. More details of these fairly straightforward calculations were given in [93].

CHAPTER XI

A Singularly Perturbed Turning Point Problem

11.1. The Problem

Differential equation problems that depend on small parameter ϵ in such a way that the order of the equation is lower for $\epsilon = 0$ than for $\epsilon \neq 0$ but remains positive are now commonly called "singular perturbation problems." The condition that the order remain positive for $\epsilon = 0$ is not a very distinguishing property of the differential equation as such. The equation

$$\epsilon u'' - 2xu' + ku = 0, \qquad k \text{ a constant}, \tag{11.1-1}$$

for instance, which will be examined closely in the next section, becomes

$$\epsilon^2 v'' - \left(x^2 - \epsilon(1 + k)\right)v = 0 \tag{11.1-2}$$

under the simple change of variables

$$u = e^{x^2/2\epsilon}v. \tag{11.1-3}$$

In the first form the order decreases from two to one, in the second form from two to zero. The particular mathematical interest of singular perturbation problem lies, rather, in the question as to what happens to the solution of *boundary value problems* as $\epsilon \to 0$. The boundary conditions

$$u(\alpha) = A, \qquad u(\beta) = B, \tag{11.1-4}$$

for instance, with α, β, A, B independent of ϵ, when joined to equation (11.1-1) produce a problem where it is natural to ask whether or not it has a solution that tends to a solution of the "reduced equation"

$$-2xu_0' + ku_0 = 0, \tag{11.1-5}$$

as $\epsilon \to 0$, and, if so, what boundary conditions that limit u_0 will satisfy.

For a wide class of linear singular perturbation problems such questions are comparatively simple to answer. For the boundary value problem (11.1-1), (11.1-4), for instance, there is no serious difficulty *as long as the interval* $\alpha \leqslant x \leqslant \beta$ *does not contain the origin.* The general theory of Chapters II and III implies then that in such an interval there exist pairs of solutions of equation (11.1-1) with the asymptotic form

$$u_1 = \hat{u}_1(x,\epsilon), \qquad u_2 = \hat{u}_2(x,\epsilon)e^{x^2/\epsilon}, \tag{11.1-6}$$

$\hat{u}_1, \hat{u}_2$ being functions with asymptotic power series expansions

$$\hat{u}_j(x,\epsilon) \sim \sum_{r=0}^{\infty} u_{jr}(x)\epsilon^r, \qquad \epsilon \to 0+, \quad j = 1,2, \tag{11.1-7}$$

uniformly valid for $\alpha \leqslant x \leqslant \beta$. These solutions are not uniquely determined by their asymptotic expansions, but to within arbitrary constant factors independent of ϵ one has

$$u_{10}(x) = x^{k/2}, \qquad u_{20}(x) = x^{(k+1)/2},$$

at least if k is independent of ϵ.

The solution which satisfies the boundary conditions (11.1-4) has the form

$$u(x,\epsilon) = c_1 u_1(x,\epsilon) + c_2 u_2(x,\epsilon) \tag{11.1-8}$$

with constants c_1, c_2 that may depend on ϵ. They can be determined from the conditions

$$c_1 u(\alpha,\epsilon) + c_2 u_2(\alpha,\epsilon) = A, \qquad c_1 u_1(\beta,\epsilon) + c_2 u_2(\beta,\epsilon) = B \tag{11.1-9}$$

by Cramer's rule. If $0 < \alpha < \beta$, one finds that

$$c_1(\epsilon) = \frac{A}{u_1(\alpha,\epsilon)} + O(e^{(\alpha^2 - \beta^2)/\epsilon}), \qquad c_2(\epsilon) = O(e^{-\beta^2/\epsilon}),$$

so that

$$u(x,\epsilon) = A \frac{u_1(x,\epsilon)}{u_1(\alpha,\epsilon)} + O(e^{(x^2 - \beta^2)/\epsilon}). \tag{11.1-10}$$

Thus, the solution of the boundary value problem possesses, for $0 < \alpha \leqslant x < \beta$, an asymptotic series expansion in powers of ϵ. Except very close to $x = \beta$, the "boundary layer correction" is asymptotically negligible. In particular, $\lim_{\epsilon \to 0} u(x,\epsilon) = u_{10}(x)$ in $\alpha \leqslant x < \beta$, and this is the solution of the reduced equation (11.1-5) that satisfies the left, but generally not the right, boundary condition.

If $\alpha < \beta < 0$, an analogous calculation shows that it is then the right-hand boundary condition that is preserved in the limit and that the boundary layer appears at the left end point.

The extension of the arguments in the example above to two-point boundary value problems for general linear analytic differential equations whose order decreases from n to m at $\epsilon = 0$ has been thoroughly studied

([84], [21]), and the results are easy to describe, provided certain conditions, which amount to the absence of turning points, are satisfied. If turning points are present, the analysis becomes, however, a very subtle problem, even when the order drops from two to one.

Only this last mentioned case, which has recently attracted much attention ([2], [48], [8], [33], [35], [79], etc.), will be dealt with here and that only under the additional assumption that the turning point has the lowest possible order. Consider, accordingly, the differential equation

$$\epsilon u'' + f(x,\epsilon)u' + g(x,\epsilon)u = 0 \tag{11.1-11}$$

on some interval on the real axis, which for simplicity will be taken as $I := [-1, 1]$. To keep the problem in the framework of this report, f and g will be assumed to be analytic functions holomorphic and real for $x \in I$, $|\epsilon| \leqslant \epsilon_0$. For large parts of the theory, but definitely not for all of it, the analyticity in x is not indispensable.

Let again the boundary conditions (11.1-4) be prescribed. They now become

$$u(-1,\epsilon) = A, \qquad u(1,\epsilon) = B. \tag{11.1-12}$$

The reduced equation is

$$f(x,0)u_0' + g(x,0)u_0 = 0. \tag{11.1-13}$$

The problem does not involve turning points as long as the decisive condition

$$f(x,0) \neq 0 \qquad \text{for} \quad x \in I \tag{11.1-14}$$

is satisfied. The arguments given above for the special equation (11.1-1) in an interval that did not contain the origin can then be adapted with very little change and with an analogous result. The end point at which the boundary layer appears depends on the sign of $f(x,0)$ on I.

If, however, the function $f(x,0)$ has a zero on I that point is a turning point according to the definition adopted in Chapter III. In the present chapter the resulting boundary value problem will be solved under the further restrictive hypothesis that the zero is of order one. The method followed is that of Sibuya in [79].

11.2. A Simple Example

In the interval I, i.e., when $\alpha = -1$, $\beta = 1$ in formula (11.1-4) the arguments that led to the asymptotic representation (11.1-10) for the solution of the boundary value problem for the equation (11.1-1) are no longer valid. It is still true that (11.1-6) and the formal series in (11.1-7) constitute formal series solutions, but these series no longer represent one and the same solution asymptotically on the whole interval I. This matter is more

conveniently analyzed for the differential equation (11.1-2) which is equivalent to (11.1-1). That equation possesses two formal series solutions of the
form

$$\hat{v}^{-}(x,\epsilon)e^{-x^2/2\epsilon}, \qquad \hat{v}^{+}(x,\epsilon)e^{x^2/2\epsilon}, \qquad (11.2\text{-}1)$$

where

$$\hat{v}^{\pm}(x,\epsilon) = \sum_{r=0}^{\infty} v_r^{\pm}(x)\epsilon^r. \qquad (11.2\text{-}2)$$

The exponential factors in (11.2-1) are unique, but there exist infinitely
many choices for the series in (11.2-2). Among them there is one for which
the $v_r^{\pm}(x)$ have the order of magnitude

$$v_r^{\pm}(x) = O(x^{-1/2-2r}), \qquad r = 0,1,\ldots, \qquad (11.2\text{-}3)$$

as $x \to \infty$. The property (11.2-3) characterizes the formal solutions uniquely
except for an arbitrary constant nonzero factor independent of ϵ. The
functions $v_r^{\pm}(x)$ can be calculated termwise.

The facts stated above were proved in Chapter VIII. The theory of
section 8.2 also supplies the following additional information. There are
four separation curves—they are straight rays in the case of the simple
equation (11.1-2)—through the turning point $x = 0$. They have directions
$-\frac{1}{4}\pi + \frac{1}{2}(j-1)\pi$ ($j \bmod 4$) and will be denoted by l_j. Let S_j be the open
quadrant between l_j and l_{j+1}. To every quadrant S_j there belongs one
solution of (11.1-2) which has as its asymptotic expansion there the one of
the two formal expressions in (11.2-1) for which the real part of the
exponent is negative in S_j, and therefore positive in S_{j-1} and S_{j+1}. The
asymptotic expansion is valid in the whole unbounded domain $\mathbb{C} - \overline{S}_{j+2}$,
which consists of three of the four sectors S_j ($\overline{S}_j$ is the closure of S_j). In any
such domain at positive distance from $\overline{S}_{j+2}$ the expansion is even uniform.
Moreover, if (11.2-3) is true, it is a doubly asymptotic expansion in the
sense of Theorem 10.4-2 so that it is unique, except for a constant
normalization factor independent of ϵ.

The general theory which has just been summarized cannot supply a
complete solution of the lateral asymptotic connection problem, and it says
nothing about the central connection problem. Fortunately, equation (11.1-
2) is essentially Weber's equation, whose solutions, known also as parabolic
cylinder functions, can be obtained completely by special methods that
supply all the desired information. To draw on that knowledge, consider
Weber's differential equation in its standard form, namely,

$$\frac{d^2z}{d\xi^2} - (\xi^2 - a)z = 0 \qquad (11.2\text{-}4)$$

with a a constant (that may depend on ϵ). It is known that one of its
solutions, $z_1(\xi,a)$, tends to zero for $\xi \to +\infty$. With an appropriate normalization factor this solution and its properties are listed in the literature, e.g.,

in [1]. Simple symmetry arguments show that all four functions

$$z_j(\xi, a) := z_1\big((-i)^{j-1}\xi, (-1)^{j-1}a\big) \qquad (j \bmod 4)$$

are solutions of (11.2-4). Therefore, the four functions

$$v_j(x, \epsilon) = z_j\left(\frac{x}{\epsilon^{1/2}}, a\right) \tag{11.2-5}$$

satisfy the differential equation

$$\epsilon^2 \frac{d^2v}{dx^2} - (x^2 - \epsilon a)v = 0, \tag{11.2-6}$$

which is (11.1-2) with

$$a = 1 + k. \tag{11.2-7}$$

The asymptotic theory of Weber's equation not only confirms the results of the general theory described above with v_j being the solution that is recessive in S_j, dominant in S_{j-1} and S_{j+1}, but beyond that, it supplies the linear relations that allow one to continue the asymptotic descriptions into the fourth sector S_{j+2}. One has, in fact, the relation $z_1 = \kappa z_2 + \lambda z_3$, i.e.,

$$v_1 = \kappa v_2 + \lambda v_3, \tag{11.2-8}$$

where

$$\kappa = 2^{-a/2} e^{i\pi(a+1)/4} \sqrt{2\pi} \,\Big/\, \Gamma\left(\frac{1-a}{2}\right), \qquad \lambda = -i e^{i\pi a/2}. \tag{11.2-9}$$

Armed with that information on the solutions of equation (11.1-2) the original boundary value problem for (11.1-1) can now be solved asymptotically. That equation admits the four particular solutions

$$u_j := v_j e^{x^2/2\epsilon} \qquad (j \bmod 4). \tag{11.2-10}$$

The v_j, given by (11.2-5), have in $\mathbb{C} - \bar{S}_{j+2}$ asymptotic expansions of the form (11.2-1), (11.2-2) that are recessive in S_j. Thus

$$v_j(x, \epsilon) = \hat{v}_j(x, \epsilon) \exp\{(-1)^j x^2/2\epsilon\},$$

where the $\hat{v}_j$ have asymptotic series $\hat{\mathbf{v}}_j$ in powers of ϵ. Hence

$$\begin{aligned}
& u_1 \sim \hat{\mathbf{v}}_1 && \text{in } \mathbb{C} - \bar{S}_3, && u_2 \sim \hat{\mathbf{v}}_2 e^{-x^2/\epsilon} && \text{in } \mathbb{C} - \bar{S}_4, \\
& u_3 \sim \hat{\mathbf{v}}_3 && \text{in } \mathbb{C} - \bar{S}_1, && u_4 \sim \hat{\mathbf{v}}_4 e^{-x^2/\epsilon} && \text{in } \mathbb{C} - \bar{S}_2.
\end{aligned} \tag{11.2-11}$$

Moreover, by (11.2-8),

$$u_1 = \kappa u_2 + \lambda u_3 \sim \kappa \hat{\mathbf{v}}_2 e^{x^2/\epsilon} + \lambda \hat{\mathbf{v}}_3 \qquad \text{in } S_3. \tag{11.2-12}$$

The solution of the boundary problem (11.1-1), (11.1-4), when it exists, can be written

$$u = c_1 u_1 + c_2 u_2, \tag{11.2-13}$$

where c_1, c_2 may depend on ϵ, k, A and B, but not on x. The routine calculation of c_1 and c_2 will be described in some detail, in order to reveal clearly the structure of the answer. One has to solve the two linear algebraic equations

$$c_1 u_1(-1,\epsilon) + c_2 u_2(-1,\epsilon) = A, \qquad c_1 u_1(1,\epsilon) + c_2 u_2(1,\epsilon) = B. \quad (11.2\text{-}14)$$

Now,

$$u_2(\pm 1,\epsilon) = \hat{v}_2(\pm 1,\epsilon)e^{1/\epsilon} \sim \hat{v}_2(\pm 1,\epsilon)e^{1/\epsilon},$$

while

$$u_1(+1,\epsilon) = \hat{v}_1(+1,\epsilon) \sim \hat{v}_1(+1,\epsilon),$$

but

$$u_1(-1,\epsilon) = \kappa u_2(-1,\epsilon) + \lambda u_3(-1,\epsilon) \sim \kappa \hat{v}_2(-1,\epsilon)e^{1/\epsilon} + \lambda \hat{v}(-1,\epsilon).$$
$$(11.2\text{-}15)$$

Inserting these formulas into (11.1-14) and applying Cramer's rule one obtains the explicit expressions

$$c_1 = \frac{\begin{vmatrix} A & \hat{v}_2(-1) \\ B & \hat{v}_2(1) \end{vmatrix}}{\kappa \hat{v}_2(-1)\hat{v}_2(1)e^{1/\epsilon} + \begin{vmatrix} \lambda \hat{v}_3(-1) & \hat{v}_2(-1) \\ \hat{v}_1(1) & \hat{v}_2(1) \end{vmatrix}},$$

$$(11.2\text{-}16)$$

$$c_2 = \frac{B\kappa \hat{v}_2(-1) + \begin{vmatrix} \lambda \hat{v}_3(-1) & A \\ \hat{v}_1(1) & B \end{vmatrix}e^{-1/\epsilon}}{\kappa \hat{v}_2(-1)\hat{v}_2(1)e^{1/\epsilon} + \begin{vmatrix} \lambda \hat{v}_3(-1) & \hat{v}_2(-1) \\ \hat{v}_1(1) & \hat{v}_2(1) \end{vmatrix}}.$$

For the sake of brevity the dependence on ϵ has not been set in evidence in the last two formulas. The solution of the boundary value problem is, then,

$$u(x,\epsilon) = c_1(\epsilon)\hat{v}_1(x,\epsilon) + c_2(\epsilon)\hat{v}_2(x,\epsilon)e^{x^2/\epsilon}, \qquad (11.2\text{-}17)$$

and also

$$u(x,\epsilon) = c_1(\epsilon)\left[\kappa \hat{v}_2(x,\epsilon)e^{x^2/\epsilon} + \lambda \hat{v}_3(x,\epsilon) \right] + c_2(\epsilon)\hat{v}_2(x,\epsilon)e^{x^2/\epsilon}. \quad (11.2\text{-}18)$$

The form (11.2-17) is suited for the asymptotic description in S_1, while (11.2-18) is asymptotically known in S_3.

Now, κ in (11.2-9) vanishes whenever $(1-a)/2$ is a negative integer which happens, by (11.2-7), if and only if k is an even positive integer. If k does *not* have such an exceptional value and is independent of ϵ, then the formulas (11.2-16) show that

$$c_j(\epsilon) = O(e^{-1/\epsilon}), \qquad j = 1,2 \quad \text{as} \quad \epsilon \to 0+$$

and it follows from (11.2-17) and (11.2-18) that $\lim_{\epsilon \to 0+} u(x,\epsilon) = 0$ in $\{x \mid -1 < x < 1, x \neq 0\}$.

Lemma 6.2-1, which says that $\hat{v}_j$ and $\hat{v}_j^{-1}$ remain bounded, as $x \to 0$, proves that u tends to zero even at $x = 0$ and, therefore, *everywhere* on $0 < x < 1$. Now assume that k *is* an even positive integer. Then $\kappa = 0$, $\lambda = 1$, $v_1 = v_3$, and one concludes from (11.2-16) that

$$c_j(\epsilon) \sim \left[\sum_{r=0}^{\infty} c_{jr}\epsilon^r \right] \exp\{(1-j)/\epsilon\} \qquad \text{as} \quad \epsilon \to +0, \quad j = 1, 2,$$

with certain constants c_{jr}, because the denominator in (11.2-16) has a nonzero limit, as $\epsilon \to 0 +$. This can be verified by inspection of the limits of $\hat{v}_j(\pm 1, \epsilon)$ at $\epsilon = 0$. Moreover, (11.2-17), (11.2-18) reduce to

$$u(x,\epsilon) = c_1(\epsilon)u_3(x,\epsilon) + O(e^{(x^2-1)/\epsilon})$$

for $-1 < x < 1$. Therefore, u possesses an asymptotic series representation

$$u(x,\epsilon) \sim \sum_{r=0}^{\infty} u_r(x)\epsilon^r \qquad \text{for} \quad -1 < x < 1, \quad \epsilon \to 0 +. \quad (11.2\text{-}19)$$

In general, the series in (11.2-19) is not an asymptotic representation of u at $x = -1$ and $x = 1$. There appear therefore boundary layers near those points.

The exceptional case that the differential equation (11.1-1) possesses a solution which tends in all of $-1 \leqslant x \leqslant 1$ to a bounded nonzero function has been called the "asymptotic resonance" case by Ackerberg and O'Malley [2].

For some applications it is important to investigate what happens if k is permitted to change with ϵ. A *necessary* condition for asymptotic resonance is, of course, that

$$\lim_{\epsilon \to 0+} k(\epsilon) = 2n, \qquad n \text{ a positive integer.} \quad (11.2\text{-}20)$$

To find sufficient conditions, observe that because of (11.2-7) and (11.2-9), κ and λ in formulas (11.2-16), (11.2-17) are functions of ϵ and that (11.2-20) implies

$$|\kappa| \leqslant \text{const.}|k(\epsilon) - 2n|, \qquad \lim_{\epsilon \to 0+} \lambda = 1, \quad (11.2\text{-}21)$$

since the Γ-function has poles of order one at the negative integers. Therefore, if $k(\epsilon)$ approaches $2n$ so fast that

$$(k(\epsilon) - 2n)e^{1/\epsilon} = o(1) \qquad \text{as} \quad \epsilon \to 0 +, \quad (11.2\text{-}22)$$

then, also,

$$\kappa e^{1/\epsilon} = o(1) \quad (11.2\text{-}23)$$

and the arguments leading to (11.2-19) remain valid. If

$$\lim_{\epsilon \to 0+} (k(\epsilon) - 2n)e^{1/\epsilon} = \infty \qquad \text{for all} \quad n \in \mathbb{N}, \quad (11.2\text{-}24)$$

then $\lim_{\epsilon \to 0+} u(x,\epsilon) = 0$. If neither (11.2-22) or (11.2-24) is true, the asymptotic behavior of $u(x,\epsilon)$ is somewhat more complicated to state. It is, for instance, possible that for a null-sequence of ϵ-values there fails to exist a solution. No serious new questions arise, however, and the matter need not be pursued in detail. The results of this section will now be restated as a theorem.

Theorem 11.2-1.   *If for some positive integer n the function $k(\epsilon)$ satisfies the relation*

$$\lim_{\epsilon \to 0+} (k(\epsilon) - 2n)e^{1/\epsilon} = 0,$$

then the real boundary value problem

$$\epsilon u'' - 2xu' + k(\epsilon)u = 0,$$

$$u(-1) = A, \qquad u(1) = B, \qquad A^2 + B^2 \neq 0$$

possesses a solution with an asymptotic expansion

$$u(x,\epsilon) \sim \sum_{r=0}^{\infty} u_r(x)\epsilon^r, \qquad u_0(x) \not\equiv 0, \quad \epsilon \to 0+,$$

valid in $-1 < x < 1$. If

$$\lim_{\epsilon \to 0+} (k(\epsilon) - 2n)e^{1/\epsilon} = \infty$$

for all positive integers n, then

$$\lim_{\epsilon \to 0+} u(x,\epsilon) \equiv 0 \qquad in \quad -1 < x < 1.$$

Corollary.   *If the boundary conditions in Theorem 11.2-1 are replaced by the conditions (11.1-4) with $\alpha < 0 < \beta$, then the factor $e^{1/\epsilon}$ in the theorem must be replaced by $e^{r_0^2/\epsilon}$, $r_0 = \max(|\alpha|, \beta)$.*

11.3. The General Case: Formal Part

The general differential equation (11.1-11) with the boundary conditions (11.1-2) will now be studied under the hypothesis that

$$f(x,0) = -xh(x), \qquad h(x) > 0 \quad in \quad -1 \leqslant x \leqslant 1. \qquad (11.3-1)$$

This is the natural generalization of the special equation (11.1-1). Other generalizations, such as the occurrence of a second- or higher-order turning point in $-1 \leqslant x \leqslant 1$ or the case that $h(x) < 0$ have been only incompletely studied, but the behavior of the solution is then quite different.

The differential equation (11.1-11) could be asymptotically solved by the general methods of Chapters V and VI, except when it comes to the determination of the asymptotic connection formulas. For those the natural approach is through the theory of Special Functions, in particular the parabolic cylinder or Weber functions. It is therefore preferable to aim directly at that theory, taking advantage of the special form of the differential equation.

To begin with, equation (11.1-11) will be transformed into one that differs little, for small ϵ, from a form of Weber's equation. To that end, one first changes the independent variable x to t by setting

$$t := \left[\int_0^x \xi h(\xi)\, d\xi \right]^{1/2}. \tag{11.3-2}$$

If the sign of the square root is fixed by stipulating that $t > 0$ for $x > 0$, then $t = t(x)$ maps the neighborhood of the interval $I = \{x \mid -1 < x < 1\}$ conformally onto a neighborhood of the interval $I^* := t(I)$. The differential equation (11.1-11) is taken into

$$\epsilon \frac{d^2 u}{dt^2} + F(t,\epsilon) \frac{du}{dt} + G(t,\epsilon) u = 0 \tag{11.3-3}$$

with

$$F = (t')^{-2}\{t'f + \epsilon t''\}, \qquad G = (t')^{-2} g. \tag{11.3-4}$$

Here, t and t'' stand for the functions of t obtained by replacing x with its expression in terms of t in $dt(x)/dx$ and $d^2 t(x)/dx^2$. Since $2tt' = -xh(x) = -f(x,0)$, by (11.3-2), the function F in (11.3-3) has the form

$$F(t,\epsilon) = -2t + \epsilon F_1(t,\epsilon). \tag{11.3-5}$$

The functions F_1 and G are holomorphic for $t \in I^*$, $0 \leqslant \epsilon \leqslant \epsilon_0$.

In analogy to the transformation (11.1-2) one next removes the term in (11.3-3) that contains the first derivative by setting

$$u = v \exp\left\{ -\frac{1}{2\epsilon} \int_0^t F(s,\epsilon)\, ds \right\} \tag{11.3-6}$$

and finds an equation of the form

$$\epsilon^2 \frac{d^2 v}{dt^2} - \left(t^2 + \epsilon R(t,\epsilon) \right) v = 0 \tag{11.3-7}$$

with

$$R = \left(t^2 - \frac{1}{4} F^2 \right) \epsilon^{-1} - \frac{1}{2} \frac{\partial F}{\partial t} + G. \tag{11.3-8}$$

Because of (11.3-5) the function F is not only holomorphic in ϵ for $\epsilon \neq 0$, but even at $\epsilon = 0$.

In order to simplify the differential equation further by the theory of Chapter V it is convenient to set

$$y := \begin{pmatrix} v \\ \epsilon \dfrac{dv}{dt} \end{pmatrix}$$

and to deal with the vectorial differential equation

$$\epsilon \frac{dy}{dt} = \left[\begin{pmatrix} 0 & 1 \\ t^2 & 0 \end{pmatrix} + \begin{pmatrix} 0 & 0 \\ R(t,\epsilon) & 0 \end{pmatrix} \epsilon \right] y, \tag{11.3-9}$$

which is equivalent to (11.3-7). A formal simplification of such a system was described in Theorem 5.2-2: There exists a formal matrix series

$$\mathbf{P}(t,\epsilon) = \sum_{r=0}^{\infty} P_r(t,\epsilon)\epsilon^r, \qquad P_0(t) \equiv I \qquad (11.3\text{-}10)$$

with coefficients holomorphic in a complex region of the t-plane which contains the interval t^* such that

$$y = \mathbf{P}(t,\epsilon)w \qquad (11.3\text{-}11)$$

takes (11.3-9) into

$$\epsilon \frac{dw}{dt} = \begin{bmatrix} 0 & 1 \\ t^2 + \epsilon \displaystyle\sum_{r=0}^{\infty} b_r \epsilon^r & 0 \end{bmatrix} w. \qquad (11.3\text{-}12)$$

The b_r are constants. The first of the two scalar equations contained in (11.3-12), i.e.,

$$\epsilon^2 \frac{d^2 w_1}{dt^2} - \left(t^2 + \epsilon \sum_{r=0}^{\infty} b_r \epsilon^r \right) w_1 = 0, \qquad (11.3\text{-}13)$$

differs from (11.1-2) only by the fact that $\sum_{r=0}^{\infty} b_r \epsilon^r$ is a formal series, not a function.

In the proof of Theorem 5.2-2 the arguments were limited to a neighborhood of the origin. A check of the reasoning there shows, however, that the coefficients P_r in (11.3-10) remain holomorphic on all of I^*.

The analytic companions of the formal theorem just quoted are Theorems 6.3-2 and 6.6-2 but even the *formal* result implies a fact about the formal solution of equation (11.1-11) which is of some interest in itself, and which has been the subject of some recent papers ([35], [48]). It turns out that when (11.3-1) is true such solutions have singularities at $x = 0$ unless certain conditions, infinite in number, are satisfied. The theorem to be stated shows that those conditions become quite simple when expressed in terms of the formal differential equation (11.3-13).

Theorem 11.3-1. *If the real differential equation*

$$\epsilon u'' + f(x,\epsilon)u' + g(x,\epsilon)u = 0 \qquad (11.3\text{-}14)$$

with

$$f(x,0) = -xh(x), \qquad h(x) > 0 \quad in \quad -1 \leqslant x \leqslant 1, \qquad (11.3\text{-}15)$$

has holomorphic coefficients for $-1 \leqslant x \leqslant 1$, $0 < \epsilon \leqslant \epsilon_0$, then it can be formally satisfied by a nontrivial power series $\sum_{r=0}^{\infty} u_r(x)\epsilon^r$ with coefficients holomorphic in $1 \leqslant x \leqslant 1$, if and only if the corresponding numbers b_r in the transformed differential equation (11.3-13) have the values

$$b_0 = -(1 + 2m), \qquad m \text{ a non-negative integer}, \quad b_r = 0, \quad r > 0. \qquad (11.3\text{-}16)$$

PROOF.   Assume that $u(x, \epsilon) := \sum_{r=0}^{\infty} u_r(x) \epsilon^r$ is a formal solution of (11.3-14). Then the expression

$$w(t, \epsilon) = P^{-1}(t, \epsilon) \begin{bmatrix} u(x(t), \epsilon) \exp\left\{ \dfrac{1}{2\epsilon} \int_0^t F(\tau, \epsilon) \, d\tau \right\} \\ \epsilon \dfrac{d}{dt} \left[u(x(t), \epsilon) \exp\left\{ \dfrac{1}{2\epsilon} \int_0^t F(\tau, \epsilon) \, d\tau \right\} \right] \end{bmatrix}$$

is a formal solution of (11.3-13) by virtue of (11.3-6) and (11.3-11). (Of course, w is not a series in powers of ϵ.) The further transformation

$$w_1 = e^{-(1/2\epsilon)t^2} p \tag{11.3-17}$$

takes (11.3-13) into

$$\epsilon \frac{d^2 p}{dt^2} - 2t \frac{dp}{dt} - \left(1 + \sum_{r=0}^{\infty} b_r \epsilon^r \right) p = 0, \tag{11.3-18}$$

which is formally solved by

$$p(t, \epsilon) := w_1(t, \epsilon) e^{(1/2\epsilon)t^2}. \tag{11.3-19}$$

Thanks to formula (11.3-5) the expression $p(t, \epsilon)$ *is a formal series in powers of ϵ with holomorphic coefficients.* Now, insert that series,

$$p(t, \epsilon) =: \sum_{r=0}^{\infty} p_r(t) \epsilon^r, \tag{11.3-20}$$

for p into (11.3-16) and calculate the p_r from the resulting series of conditions:

$$2t \frac{dp_0}{dt} + (1 + b_0) p_0 = 0,$$

$$2t \frac{dp_r}{dt} + (1 + b_0) p_r = \frac{d^2 p_{r-1}}{dt^2} - p_{r-1} b_1 - p_{r-2} b_2 - \cdots - p_1 b_{r-1}, \qquad r > 0.$$

One finds, first, that $p_0(t) = \text{const.} \, t^{-(1+b_0)/2}$. Hence, $(1 + b_0)/2$ must be a nonpositive integer if p_0 is to be holomorphic at $t = 0$, say, $(1 + b_0)/2 = -m$, i.e., $b_0 = -(1 + 2m)$, m a non-negative integer. Then $p_1(t) = \text{const.} \, t^m - \frac{1}{2} b_1 t^m \log t - m(m-1) t^{m-1}$, which is holomorphic at $t = 0$ if and only if $b_1 = 0$. The complete induction argument which shows that all p_r are holomorphic on I^* if, and only if, (11.3-16) is true, is straightforward.

   Conversely, if (11.3-16) holds, then the p_r can be calculated as described, and (11.3-19) then defines $u(x, \epsilon)$ as a formal series with coefficients holomorphic on $-1 \leqslant x \leqslant 1$ that satisfies (11.3-14). Thus, the theorem is proved.

   The conditions (11.3-16), transformed into relations among the coefficients of the original differential equation (11.3-14) are sometimes called

"Matkowksy's conditions," because he described an algorithm for finding these conditions and the series $\mathbf{u}$ in [48]. It is natural to conjecture that this series is an asymptotic representation of a solution of the differential equation and that, therefore, Matkowsky's conditions are not only necessary but sufficient for the existence of solutions with nontrivial asymptotic power series. Beyond that it seems plausible that if the boundary value problem (11.1-11), (11.3-1), (11.1-12) has *any* solution that remains bounded in $[-1, 1]$ and does not have the zero function as its limit for $\epsilon \to 0 +$, it must have such an asymptotic series expansion. This matter will now be investigated.

11.4. The General Case: Analytic Part

Corresponding to the formal Theorem 5.2-2 invoked in the preceding section there exists the strong analytic result in Theorem 6.7-1, restated here, for convenience, in the notation of this chapter.

Theorem 11.4-1. *There exists a matrix-valued function $P(t, \epsilon)$ holomorphic for $t \in I^*$, $\epsilon \in S := \{\epsilon / 0 < |\epsilon| \leqslant \epsilon_0, |\arg \epsilon| < \rho\}$ which has the series (11.3-10) as its uniformly asymptotic expansion in that region and has the property that the transformation*

$$y = Pw$$

takes the differential equation (11.3-9) into

$$\epsilon \frac{dw}{dt} = \begin{pmatrix} 0 & 1 \\ t^2 + \epsilon b(\epsilon) & 0 \end{pmatrix} w, \tag{11.4-1}$$

where b is independent of t, and

$$b(\epsilon) \sim \sum_{r=0}^{\infty} b_r \epsilon^r \qquad as \quad \epsilon \to 0 \quad in \ S. \tag{11.4-2}$$

For the aim of this section, which is to investigate whether the Matkowsky conditions are *sufficient* for the existence of solutions of the boundary value problem (11.1-11), (11.1-12), (11.3-1) whose limit, as $\epsilon \to 0$, is not zero, it may be assumed that (11.4-2) has the simple form $b(\epsilon) \sim q$, q some negative odd integer, as $\epsilon \to 0$ in S. In more explicit notation this means that

$$b(\epsilon) = q + \delta(\epsilon), \qquad q \text{ a negative odd integer, with} \tag{11.4-3}$$

$$\delta(\epsilon) \sim 0 \qquad \text{for} \quad \epsilon \to 0 \quad \text{in } S. \tag{11.4-4}$$

If (11.4-3) is satisfied, the equation for the first component w_1 of the vector w in (11.4-1) is

$$\epsilon^2 \frac{d^2u}{dt^2} - (t^2 + \epsilon(q + \delta(\epsilon))u_1 = 0, \qquad (11.4\text{-}5)$$

which, except for the notation, is identical with (11.1-2) for $k = k(\epsilon)$,

$$k(\epsilon) = q - 1 + \delta(\epsilon), \qquad (11.4\text{-}6)$$

or, if one sets

$$q = 1 - 2n, \qquad n \text{ a positive integer,}$$

with

$$k(\epsilon) - 2n = \delta(\epsilon).$$

The condition of the corollary to Theorem 11.2-1 reads, now

$$\lim_{\epsilon \to 0+} \delta(\epsilon) e^{r_0^2/\epsilon} = 0. \qquad (11.4\text{-}7)$$

If this condition is satisfied, Theorem 11.2-1 proves that the formal series solution of (11.1-11) with (11.3-1)—which exists by Theorem 11.3-1, thanks to the assumptions made on the b_r, $r = 0, 1, \ldots$, —is asymptotic to a true solution, and this is the result conjectured before.

Unfortunately, the proof of Theorem 11.4-1 does not contain any simple way for replacing (11.4-4) with the stronger statement (11.4-7). In [77], Sibuya has given a subtle, but long, argument resulting in the following theorem.

Theorem 11.4-2. *Assume that the coefficients f and g of the differential equation (11.1-11) are holomorphic for $x \in D$, $|\epsilon| \le \epsilon_0$, and that $f(x, 0)$ has the form (11.3-1). Then the differential equation possesses solutions with an asymptotic expansion*

$$u(x, \epsilon) \sim \sum_{r=0}^{\infty} u_r(x)\epsilon^r, \qquad u_0(x) \not\equiv 0, \quad -1 < x < 1,$$

as $\epsilon \to 0+$, provided

  (i) *The Matkowsky conditions (11.3-16) on the transformed differential equation (11.3-12) are satisfied;*
 (ii) *the image $t(D)$ of D under the mapping (11.3-2) contains the disk*

$$|t| \le \max\{t(-1), t(+1)\}.$$

In two as yet unpublished papers David Ching-her Lin has shown that condition (ii) is not necessary. On the other hand, it is not difficult to see that for data that are infinitely differentiable but not analytic the Matkowsky conditions are *not* sufficient for asymptotic resonance.

The principal tool for the proof of Theorem 11.4-2 is a result, of interest in itself, concerning the gap between the statement that a function is asymptotic to zero and the statement that it is exponentially small. To shorten the descriptions, a function $\delta(\epsilon)$ holomorphic in a sector $S := \{\epsilon \,|\, 0 < |\epsilon| \leqslant \epsilon_0, \alpha < \arg \epsilon < \beta\}$ will be called "flat in S," if it is asymptotic to zero, as $\epsilon \to 0$ in S. The letters $c_0, c_1, \ldots,$ below, indicate certain positive constants, independent of ϵ.

Theorem 11.4-3. *Let*

$$S_\mu = \{\epsilon \,|\, 0 < |\epsilon| \leqslant \epsilon_0, \alpha_\mu < \arg \epsilon < \beta_\mu\}, \qquad \mu = 1, 2, \ldots, N,$$

be sectors whose union covers the punctured disk $0 < |\epsilon| \leqslant \epsilon_0$, *and let* δ_μ, $\mu = 1, 2, \ldots, N$, *be functions of* ϵ *each of which is flat in* S_μ. *If*

$$|\delta_\mu(\epsilon) - \delta_\nu(\epsilon)| \leqslant c_0 e^{-c_1/|\epsilon|} \qquad in \quad S_\mu \cap S_\nu, \tag{11.4-8}$$

whenever $S_\mu \cap S_\nu \neq \emptyset$, *then*

$$|\delta_\mu(\epsilon)| \leqslant c_2 e^{-c_1/|\epsilon|} \qquad in \; S_\mu, \quad \mu = 1, 2, \ldots, N. \tag{11.4-9}$$

Since this theorem is not directly concerned with differential equations it will not be proved here. Proofs can be found in [67] and in [79].

To see the relevance of that Theorem 11.4-3 to the problem at hand, observe that, thanks to the assumption that the coefficients of the differential equation (11.1-11) are holomorphic in ϵ in a full disk, not only in a sector, Theorem 11.4-1 is valid in (possibly narrow) sectors of the ϵ-place that contain any preassigned direction $\arg \epsilon = \vartheta$. To see that, it suffices to perform the changes of variables

$$\epsilon = \epsilon^* e^{-i\vartheta}, \qquad t = t^* e^{-i\vartheta/3}, \qquad y = \begin{pmatrix} 1 & 0 \\ 0 & e^{-i\vartheta/3} \end{pmatrix} y^*$$

in the differential equation (11.3-9) and to apply Theorem 11.4-1 to the new differential equation. By returning to the old variables one obtains the same result as in Theorem 11.4-1 but with asymptotic expansions valid in a sector S_ϑ centered as $\arg \epsilon = \vartheta$. The functions P and b, there, must be replaced by functions P_ϑ, b_ϑ. Moreover, $\delta(\epsilon)$ in (11.4-3) must be replaced by $\delta_\vartheta(\epsilon)$, which is flat in S_ϑ.

Thus, there is now a family of differential equations

$$\epsilon \frac{dw}{dt} = \begin{pmatrix} 0 & 1 \\ t^2 + \epsilon(q + \delta_\vartheta(\epsilon)) & 0 \end{pmatrix} w, \tag{11.4-10}$$

depending on ϑ, such that to every fundamental matrix solution W_ϑ there corresponds a solution

$$Y_\vartheta = P_\vartheta W_\vartheta \tag{11.4-11}$$

of the differential equation (11.3-9). This latter differential equation—and this is important—does not depend on ϑ.

The principal hypothesis of Theorem 11.4-3, namely, the inequality (11.4-8), will now be shown to be true for the functions $\delta_\vartheta(\epsilon)$ in (11.4-10) by studying the connection formulas for the corresponding solutions Y_ϑ of (11.3-9). In preparation of that calculation the connection formulas for Weber's equation introduced in scalar form in Section 11.2 must be recast in matrix notation: Let $z(\xi, a)$ be the particular solution of Weber's equation introduced there and define

$$z_j(\xi, a) := z\left((-i)^{j-1}\xi, (-1)^{j-1}a\right) \qquad (j \bmod 4)$$

$$Z_j(\xi, a) := \begin{bmatrix} z_j(\xi, a) & z_{j+1}(\xi, a) \\ \dfrac{dz_j(\xi, a)}{d\xi} & \dfrac{dz_{j+1}(\xi, a)}{d\xi} \end{bmatrix},$$

then

$$Z_j(\xi, a) = Z_{j-1}(\xi, a)C\left((-1)^{j+1}a\right) \tag{11.4-12}$$

with

$$C(a) = \begin{pmatrix} \kappa(a) & 1 \\ \lambda(a) & 0 \end{pmatrix}. \tag{11.4-13}$$

The matrices $Z = Z_j(\xi, a)$ satisfy

$$\frac{dZ}{d\xi} = \begin{pmatrix} 0 & 1 \\ \xi^2 - a & 0 \end{pmatrix} Z;$$

hence, four matrix solutions of (11.4-10) are given by

$$W_{j\vartheta}(t, \epsilon) := \begin{pmatrix} 1 & 0 \\ 0 & \epsilon^{1/2} \end{pmatrix} Z_j\left(\frac{t}{\epsilon^{1/2}}, -(q + \delta_\vartheta(\epsilon))\right), \tag{11.4-14}$$

as can be verified directly. Four corresponding solutions of (11.3-9) are then

$$Y_{j\vartheta} := P_\vartheta W_{j\vartheta}. \tag{11.4-15}$$

They are connected, thanks to (11.4-11) and (11.4-12), by the formula

$$Y_{j\vartheta} = Y_{j+1,\vartheta} C_{\vartheta j}, \tag{11.4-16}$$

where

$$C_{\vartheta j} := C\left((-1)^{j+1}(q + \delta_\vartheta(\epsilon))\right). \tag{11.4-17}$$

Since the differential equation (11.3-9) does not depend on ϑ, any two solutions $Y_{j\alpha}, Y_{j\beta}$ must be connected by a formula

$$Y_{j\alpha} = Y_{j\beta} K_j \qquad (j \bmod 4), \tag{11.4-18}$$

with K_j independent of t. Of course, K_j depends on α and β, but this dependence need not be exhibited explicitly, because α and β will be kept fixed for the remainder of this section. They must be chosen so that $S_\alpha \cap S_\beta \neq \emptyset$. Combining (11.4-16) and (11.4-18) one sees that the various connection matrices are related by the equality

$$K_{j+1} = C_{\beta j} K_j C_{\alpha j}^{-1}. \tag{11.4-19}$$

Some information on the matrices K_j can be obtained by applying the asymptotic formulas for Weber's functions to the representation

$$K_j = W_{j\beta}^{-1} P_\beta^{-1} P_\alpha W_{j\alpha}, \tag{11.4-20}$$

which follows from (11.4-15) and (11.4-18). These asymptotic formulas (obtainable from the general asymptotic theory, as well as from a specialized study of the properties of Weber's equation), can be written, in matrix form, as follows:

$$Z_j(\xi, a) = \hat{Z}_j(\xi, a) e^{Q_j(\xi)}, \tag{11.4-21}$$

where

$$Q_j(\xi) = (-1)^j \begin{pmatrix} 1 & 0 \\ 0 & -1 \end{pmatrix} \xi^2 / 2 \tag{11.4-22}$$

and

$$\hat{Z}_j(\xi, a) \sim \sum_{r=1}^{\infty} Z_{jr}(a) \xi^{-r} \qquad \text{as} \quad \xi \to \infty \tag{11.4-23}$$

in the sector

$$\left| \arg \xi - \tfrac{1}{2} j\pi \right| < \tfrac{3}{4} \pi.$$

The functions Z_{jr} are entire. From (11.4-14) and (11.4-21) one obtains, then, asymptotic expansions for the $W_{j\vartheta}$, valid for $t \to \infty$, as well as for $\epsilon \to 0$, provided $t \in t(D)$ and

$$\left| \arg \frac{t}{\epsilon^{1/2}} - j \frac{\pi}{2} \right| < \frac{\pi}{2} . \tag{11.4-24}$$

One has, thus,

$$W_{j\vartheta}(t, \epsilon) = \hat{W}_{j\vartheta}(t, \epsilon) e^{Q_j(t\epsilon^{-1/2})}$$

with

$$\hat{W}_{j\alpha}(t, \epsilon) \sim \hat{W}_{j\beta}(t, \epsilon),$$

as long as (11.4-24) is true.

Now require, in addition, that $\epsilon \in S_\alpha \cap S_\beta$, then P_α and P_β also have the same asymptotic series in powers of ϵ. Therefore, K_j in (11.4-20) admits an asymptotic representation

$$K_j = I + e^{-Q(t\epsilon^{-1/2})} \Omega_j(t, \epsilon) e^{Q(t\epsilon^{-1/2})} \tag{11.4-25}$$

with

$$\Omega_j(t, \epsilon) \sim 0, \tag{11.4-26}$$

for $t \in t(D)$, $\epsilon \in S_\alpha \cap S_\beta \neq \emptyset$, and if (11.4-24) is satisfied.

In (11.4-19) the matrices $C_{\beta j}$, $C_{\alpha j}$ are known functions of the quantities $\delta_\alpha(\epsilon)$, $\delta_\beta(\epsilon)$. The formulas (11.4-25) contain asymptotic information on the matrices K_j, K_{j+1} in (11.4-19). By combining these formulas it is possible to improve the known relation $\delta_\beta(\epsilon) - \delta_\alpha(\epsilon) \sim 0$.

To begin with, no matter how narrow the sector $S_\alpha \cap S_\beta$, there are values of $t \in t(D)$ and of $\epsilon \in S_\alpha \cap S_\beta$ for which (11.4-24) is true and $t\epsilon^{-1/2}$ is real, as well as other such values for which $t\epsilon^{-1/2}$ is a pure imaginary number. Choosing such values with $|t| = r$, one obtains on the entries $k_j^{\rho\sigma}$ of the matrix K_j the information

$$k_j^{11} \sim 1, \qquad k_j^{22} \sim 1, \tag{11.4-27}$$

$$|k_j^{\rho\sigma}| \leqslant ke^{-r^2/|\epsilon|}, \qquad \rho \neq \sigma \quad \rho,\sigma = 1,2. \tag{11.4-28}$$

Here k is some constant independent of ϵ, and r can be chosen—by hypothesis (ii) of Theorem 11.4-2—so that the disk $|t| \leqslant r$ contains the I^* in its interior, but is contained in $t(D)$. [I^* was defined after formula (11.3-2).]

To shorten the expressions for the calculations that follow, the notation

$$a(\epsilon) \approx b(\epsilon) \qquad \text{in } S$$

will be introduced to indicate that

$$[a(\epsilon) - b(\epsilon)]e^{r^2/|\epsilon|}$$

remains bounded, as $\epsilon \to 0$ in the sector S.

Now, the relation

$$K_{j+1}C_{\alpha j} - C_{\beta j}K_j = 0,$$

which is equivalent to (11.4-19), implies

$$\begin{bmatrix} k_{j+1}^{11} & 0 \\ 0 & k_{j+1}^{22} \end{bmatrix}\begin{bmatrix} \kappa_{\alpha j} & 1 \\ \lambda_{\alpha j} & 0 \end{bmatrix} - \begin{bmatrix} \kappa_{\beta j} & 1 \\ \lambda_{\beta j} & 0 \end{bmatrix}\begin{bmatrix} k_j^{11} & 0 \\ 0 & k_j^{22} \end{bmatrix} \approx 0$$

in $S_\alpha \cap S_\beta$, in consequence of (11.4-27), (11.4-28). Therefore, in particular, one sees from the first row of this matrix relation that

$$k_{j+1}^{11}\kappa_{\alpha j} - k_j^{11}\kappa_{\beta j} \approx 0, \qquad k_{j+1}^{11} - k_j^{22} \approx 0,$$

and, thus, after elimination of k_{j+1}^{11},

$$(k_j^{22} - k_j^{11})\kappa_{\alpha j} + \kappa_j^{11}(\kappa_{\alpha j} - \kappa_{\beta j}) \approx 0. \tag{11.4-29}$$

On the other hand, (11.4-19) is also equivalent—with j replaced by $j - 1$—to

$$C_{\beta,j-1}^{-1}K_j - K_{j-1}C_{\alpha,j-1}^{-1} = 0,$$

i.e.,

$$\begin{bmatrix} 0 & \lambda_{\beta,j-1}^{-1} \\ 1 & -\kappa_{\beta,j-1}\lambda_{\beta,j-1}^{-1} \end{bmatrix}\begin{bmatrix} k_j^{11} & 0 \\ 0 & k_j^{22} \end{bmatrix} - \begin{bmatrix} k_{j-1}^{11} & 0 \\ 0 & k_{j-1} \end{bmatrix}\begin{bmatrix} 0 & \lambda_{\alpha,j-1}^{-1} \\ 1 & -\kappa_{\alpha,j-1}\lambda_{\alpha,j-1}^{-1} \end{bmatrix} \approx 0.$$

The last row of this matrix relation yields

$$k_j^{11} - k_{j-1}^{22} \approx 0, \qquad \frac{\kappa_{\beta,j-1}}{\lambda_{\beta,j-1}}k_j^{22} - \frac{\kappa_{\alpha,j-1}}{\lambda_{\alpha,j-1}}k_{j-1}^{22} \approx 0.$$

Eliminating k_{j-1}^{22} and introducing the abbreviation

$$\mu_{\vartheta j} := \kappa_{\vartheta j}/\lambda_{\vartheta j} \qquad (11.4\text{-}30)$$

one then has,

$$\mu_{\beta,j-1}k_j^{22} - \mu_{\alpha,j-1}k_j^{11} \approx 0,$$

or

$$\mu_{\beta,j-1}\big(k_j^{22} - k_j^{11}\big) + k_j^{11}\big(\mu_{\beta,j-1} - \mu_{\alpha,j-1}\big) \approx 0. \qquad (11.4\text{-}31)$$

The derivative of $\kappa(a)$ is bounded and does not vanish at $a = \pm q$, as can be seen from (11.2-9). Therefore, the same is true of the derivative of $\mu_{\alpha,j-1}$, and the Theorem of the Mean implies that

$$0 < c_2 \leqslant |(\mu_{\beta,j-1} - \mu_{\alpha,j-1})/(\delta_\beta - \delta_\alpha)| \leqslant c_3 < \infty, \qquad (11.4\text{-}32)$$

for $\epsilon \in S_\alpha \cap S_\beta$ and $|\epsilon|$ sufficiently small. Similarly,

$$0 < c_4 \leqslant |(\kappa_{\beta j} - \kappa_{\alpha j})/(\delta_\beta - \delta_\alpha)| \leqslant c_5 < \infty. \qquad (11.4\text{-}33)$$

Now, choose j odd. Then

$$\lim_{\epsilon \to 0} \mu_{\beta,j-1} = 0, \qquad \lim_{\epsilon \to 0} \kappa_{\alpha,j} \neq 0, \qquad (11.4\text{-}34)$$

because of (11.2-9), (11.4-13), (11.4-17), (11.4-30) and the fact that $\Gamma^{-1}(\tfrac{1}{2}(1-a)) = 0$ whenever a is an odd positive integer. Here, it must be remembered that q is a negative odd integer. Combining (11.4-32) with (11.4-31) leads to the inequality

$$|\delta_\beta - \delta_\alpha| \leqslant c_2^{-1}|k_j^{11}|^{-1}|\mu_{\beta,j-1}|\,|k_j^{22} - k_j^{11}| + c_6 e^{-r^2/|\epsilon|} \qquad (11.4\text{-}35)$$

and combining (11.4-29) with (11.4-33) implies

$$|k_j^{22} - k_j^{11}| \leqslant |k_j^{11}|\,|\kappa_{\alpha j}|^{-1}c_5|\delta_\beta - \delta_\alpha| + c_7 e^{-r^2/|\epsilon|} \qquad (11.4\text{-}36)$$

Insertion of (11.4-36) into (11.4-35) and reference to (11.4-34) produces, at long last, the desired inequality

$$|\delta_\beta - \delta_\alpha| \leqslant c_8 e^{-r^2/|\epsilon|},$$

for $\epsilon \in S_\alpha \cap S_\beta$ and sufficiently small.

CHAPTER XII

Appendix: Some Linear Algebra for Holomorphic Matrices

12.1. Vectors and Matrices of Holomorphic Functions

The theory of linear ordinary differential equations involves a considerable amount of linear algebra. If, as in this book, the differential equations are analytic, it is essential to know which of the common algebraic operations can be carried out in the ring of functions of one complex variable that are holomorphic in some given region. In this section, some useful general theorems, less simple than their analogs for constant matrices, will be proved.

First some notation and terminology: The class of matrices of n rows and m columns whose entries are functions of a complex variable x holomorphic for x in a region $\Omega \in \mathbb{C}$ will be denoted by $H_{mn}(\Omega)$. If the region Ω is clearly understood it may be omitted from the notation and I shall then write simply H_{mn}. A set of p vectors $v_j \in H_{n1}, j = 1, 2, \ldots, p$, will be called *independent* in Ω if the constant vectors $v_j(x), j = 1, 2, \ldots, p$, are linearly independent over the field $\mathbb{C}$ at all points x of Ω. Such a set of vectors will be called *dependent* if the $v_j(x)$ are linearly dependent at all points of Ω. Observe that with this terminology there are sets of vectors in H_{n1} that are neither dependent nor independent. A set of n vectors in H_{n1} which are independent in Ω may be called a "holomorphic basis in Ω."

The concept of the greatest common divisor of a finite set of functions, familiar for polynomials, can be extended to the class $H_{11}(\Omega)$ and turns out to be useful in this section. A common divisor of p functions $f_j \in H_{11}(\Omega)$, $j = 1, 2, \ldots, p$, not all identically zero, is by definition, a function $g \in H_{11}(\Omega)$ such that the functions f_j/g are in $H_{11}(\Omega)$. If the functions f_j/g have no common zeros in Ω, then g is called a "greatest common divisor" of $f_1, f_2, \ldots, f_p$ in Ω. It can be proved with the help of the Weierstrass Factorization Theorem and the Mittag Leffler Theorem (see, e.g., [69], pp. 297–298, 308–309) that such functions g exist, and that there exist functions $h_j \in H_{11}(\Omega), j = 1, 2, \ldots, p$, so that

$$\sum_{j=1}^{n} f_j h_j = g,$$

if g is a greatest common divisor of the f_j in Ω. The fact just stated is useful for the proof of the lemma below.

Lemma 12.1-1. *If $v \in H_{n1}(\Omega)$ and $v(x)$ is nowhere zero in Ω, then there exist $n - 1$ vectors $w_j \in H_{n1}(\Omega), j = 1, 2, \ldots, n - 1$, such that the set $v, w_1, w_2, \ldots, w_{n-1}$ is a holomorphic basis in Ω.*

PROOF. (The essence of this proof is due to Walter Rudin.) The proof proceeds by induction with respect to the dimension n. For $n = 1$ the statement is vacuously true. To explain clearly the structure of the argument, the case $n = 2$ will be discussed explicitly in a way that brings out the inductive procedure. Let $v = (v^{(1)}, v^{(2)})^T$, and let $\phi \in H_{11}$ be such that $v^{(1)} = \phi \tilde{v}^{(1)}$ with $\tilde{v}^{(1)} \in H_{11}$ and $\tilde{v}^{(1)}(x)$ nowhere zero in Ω. By assumption the greatest common divisor of ϕ and $v^{(2)}$ is 1. Therefore, there exist two functions $\alpha, \beta \in H_{11}$ so that

$$\phi \alpha + v^{(2)} \beta = 1. \tag{12.1-1}$$

Now a vector $w_1 = (w_1^{(1)}, w_1^{(2)})^T$ must be found with which

$$\det \begin{pmatrix} v^{(1)} & w_1^{(1)} \\ v^{(2)} & w_1^{(2)} \end{pmatrix} = \phi \tilde{v}^{(1)} w_1^{(2)} - v^{(2)} w_1^{(1)}$$

is nowhere zero in Ω. By (12.1-1) this will be the case if

$$w_1^{(1)} = -\beta, \qquad w_2^{(1)} = \alpha / \tilde{v}^{(1)}.$$

Next, assume that the lemma is true for some n, say for $n = m$, and let $v = (v^{(1)}, v^{(2)}, \ldots, v^{(m+1)})^T$ be the given vector in $H_{m+1,1}(\Omega)$. If ϕ is a greatest common divisor of $v^{(1)}, v^{(2)}, \ldots, v^{(m)}$, then the vector

$$\tilde{v} := (v^{(1)}/\phi, v^{(2)}/\phi, \ldots, v^{(m)}/\phi)^T \in H_{m1},$$

is nowhere zero in Ω. By the inductive hypothesis there exist $m - 1$ vectors

$$\tilde{w}_j \in H_{m1}, \qquad j = 1, 2, \ldots, m - 1,$$

for which the matrix

$$\tilde{M}_m = \{\tilde{v}, \tilde{w}_1, \tilde{w}_2, \ldots, \tilde{w}_{m-1}\} \in H_{mm}$$

formed by the indicated column vectors has nowhere in Ω a zero determinant. Set

$$M_m := \{\phi \tilde{v}, \tilde{w}_1, \ldots, \tilde{w}_{m-1}\},$$

then

$$\det M_m = \phi \det \tilde{M}_m .$$

By hypothesis, v has no zeros in Ω, therefore, α, β in H_{11} can be found such that

$$\phi \alpha + v^{(m+1)} \beta = 1.$$

Now enlarge the $m \times m$ matrix M_m to a matrix $M_{m+1} \in H_{m+1,m+1}$ by adjoining to it an $(m+1)$st column $u = (u^{(1)}, u^{(2)} \ldots, u^{(m+1)}) \in H_{m+1,1}$ still to be determined and the $(m+1)$st row $(v^{(m+1)}, 0, \ldots, 0, u^{(m+1)})$, as indicated in the formula below:

$$M_{m+1} := \begin{bmatrix} & & & u(1) \\ & M_m & & u(2) \\ & & & \vdots \\ v^{(m+1)}, 0 & \cdots & 0, u^{(m+1)} \end{bmatrix}.$$

Its first column is v and its determinant is

$$\det M_{m+1} = (-1)^{m+1} N_m + \phi \det \tilde{M}_m u^{(m+1)}, \qquad (12.1\text{-}2)$$

where N_m is the determinant of the $m \times m$ matrix obtained from M_m by canceling its first column and adjoining $(u^{(1)}, u^{(2)}, \ldots, u^{(m)})^T$ as its last column. Now choose

$$u^{(j)} = \tilde{v}^{(j)} \beta, \quad j = 1, 2, \ldots, m, \quad u^{(m+1)} = \alpha,$$

then

$$N_m = (-1)^m \det \tilde{M}_m \cdot \beta,$$

and (12.1-2) becomes

$$\det M_{m+1} = \det \tilde{M}_m (v^{(m+1)} \beta + \phi \alpha),$$

which is nowhere zero. The $m + 1$ columns of M_{m+1} are the desired basis.

Theorem 12.1-1. *Let $v_j \in H_{n1}(\Omega)$, $j = 1, 2, \ldots, p < n$, be p given vectors independent in Ω. Then there exist $n - p$ additional vectors $w_k \in H_{n1}(\Omega)$, $k = 1, 2, \ldots, n - p$, so that the vectors $v_1(x), v_2(x), \ldots, v_p(x), w_1(x), w_2(x), \ldots, w_{n-p}(x)$ are a basis of $\mathbb{C}$ for all $x \in \Omega$.*

PROOF. Lemma 12.1-1 states the theorem for $p = 1$. Assume it to be true for $p = m < n - 1$, then it suffices to show it holds for $p = m + 1$, as well. Let the independent vectors $v_j \in H_{n1}, j = 1, 2, \ldots, m + 1$, be given. By the inductive hypothesis there exist $n - m$ vectors $u_k \in H_{n1}$, $k = 1, 2, \ldots, n - m$, and n scalar functions $a^{(j)}, b^{(k)}, j = 1, 2, \ldots, m; k = 1, 2, \ldots, n - m$, such that vectors $v_1, v_2, \ldots v_n, u_1, u_2, \ldots, u_{n-m}$ form a basis and that

$$v_{m+1} = \sum_{j=1}^{m} a^{(j)} v_j + \sum_{k=1}^{n-m} b_1^{(k)} u_k .$$

These coefficients $a^{(j)}, b_1^{(k)}$ can be calculated by Cramer's rule and are therefore in $H_{11}(\Omega)$. The row vector

$$b_1 = \left(b_1^{(1)}, b_1^{(2)}, \ldots, b_1^{(n-m)}\right) \in H_{1,n-m}(\Omega)$$

is nowhere zero in Ω, because the vectors $v_1, v_2, \ldots, v_{m+1}$ are, by hypothesis, independent. By Lemma 12.1-1 one can adjoin to the vector b_1 a set of $n - m - 1$ row vectors

$$b_j = \left(b_j^{(1)}, b_j^{(2)}, \ldots, b_j^{(n-m)}\right) \in H_{1,n-m},$$

$j = 2, 3, \ldots, n - m$, so that the vectors $b_1, b_2, \ldots, b_{n-m}$ are independent in Ω. The vectors

$$v_1, v_2, \ldots, v_{m+1}, \qquad w_1 = \sum_{k=1}^{n-m} b_2^{(k)} u_k,$$

$$w_2 = \sum_{k=1}^{n-m} b_3^{(k)} u_k, \ldots, \qquad w_{n-m-1} = \sum_{k=1}^{n-m} b_{n-m}^{(k)} u_k$$

then form the desired basis, because the components of these vectors with respect to the basis $v_1, v_2, \ldots, v_m, u_1, \ldots, u_{n-m}$ form the matrix

$$
\begin{bmatrix}
 & & I_m & & & & 0 & \\
a^{(1)} & a^{(2)} & \cdots & a^{(m)} & b_1^{(1)} & b_1^{(2)} & \cdots & b_1^{(n-m)} \\
0 & 0 & \cdots & 0 & b_2^{(1)} & b_2^{(2)} & \cdots & b_2^{(n-m)} \\
0 & 0 & \cdots & 0 & b_2^{(1)} & b_3^{(2)} & \cdots & b_3^{(n-m)} \\
\vdots & & \vdots & & \vdots & & & \vdots \\
0 & 0 & \cdots & 0 & b_{n-m}^{(1)} & b_{n-m}^{2} & \cdots & b_{n-m}^{(n-m)}
\end{bmatrix}
$$

(I_m is the identity matrix of order m), and the $n \times n$ matrix in this formula is everywhere invertible in Ω.

Theorem 12.1-2. *Let $A \in H_{nn}(\Omega)$. If the rank of $A(x)$ does not exceed r at any point $x \in \Omega$, then the equation*

$$Av = 0 \tag{12.1-3}$$

for v can be satisfied by at least $n - r$ vectors $v_j \in H_{n1}, j = 1, 2, \ldots, n - r$, which are independent in Ω.

PROOF. Let $s \leqslant r$ be the largest order of a minor of A that is not identically zero. If $s = n$, the theorem is trivially true. Assume that $s < n$. Then there is a permutation matrix P such that the leading $s \times s$ minor of $P^{-1}AP$ is not identically zero. If v is a solution of (12.1-3), then $w = P^{-1}v$ is a solution of $(P^{-1}AP)w = 0$. Hence, no generality is lost by assuming that the leading $s \times s$ minor of A itself is not identically zero. Write A in the partitioned form

$$A = \begin{pmatrix} A_{11} & A_{12} \\ A_{21} & A_{22} \end{pmatrix}$$

with A_{11} having dimension $s \times s$. From the identity

$$\begin{pmatrix} I_s & 0 \\ -A_{21}A_{11}^{-1} & I_{n-s} \end{pmatrix} \begin{pmatrix} A_{11} & A_{12} \\ A_{21} & A_{22} \end{pmatrix}$$

$$= \begin{pmatrix} A_{11} & A_{12} \\ 0 & -A_{21}A_{11}^{-1}A_{12} + A_{22} \end{pmatrix} \qquad (12.1\text{-}4)$$

it follows that

$$-A_{21}A_{11}^{-1}A_{12} + A_{22} = 0, \qquad (12.1\text{-}5)$$

since otherwise the right member in (12.1-4) would have rank greater than s at some points of Ω, while the left member has everywhere at most rank s. The identity

$$A \begin{pmatrix} A_{11}^{-1}A_{12} \\ -I_{n-s} \end{pmatrix} = 0,$$

which is an immediate consequence of (12.1-5), exhibits $n - s$ pointwise independent vector solutions of (12.1-3). These solutions may, however, have poles at the zeros of $\det A_{11}$. To each of these solutions there corresponds a nowhere zero holomorphic solution obtained by multipling it with a suitable function in $H_{11}(\Omega)$ which cancels these poles. After the multiplication the solutions may, of course, cease to be independent in Ω. Thus, only the existence of *one* solution independent (i.e., not zero anywhere) in Ω has been established, so far.

To complete the proof by induction, assume the theorem to be true for $r = n - 1, n - 2, \ldots, n - (m - 1)$, where $2 \leqslant m < n$. By the inductive hypothesis there exist $n - (n - m + 1) = m - 1$ independent solutions in $H_{n1}(\Omega)$ for (12.1-3), say, $v_1, v_2, \ldots, v_{m-1}$. By Theorem 12.1-1 one can add to them $n - m + 1$ holomorphic vectors $w_1, w_2, \ldots, w_{n-m+1}$ such that the matrix

$$T := (v_1, v_2, \ldots, v_{m-1}, w_1, w_2, \ldots, w_{n-m+1})$$

is in H_{nn} and is pointwise invertible. The matrix

$$B := T^{-1}AT = (0, 0, \ldots, 0, T^{-1}Aw_1, T^{-1}Aw_1, \ldots, T^{-1}Aw_{n-m+1}) \qquad (12.1\text{-}6)$$

has the same rank as A in Ω. The equation

$$Bu = 0 \qquad (12.1\text{-}7)$$

has the $m - 1$ independent solutions

$$u_j = e_j, \qquad j = 1, 2, \ldots, m - 1, \tag{12.1-8}$$

where $e_j = (0, \ldots, 0, 1, 0, \ldots, 0)^T$, with the number 1 in the jth position. Because of the hypothesis that the rank of $A(x)$ does not exceed $n - m$, the same is true for the rank of $B(x)$. Hence, any $n - m + 1$ rows of B are dependent for all $x \in \Omega$. Let $\tilde{B}$ be the particular submatrix of order $(n - m + 1) \times (n - m + 1)$ of B which is formed by its last $n - m + 1$ rows and columns. Since the rank of $\tilde{B}$ is less than $n - m + 1$, there exists a column vector $\tilde{u}$ of dimension $n - m + 1$, holomorphic and nowhere zero in Ω, such that $\tilde{B}\tilde{u} = 0$. The n-dimensional vector

$$u_m = (0, 0, \ldots, 0, \tilde{u}^T)^T \in H_{n1}(\Omega)$$

is then a solution of $Bu = 0$ independent of the solutions u_j in (12.1-8). Finally, $Bu_m = 0$ implies $T^{-1}ATu_m = 0$, so that $v_m := Tu_m$ is a solution of (12.1-3) independent of the solutions $v_j = Tu_j$, $j = 1, 2, \ldots, m - 1$. This completes the proof of Theorem 12.1-2. A local version of this theorem was proved in [87].

If, in particular, the rank of $A(x)$ is constant and equal to r in Ω, equation (12.1-3) has $n - r$ independent solutions but not more.

Lemma 12.1-2. *Let $A \in H_{nn}(\Omega)$ have constant rank $r = n - p$. If the q independent vectors $v_j \in H_{n1}(\Omega)$, $j = 1, 2, \ldots, q$, with $q < p$ have all their values in the nullspace $N(x)$ of $A(x)$, then one can find $p - q$ additional vectors $w_k \in H_n(\Omega)$, $k = 1, 2, \ldots, p - q$, so that the vectors $v_j(x), w_k(x)$, $j = 1, 2, \ldots, q$, $k = 1, 2, \ldots, p - q$, span $N(x)$ for all $x \in \Omega$.*

PROOF. By Theorem 12.1-2 there exist p vectors $u_\mu \in H_{n1}$, $\mu = 1, 2, \ldots, p$, so that $\{u_\mu(x)\}$, $\mu = 1, 2, \ldots, p$, is a basis for $N(x)$, and by Theorem 12.1-1 one can adjoin additional vectors $u_\mu \in H_{n1}$, $\mu = p + 1, p + 2, \ldots, n$, so as to obtain a holomorphic basis of $\mathbb{C}^n$ for each point $x \in \Omega$. The vectors v_j, $j = 1, 2, \ldots, q$, have unique representations

$$v_j = \sum_{\mu=1}^{n} a_{j\mu} u_\mu, \qquad j = 1, 2, \ldots, q,$$

in which $a_{j\mu} = 0$ for $\mu > p$. For each j this formula represents a set of n equations for $a_{j\mu}$ $\mu = 1, 2, \ldots, n$, which can be solved by Cramer's rule. Therefore, all $a_{j\mu}$ are in $H_{11}(\Omega)$. The set of q vectors

$$(a_{j1}, a_{j2}, \ldots, a_{jp})^T \in H_{p1}(\Omega), \qquad j = 1, 2, \ldots, q,$$

is independent, because the v_j, $j = 1, 2, \ldots, q$, are independent. It can be completed to a set of p independent vectors in H_{p1} (Theorem 12.1-1). If the full set of these p vectors is denoted by $(a_{j1}, a_{j2}, \ldots, a_{jp})^T$, $j = 1, 2, \ldots, p$, the vectors

$$w^{(k)} = \sum_{\mu=1}^{p} a_{j\mu} u_\mu, \qquad j = q + 1, q + 2, \ldots, p - q,$$

have the property required in the thoerem.

12.2. Reduction to Jordan Form

Every constant matrix with entries in the field of complex numbers is similar, in this field, to its so-called Jordan matrix, which is unique except for the order of its blocks. If A is a matrix in $H_{nn}(\Omega)$, it may or may not be the case that the matrix valued function J_A which for each $x \in \Omega$ is the Jordan matrix $J_A(x)$ of $A(x)$ is also holomorphic. Even if $J_A \in H_{nn}(\Omega)$, it is not always true that A is holomorphically similar to J_A, i.e., that a matrix $T \in H_{nn}(\Omega)$ can be found which is invertible in Ω and with which

$$T^{-1}AT \in J_A \qquad \text{in } \Omega.$$

Examples will be given later in this section. The aim of this section is the derivation of sufficient criteria for a matrix in $H_{nn}(\Omega)$ to be holomorphically similar to its Jordan form. The conditions to be given are almost, but not quite, necessary.

Theorem 12.2-1. *If all eigenvalues of $A \in H_{nn}(\Omega)$ are in $H_{11}(\Omega)$ and if all eigenvalues of A that are equal at one point of Ω are identically equal, then A is in Ω holomorphically similar to a direct sum of matrices each of which has only one distinct eigenvalue.*

PROOF. Let $\lambda_1, \lambda_2, \ldots, \lambda_p$ be the distinct eigenvalues of A in Ω. By the conditions on Ω, they are in $H_{11}(\Omega)$. It follows from Theorem 12.1-2 that the equation $(A - \lambda_1 I)v = 0$ has at least one vector solution $v = v_1 \in H_{n1}(\Omega)$ which is nowhere zero in Ω, and by Theorem 12.1-1 there exist $n - 1$ vectors $w_j \in H_{n1}, j = 1, 2, \ldots, n - 1$, so that $v_1, w_1, w_2, \ldots, w_{n-1}$ are independent in Ω. Let $T_1 \in H_{nn}$ be the matrix with columns $v_1, w_1, w_2, \ldots, w_{n-1}$. Then $B := T_1^{-1}AT_1$ has the form

$$B = \begin{bmatrix} \lambda_1 & B_{21} \\ 0 & \\ \vdots & B_{22} \\ 0 & \end{bmatrix}$$

with $B \in H_{nn}$, $B_{21} \in H_{1,n-1}$, $B_{22} \in H_{n-1,n-1}$. The matrix B_{22} has the same eigenvalues as A, except that λ_1 occurs with multiplicity one less than in B.

A second similarity transformation, this time with a matrix

$$T_2 = \begin{pmatrix} 1 & 0 \\ 0 & S_{22} \end{pmatrix},$$

where $S_{22} \in H_{n-1,n-1}(\Omega)$, subjects B_{22} to a similarity transformation with the matrix S_{22}. The previous argument can now be repeated, and after n iterations the matrix A will have been transformed into an upper triangular matrix in $H_{nn}(\Omega)$. For constant matrices the preceding argument is quite familiar, but Theorem 12.1-1 was needed to extend it to matrices in $H_{nn}(\Omega)$.

For notational convenience it will now be assumed that the original matrix A has already this triangular form. If the distinct eigenvalues $\lambda_1, \lambda_2, \ldots, \lambda_p$ have multiplicities $m_1, m_2, \ldots, m_p$ which, by assumption, are constant in Ω, then A is partitioned into p^2 submatrices A_{jk} of orders $m_j \times m_k$, respectively, so that $A_{jk} \equiv 0$ for $j > k$ and that A_{jj} is upper triangular.

It remains to be shown that the matrices A_{jk}, $j < k$ can be made identically zero by suitable further similarity transformations. To annul A_{12} in this manner, partition A in the coarser form

$$A = \begin{pmatrix} A_{11} & A_{12} \\ 0 & C_{22} \end{pmatrix},$$

where

$$C_{22} = \begin{bmatrix} A_{21} & A_{22} & \cdots & A_{2p} \\ A_{31} & A_{32} & \cdots & A_{2p} \\ \cdots & & \cdots & \\ A_{p1} & A_{p2} & \cdots & A_{pp} \end{bmatrix}.$$

The dimensions of C_{22} are $(n - m_1) \times (n - m_1)$. Now take

$$T = \begin{pmatrix} I_{m_1} & T_{12} \\ 0 & I_{n-m_1} \end{pmatrix}$$

with T_{12} still to be chosen. Then

$$T^{-1}AT = \begin{pmatrix} A_{11} & A_{11}T_{12} - T_{12}C_{22} + A_{12} \\ 0 & C_{22} \end{pmatrix}.$$

The rectangular matrix T_{12} must satisfy the equation for X,

$$XC_{22} - A_{11}X = A_{12}. \tag{12.2-1}$$

It is a fact of elementary matrix theory that the operator on X in the left member has only the nullspace zero, if, as in this case, C_{22} and A_{11} have no common eigenvalue. Therefore, equation (12.2-1) has a unique solution, which can be calculated by Cramer's rule and is therefore holomorphic in Ω. That argument can be repeated, with C_{22} taking the place of A. After p such transformations the proof of Theorem 12.2-1 is at hand.

EXAMPLE.   Let

$$A(x) = \begin{bmatrix} x & 1 & a(x) \\ 0 & x & b(x) \\ 0 & 0 & 0 \end{bmatrix} \tag{12.2-2}$$

with $a, b \in H_{11}(\Omega)$. The hypotheses of Theorem 12.2-1 are satisfied if, and only if, Ω does not contain the origin. A short calculation shows that the most general matrix T such that, for $x \neq 0$,

$$T^{-1}(x)A(x)T(x) = \begin{bmatrix} x & 1 & 0 \\ 0 & x & 0 \\ 0 & 0 & 0 \end{bmatrix} \tag{12.2-3}$$

in Ω is

$$T(x) = \begin{bmatrix} \alpha(x) & \beta(x) & -a(x)\gamma(x)x^{-1} + b(x)\gamma(x)x^{-2} \\ 0 & \alpha(x) & -b(x)\gamma(x)x^{-1} \\ 0 & 0 & \gamma(x) \end{bmatrix}.$$

The scalar functions α, β, γ are arbitrary, so far, except that α and γ must not vanish anywhere in Ω. If they are holomorphic, so is T, provided zero is not a point of Ω. If $0 \in \Omega$, then T has a pole at $x = 0$ unless the conditions

$$b(0) = 0, \qquad \left(\frac{db}{dx} \right)_{x=0} = a(0) \tag{12.2-4}$$

are satisfied. Therefore, A is, in general not holomorphically similar to the block-diagonal matrix in (12.2-3) in regions that contain the origin. If the conditions (12.2-4) are not satisfied two possibilities arise: There may exist a constant invertible matrix S such that

$$S^{-1}A(0)S = \begin{bmatrix} 0 & 1 & 0 \\ 0 & 0 & 0 \\ 0 & 0 & 0 \end{bmatrix}.$$

Then A is pointwise similar to the block-diagonal matrix in (12.2-3), even in regions containing $x = 0$. This happens if and only if $b(0) = 0$, as is easily verified. Thus, there do exist cases in which the Jordan matrix $J_A(x)$ is holomorphic in some region but, nevertheless, not holomorphically similar to all holomorphic matrices to which it is pointwise similar.

Finally, if $b(0) \neq 0$ in (12.2-2), then the Jordan form of $A(0)$ is

$$J_A(0) = \begin{bmatrix} 0 & 1 & 0 \\ 0 & 0 & 1 \\ 0 & 0 & 0 \end{bmatrix}$$

so that J_A is then not holomorphic at $x = 0$.

Theorem 12.2-2. *A matrix $A \in H_{nn}(\Omega)$ is in Ω holomorphically similar to its*

Jordan matrix J_A if

(i) $J_A \in H_{nn}(\Omega)$,
   and if
(ii) *eigenvalues of $A(x)$ that are equal at one point of Ω are identically equal.*

Condition (i) is clearly also necessary. It can also be restated by saying that the elementary divisors of A must have constant degrees in Ω and that the eigenvalues of A have no branch points in Ω. The example after Theorem 12.2-1 shows that condition (ii) is not necessary but almost necessary in the sense that matrices that are holomorphically similar to their Jordan matrix although (ii) is not true are exceptional.

PROOF OF THEOREM 12.2-2.   By virtue of Theorem 12.2-1 no generality is lost if from now on A is assumed to have only one distinct eigenvalue λ. Since

$$T^{-1}(A - \lambda I)T = T^{-1}AT - \lambda I,$$

a similarity transformation that takes A into its Jordan form does the same to $A - \lambda I$. It therefore suffices to study nilpotent matrices A.

Thanks to the results of Section 12.1, most proofs of the reduction of constant nilpotent matrices to Jordan form carry literally over to matrices in $H_{nn}(\Omega)$. The arguments given here are those in the account by Hermann Weyl in [90], pp. 97–99. The description will therefore be held brief, with emphasis on the points where the results of Section 12.1 are needed.

Let $w_j \in H_{n1}(\Omega)$, $j = 1, 2, \ldots, q$, be given vectors. Following Weyl's terminology a set of vectors $v_k \in H_{n1}(\Omega)$, $k = 1, 2, \ldots, p$, will be called "independent modulo $w_1, w_2, \ldots, w_q$ in Ω", if for every $x \in \Omega$ the span of the vectors $w_j(x)$, $j = 1, 2, \ldots, q$, over the field $\mathbb{C}$ has only the zero vector in common with the span of the vectors v_k.

There is an integer $l < n$ such that $A^l = 0$ but $A^{l-1} \neq 0$. Denote the nullspace of $A^{l-j}(x)$ by N_j. It depends on x but its dimension does not, because it is also the dimension of the nullspace of J_A^{l-j}, and the Jordan matrix J_A of the nilpotent matrix A is constant. Clearly, $v \in N_j$ implies $Av \in N_{j-1}$ and, therefore,

$$N_0 \supset N_1 \supset \cdots \supset N_l = 0.$$

Let m_j be the dimension of N_j. Then $n = m_0 > m_1 \cdots > m_l = 0$.

By Theorem 12.1-2 each nullspace N_j has a holomorphic basis, say, $\{u_{j1}, u_{j2}, \ldots, u_{jm_j}\}$ or, in abbreviated notation, $\{u_{jk}\}_{m_j}$. The basis $\{u_{1k}\}_{m_1}$ of N_1 can be completed to a basis of $N_0 = \mathbb{C}^n$ by adjoining $r_0 = m_0 - m_1$ vectors $\{v_{0k}\}_{r_0}$. This follows from Theorem 12.1-1. The vectors $\{v_{0k}\}_{r_0}$ are not only independent but even independent modulo $\{u_{1k}\}_{m_1}$. More generally, on the strength of Lemma 12.1-2 the basis $\{u_{jk}\}_{m_j}$ of N_j can be completed to a basis of N_{j-1} by adjoining $r_{j-1} = m_{j-1} - m_j$ holomorphic vectors $\{v_{j-1,k}\}_{r_{j-1}}$. One has

$$\sum_{j=0}^{l-1} r_j = n.$$

The total set $v_{\mu\nu}$ of these vectors is a basis of $\mathbb{C}^n$ at every $x \in \Omega$, because each group is independent modulo the previous group.

Each set $\{v_{jk}\}_{r_j}, j = 0, 1, \ldots, l-1$, will now be replaced by an equivalent set, as follows: The set $\{v_{0k}\}_{r_0}$ is not changed. The set $\{v_{1k}\}_{r_1}$ is replaced by the vectors $\{Av_{0k}\}_{r_0}$ together with $s_1 = r_1 - r_0$ vectors $\{z_{1k}\}_{s_1}$. This is possible, because the vectors $\{Av_{0k}\}_{r_0}$ in N_1 are independent modulo N_2. In fact, if for some $x \in \Omega$ one has constants $c_k, k = 1, 2, \ldots, r_0$ not all zero, with

$$\sum_{k=1}^{r_0} c_k Av_{0k} \in N_2,$$

i.e., with

$$A^{l-2}\left[A \sum_{k=1}^{r_0} c_k v_{0k} \right] = 0,$$

then

$$\sum_{k=0}^{r_0} c_k v_{0k} \in N_1,$$

contrary to the construction of $\{v_{0k}\}_{r_0}$.

This argument shows, incidentally, that $r_1 > r_0$. Continuing in this fashion and defining $s_j = r_j - r_{j-1}$ one obtains the following replacements: $\{v_{1k}\}_{r_1}$ by $\{Av_{0k}\}_{r_0} \cup \{z_{1k}\}_{s_1}$; $\{v_{2k}\}_{r_2}$ by $\{Av_{0k}\}_{r_0} \cup \{Az_{2k}\}_{s_1} \cup \{z_{2k}\}_{s_2}; \ldots; \{v_{jk}\}_{r_j}$ by $\{A^j v_{0k}\}_{r_0} \cup \{A^{j-1}z_{1k}\}_{s_1} \cup \cdots \cup \{z_{jk}\}_{s_j}$, etc., for $j = 1, 2, \ldots, l-1$. The n vectors

$$\{A^\mu v_{0k}\}_{r_0}, \qquad \mu = 0, 1, \ldots, l-1;$$

$$\{A^\mu z_{1k}\}_{s_1}, \qquad \mu = 0, 1, \ldots, l-2;$$

$$\vdots$$

$$\{A^\mu z_{jk}\}_{s_j}, \qquad \mu = 0, 1, \ldots, l-j-1; \qquad (12.2\text{-}5)$$

$$\vdots$$

$$\{z_{l-1,k}\}_{s_{l-1}}$$

form, for each x in Ω, a basis of $\mathbb{C}^n$.

The vectors above serve as the columns of the invertible matrix $T \in H_{nn}(\Omega)$ with which one has

$$AT = TJ_A,$$

provided the vectors are taken in the order in which they appear in the list (12.2-5). Observe, here, that the vectors $\{z_{jk}\}_{s_j}$ are in N_j, so that $\{A^{l-1}z_{jk}\}_{s_j} = 0$.

12.3. General Holomorphic Block Diagonalization

In Section 12.2 it was decisive that the multiplicity of the eigenvalues of $A(x)$ did not change in Ω. In this section a useful simplification of holomorphic matrices by means of holomorphic similarity transformations will be proved under the weaker assumption below.

Hypothesis XII-1. *The characteristic polynomial ϕ of the matrix $A \in H_{nn}(\Omega)$ can be factored into two monic polynomials of positive degree with holomorphic coefficients in Ω,*

$$\phi = \phi_1 \phi_2 \tag{12.3-1}$$

so that for no $x \in \Omega$ the polynomials ϕ_1 and ϕ_2 have a zero in common.

Theorem 12.3-1. *If Hypothesis XII-1 is satisfied, then the matrix $A \in H_{nn}(\Omega)$ is in Ω holomorphically similar to the direct sum of two matrices whose characteristic polynomials are ϕ_1 and ϕ_2, respectively.*

To illustrate the difference between Theorem 12.3-1 and Theorem 12.2-1, let

$$A(x) = \begin{bmatrix} 0 & 1 & 1 \\ x & 0 & 1 \\ 0 & 0 & x-2 \end{bmatrix}$$

and take $\Omega = \{x \,|\, |x| < 1\}$. Then $\phi(x,\lambda) = \phi_1(x,\lambda)\phi_2(x,\lambda) = (\lambda^2 - x)(\lambda - x + 2)$, and the zeros of ϕ_1 differ from the zero of ϕ_2 for all $x \in \Omega$. Theorem 12.3-1 says that $A(x)$ can be holomorphically split in Ω into a matrix of order 2 with the eigenvalues $\sqrt{x}$, $-\sqrt{x}$ and a matrix of order one with eigenvalue $x - 2$. Theorem 12.2-1, on the other hand, is not applicable in this region, because the eigenvalues $\sqrt{x}$, $-\sqrt{x}$ coincide at $x = 0$ without being identically equal. Moreover, the Jordan matrix J_A is not holomorphic at $x = 0$.

Theorem 12.3-1 is a special case of a result proved by Gingold in [16]. The more general theorem is valid not only for similarity transformations within the class of holomorphic function but also for matrices in $C^N(\Omega)$ when Ω is an interval of the real axis and N some non-negative integer.

The proof to be given here is different. It is based on the theorem in Section 12.1 and is more elementary than Gingold's. Extensions to certain rings of nonholomorphic functions should be possible.

PROOF OF THEOREM 12.3-1. Two standard results from algebra will be needed. One is the Cayley–Hamilton theorem which implies that

$$0 = \phi(A(x), x) = \phi_1(A(x), x)\phi_2(A(x), x), \tag{12.3-2}$$

identically for $x \in \Omega$. The other one is the fact that the relatively prime polynomials $\phi_1(\lambda, x), \phi_2(\lambda, x)$ satisfy for all $x \in \Omega$ a relation of the form

$$\phi_1(\lambda, x) g_1(\lambda, x) + \phi_2(\lambda, x) g_2(\lambda, x) = 1, \qquad (12.3\text{-}3)$$

where $g_1(\lambda, x), g_2(\lambda, x)$ are polynomials in λ. The relation (12.3-3) can be proved by successive divisions according to the Euclidean algorithm, applied to ϕ_1 and ϕ_2. A check of the operations involved in that calculation shows that g_1 and g_2 have coefficients in $H_{11}(\Omega)$. The polynomial identity (12.3-3) remains true if λ is replaced by $A(x)$:

$$\phi_1(A(x), x) g_1(A(x), x) + \phi_2(A(x), x) g_2(A(x), x) = 1. \quad (12.3\text{-}4)$$

Let $N_j(x)$ be the null spaces of the matrices $\phi_j(A(x), x), j = 1, 2$, and denote their dimensions by $n_j(x)$. If $v(x) \in N_1(x) \cap N_2(x)$, multiplication of the last identity by $v(x)$ and a change of the order of multiplication shows that $v(x) = 0$. Hence N_1 and N_2 have only the zero vector in common. If $v(x)$ is *any* vector in $\mathbb{C}^n$, the same multiplication and reference to (12.3-2) shows that every vector is the sum of a vector in N_2 and one in N_1. Together, the two last facts imply that

$$n_1(x) + n_2(x) = n. \qquad (12.3\text{-}5)$$

Now, the rank of the matrix $\phi_1(A(x), x)$ is the maximal order of its nonzero minors. These minors are holomorphic, therefore their zeros are a discrete set. Except at these zeros, the rank is constant, and at the zeros it may *decrease* but not increase. The same is true for $n_2(x)$, so that (12.3-5) is possible only if $n_1(x)$ and $n_2(x)$ are constant in Ω. From now on, the notation will be simplified by writing n_1, n_2 for those dimensions. Also, ϕ_1, ϕ_2 will denote the functions in $H_{nn}(\Omega)$ with values $\phi_j(A(x), x), j = 1, 2$. Observe that $N_1(x), N_2(x)$ generally change with x.

By Theorem 12.1-2 there exist n_1 vectors $v_\mu \in H_{n1}(\Omega), \mu = 1, \ldots, n_1$, which span N_1 and n_2 vectors $w_\nu \in H_{n1}(\Omega)$ that span N_2. The matrix

$$T = \{v_1, v_2, \ldots, v_{n1}, w_1, w_2, \ldots, w_{n2}\}$$

as well as its inverse is in $H_{nn}(\Omega)$. It will now be shown that

$$T^{-1}AT = \begin{pmatrix} B & 0 \\ 0 & C \end{pmatrix}$$

with $B \in H_{n_1 n_1}(\Omega), C \in H_{n_2 n_2}(\Omega)$. This is a consequence of the immediately verifiable fact that A maps N_1 and N_2 into themselves, so that

$$AT = \{Av_1, \ldots, Av_{n_1}, Aw_1, \ldots, A_{n_2}\}$$

with

$$Av_\rho = \sum_{\mu=1}^{n_1} b_{\rho\mu} v_\mu, \qquad \rho = 1, 2, \ldots, n_1,$$

$$Aw_\sigma = \sum_{\nu=1}^{n_2} c_{\sigma\nu} w_\nu, \qquad \rho = 1, 2, \ldots, n_2.$$

Also, since $T^{-1}T = I$, one has

$$T^{-1}v_\mu = e_\mu, \qquad \mu = 1, 2, \ldots, n_1,$$

$$T^{-1}w_\nu = e_{n_1+\nu}, \qquad \nu = 1, 2, \ldots, n_2,$$

where e_α, $\alpha = 1, 2, \ldots, n$, are the unit vectors with 1 in the αth position and zeros elsewhere. Therefore,

$$T^{-1}AT = \begin{bmatrix}
b_{11} & b_{21} & \cdots & b_{n_1 1} & 0 & 0 & \cdots & & 0 \\
b_{12} & b_{22} & & b_{n_1 2} & 0 & 0 & \cdots & & 0 \\
\vdots & & & \vdots & \vdots & & & & \\
b_{1n_1} & b_{2n_1} & & b_{n_1 n_1} & & 0 & 0 & \cdots & 0 \\
0 & 0 & & 0 & c_{11} & c_{21} & \cdots & & c_{n_1 1} \\
& & & & c_{12} & c_{22} & \cdots & & c_{n_2 2} \\
\vdots & \vdots & & \vdots & \vdots & \vdots & & & \vdots \\
0 & 0 & \cdots & 0 & c_{1n_2} & c_{2n_2} & \cdots & & c_{n_2 n_2}
\end{bmatrix}.$$

Since T, T^{-1} and A are in $H_{nn}(\Omega)$, so are B and C.

12.4. Holomorphic Transformation of Matrices into Arnold's Form

On the basis of Theorem 12.3-1 every matrix $A \in H_{nn}(\Omega)$ is holomorphically similar to a direct sum of holomorphic matrices whose characteristic polynomial is not the product of two monic polynomials with coefficients holomorphic and relatively prime at every point $x \in \Omega$. Although this theorem is global in characters its strength depends on Ω. For a larger Ω some factorizations may no longer be possible. To give an example: Let $\phi(\lambda, x) = (\lambda^2 - (x-1))(\lambda^2 - (x+1))$. If Ω contains neither of the points $x = 1$, $x = -1$, then ϕ can be factored into four linear polynomials; if it contains one, but not the other of these points, holomorphic factorization into three polynomials is possible, but in regions which contain $x = 1$ and $x = -1$ the two holomorphic quadratic polynomials exhibited above are the only irreducible factors.

The further simplifications to be discussed below are local in character. From now on, Ω will be a, possibly small, neighborhood of a point, say $x = 0$.

The eigenvalues of $A(x)$ are the values of certain analytic functions. By choosing Ω small enough it can be guaranteed that it contains no branch points of these functions except possibly at $x = 0$. The functions λ_j so defined are locally holomorphic in Ω except possibly at $x = 0$ where they have finite limits. If more than one number is the limit at $x = 0$ of eigenvalues, the characteristic polynomial can be factored into at least two monic polynomials with holomorphic coefficients without common zeros in Ω. Theorem 12.3-1 then applies, and the problem of further simplifications of the matrix can be reduced to the case that all eigenvalues of $A(0)$ are equal. If $\lambda_1(0)$ is that eigenvalue, $A(0) - \lambda_1(0)I$ is nilpotent. Any similarity transformation that simplifies $A(x) - \lambda_1(0)I$ leads to a corresponding simpler form for $A(x) = A(x) - \lambda_1(0)I + \lambda_1(0)I$. Another natural preliminary simplification is to transform $A(0)$ into its Jordan form. The remaining problem is then that of simplifying, by holomorphic similarity transformations, matrices $A \in H_{nn}(\Omega)$ that satisfy the following assumption.

Hypothesis XII-2 (nonrestrictive). $A(0)$ *is nilpotent and in Jordan form.*

In general, such a matrix does not satisfy the conditions of Theorem 12.2-1 or 12.3-1. The further simplification to be described below is due to Arnold [3]. Arnold's arguments are best understood in a geometric terminology: An $n \times n$ constant matrix M with entries in $\mathbb{C}$ can be interpreted as a point in $\mathbb{C}^{n^2}$. The set of all constant matrices similar to M is a manifold in $\mathbb{C}^{n^2}$, which is often called the *orbit* of M. The dimension $l(M)$ of this manifold may be zero as, e.g., for $M = I$, and it is never as big as n, so that $0 \le l(M) < n^2$. Now assume that, instead of being constant, M is a certain holomorphic function, not of the scalar variable x, but of an m-dimensional vector variable $y \in \mathbb{C}^m$. The domain of y is to be a neighborhood $N \in \mathbb{C}^m$ of $y = 0$. Arnold calls such a function a *deformation* of $M(0)$. The introduction of y is an intermediate device, convenient even for the analysis of matrices that are functions of one variable only, as will be seen presently. The deformation of $M(0)$ is also a manifold in $\mathbb{C}^{n^2}$. Of particular interest is the case that the vectors tangent to the orbit of $M(0)$ and vectors tangent at $y = 0$ to the deformation M together span all of $\mathbb{C}^{n^2}$. When this is true the orbit and the deformation are said to be *transversal* to each other.

Arnold's method consists in choosing a deformation M that is as simple in structure as possible, but which has the property that every matrix $K \in \mathbb{C}^{n^2}$ close enough to $M(0)$ is holomorphically similar to a matrix $M(y)$ for some y near zero. Since later K will be a given matrix $A \in H_{nn}(\Omega)$ it will also be required that the transformation as well as the y that corresponds to a given K be holomorphic in the entries of K. A deformation which has this property at $y = 0$ is called *versal* by Arnold. Here is a more precise wording of this definition.

Definition 12.4-1. A deformation M of $M(0)$ is called versal at $y = 0$, if the equation

$$CM(y)C^{-1} = K \qquad (12.4-1)$$

can be satisfied, by a proper choice of $C \in \mathbb{C}^{n^2}$ and of $y \in \mathbb{C}^m$, for every matrix $K \in C^{n^2}$ with $|K - M(0)| < \delta$ (δ is a certain number) and if C and y are holomorphic functions of the entries of K at $K = M(0)$ with the values $C = I, y = 0$ when $K = M(0)$.

It turns out that being transversal and being versal are equivalent properties of a deformation N. To show that, a simple variant of the implicit function theorem is needed, which will now be proved.

Lemma 12.4-1. *Let f be a mapping of a region $D \subset C^p$ into C^q, i.e., $f(z) = w$ for $z \in D$, $w \in C^q$. Assume that f is holomorphic in D and that $p > q$. Then the following statements are equivalent:*

(i) *The Jacobian matrix $\partial f / \partial z$ has maximal rank q at $z = z^0 \in D$;*
(ii) *the mapping f has a local inverse, i.e., there exists a holomorphic mapping χ that takes the neighborhood $|w - f(z^0)| < \eta$ of $w^0 = f(z^0)$ into C^p such that*

$$f(\chi(w)) = w \qquad for \quad |w - w^0| < \eta, \tag{12.4-2}$$

and that $\chi(w^0) = z^0$ (χ need not be unique).

PROOF. (i) *implies* (ii): By hypothesis, the matrix $\partial f / \partial z$ of q rows and p columns has at least one minor of order q that is not zero at $z = z^0$. Assume, without loss of generality, that this is the minor formed by the first q columns of $\partial f / \partial z$, then f defines a mapping of the q-dimensional subset of D defined by the equations $z_j = z_j^0, j = q + 1, q + 2, \ldots, p$, into C^q. By the standard implicit function theorem this mapping has a unique inverse χ with the properties described in (ii).

(ii) *implies* (i): Differentiation of the identity 12.4-2 with respect to the components of the vector w yields the matrix identity

$$\frac{\partial f}{\partial z} \frac{\partial \chi}{\partial w} = I_q,$$

where I_q is the identity matrix of order q. Hence $\partial f / \partial z$ cannot have rank less than q near $z = z^0$, as was to be proved.

Theorem 12.4-1. *A deformation M of $M(0)$ is versal at $y = 0$ if and only if it is transversal to its orbit there.*

PROOF. Equation (12.4-1) is a case of the equation $f(z) = w$ if the following identifications are made: z is the vector of dimension $n^2 + m$ consisting of the entries of C and the components of y. The point z^0 is the vector corresponding to $C = I, y = 0$. Then n^2-dimensional vector w is the matrix K, with w^0 corresponding to $M(0)$. The left member of (12.4-1) defines the function f. Instead of exhibiting explicitly the entries of the $n^2 \times (n^2 + m)$ matrix $(\partial f / \partial z)_{z = z^0}$, the differential $CM(y)C^{-1}$ at $C = I, y = 0$ will be written down. One verifies readily that it is

$$dC \cdot M(0) - M(0) \cdot dC + \sum_{\mu=1}^{m} \frac{\partial M}{\partial y_\mu}\bigg|_{y=0} dy_\mu. \qquad (12.4\text{-}3)$$

This expression represents n^2 linear forms in the $n^2 + m$ scalar variables consisting of the n^2 entries of the matrix dC and the m variables dy_μ, $\mu = 1, 2, \ldots, m$. The matrix of n^2 rows and $n^2 + m$ columns formed by the coefficients of these linear forms is the Jacobian matrix $(\partial f/\partial z)_{z=z_0}$. Lemma 12.4-1 says that if the matrix has its maximally possible rank, namely, n^2, and only then, can equation (12.4-3) be satisfied by a matrix C and a vector y which are holomorphic function of the entries of K, provided K is close to $M(0)$. Thus, M is then versal at $y = 0$. On the other hand (12.4-1) can be interpreted geometrically: $dC \cdot M(0) - M(0) \cdot dC$ represents vectors in $\mathbb{C}^{n^2}$ parallel to the tangent space of the orbit of $M(0)$, and

$$\sum_{\mu=1}^{m} \frac{\partial M}{\partial y_\mu}\bigg|_{y=0} dy_\mu,$$

is a family of vectors in the tangent space of the deformation at $y = 0$. The fact that every vector in $\mathbb{C}^{n^2}$ is a sum of two such vectors proves that the two manifolds are transversal to each other at $y = 0$, as was to be proved.

Before applying Theorem 12.4-1 to the problem of simplifying matrices $A \in H_{nn}(\Omega)$, a few examples are in order.

EXAMPLE 1

$$M(y) = \begin{pmatrix} 0 & 1 \\ y_1 & y_2 \end{pmatrix}.$$

If one writes $c = \left[\begin{smallmatrix} c_{11} & c_{21} \\ c_{12} & c_{22} \end{smallmatrix}\right]$, formula (12.4-3) is

$$\begin{pmatrix} -dc_{21} & dc_{11} - dc_{22} \\ dy_1 & dc_{21} + dy_2 \end{pmatrix}$$

for this matrix. The matrix of this set of four linear forms of the six scalar variables dc_{jk}, dy_j, $j,k = 1,2$, is, for one ordering of these variables and forms,

$$\begin{bmatrix} 0 & 0 & -1 & 0 & 0 & 0 \\ 1 & 0 & 0 & -1 & 0 & 0 \\ 0 & 0 & 0 & 0 & 1 & 0 \\ 0 & 0 & 1 & 0 & 0 & 1 \end{bmatrix}.$$

The rank is four, which is maximal. Therefore, M is transversal to its orbit at $y = 0$ and, therefore, versal as well.

EXAMPLE 2

$$M(y) = \begin{pmatrix} y_1 & 1 + y_3 \\ 0 & y_2 \end{pmatrix}.$$

A calculation analogous to that for Example 1 shows that the matrix (12.4-3) is

$$\begin{pmatrix} -dc_{21} + dy_1 & dc_{11} - dc_{22} + dy_3 \\ 0 & c_{21} + dy_2 \end{pmatrix}$$

and the matrix of these four forms in seven variables has one row that is zero, so that its rank is at most three. Hence, in this example the matrix M is not transversal to its orbit at $y = 0$ and, therefore, cannot be versal there.

This example also illustrates the fact that the condition of versality must be interpreted carefully. *Every* 2×2 matrix K close to $M(0)$ is similar to the matrix $M(y)$ for some values of y_1, y_2, y_3. That property is, however not enough. The dependence of C and y on K must be holomorphic. For

$$K = \begin{pmatrix} 0 & 1 \\ \alpha & 0 \end{pmatrix},$$

for instance, one must have $y_1 = \alpha^{1/2}$, $y_2 = -\alpha^{1/2}$, $y_3 = 0$, since K and $M(y)$ always have the same eigenvalues. The function $\alpha^{1/2}$ is not holomorphic at $\alpha = 0$.

EXAMPLE 3

$$M(y) = \begin{pmatrix} 1 + y_1 & 0 \\ 0 & y_2 \end{pmatrix}.$$

To show that this M is versal one has to solve the equation $C^{-1}KC = M(y)$ for C and y. If there is a solution, the eigenvalues of K are $\lambda_1 = 1 + y_1$, $\lambda_2 = y_2$. For K close to $M(0)$, λ_1 and λ_2 are distinct and, therefore, holomorphic in the entries of K. One must therefore have $y_1 = \lambda_1 - 1$, $y_2 = \lambda_2$, and the matrix

$$C = \begin{pmatrix} 1 & 1 \\ \lambda_1 & \lambda_2 \end{pmatrix}$$

then established the desired similarity transformation of K into $M(y)$. Observe that in this case y is uniquely determined but C, of course is not. A versal deformation for which K determines y uniquely is called "universal" by Arnold.

The foregoing concepts and results will now be applied to the given matrix $A \in H_{nn}(\Omega)$, which will play the role of K in formula (12.4-1). If M is a versal deformation at $y = 0$, the matrices C and the vector y are then holomorphic in x at $x = 0$, and $y = 0$ for $x = 0$. The task is to construct a particularly simple deformation M, simpler, in general, than the given matrix A. The deformation M must coincide with A at $x = y = 0$, i.e., $M(0) = A(0)$.

For the construction of a simple versal deformation M of $A(0) = M(0)$ a knowledge of the rank of the differential form $dC M(0) - M(0) dC$ in (12.4-3) is needed. Fortunately, this is a standard result of matrix theory, which can be found, e.g., in [15] Ch. VIII, §2 and will be stated here without proof.

Lemma 12.4-2. *Assume that $A(0)$ is in Jordan form and nilpotent and that, in addition, the diagonal blocks of $A(0)$ corresponding to its elementary divisors are arranged in order of decreasing size: $\mu_1 \geqslant \mu_2 \geqslant \cdots \geqslant \mu_p$. Then the nullspace of the mapping*

$$L(X) := XA(0) - A(0)X \qquad (12.4\text{-}4)$$

has dimension

$$d = \mu_1 + 3\mu_2 + \cdots + (2p - 1)\mu_p.$$

It follows from this lemma that the part $dC\,M(0) - M(0)\,dC$ of (12.4-3) has rank $n^2 - d$ when, as is assumed here, $M(0) = A(0)$ and $A(0)$ satisfies Hypothesis XII-2. Hence, to be versal M must be constructed so that the addition of the summation term in (12.4-3) to $dC \cdot M(0) - M(0) \cdot dC$ increases the rank from $n^2 - d$ to n^2. As only the derivatives $(\partial M / \partial y_\mu)_{y=0}$ matter it will suffice to consider deformations M that are linear in y.

Thus, m in (12.4-3) will have to be equal to d, and the constant $n \times n$ matrices $(\partial M / \partial y_\mu)|_{y=0}$ must be independent, as well as independent of the range of the operator L defined in (12.4-4). To construct a basis for the range of L another fact that can be found in Gantmacher's book [15] will be used. There, the following basis of matrices for the nullspace of L is introduced: they have nonzero entries in only one of the p^2 rectangular blocks in the partition of $A(0)$ generated by its Jordan blocks. These entries have the value 1 and will fill a slanting segment parallel to the main diagonal of $A(0)$. Also, the end points of this segment of 1's must lie on those two sides of the rectangular block which meet at its upper right-hand vertex. All such matrices belong to the basis (see Fig. 12.1).

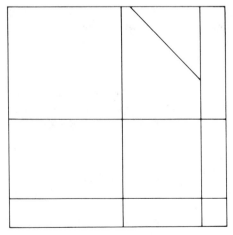

Figure 12.1. Illustration for the construction of a basis for the nullspace of L.

One easily constructed set of d vectors independent of the range of L consists of vectors orthogonal to that range. The term "orthogonal" requires that an inner product be introduced in the space of all $n \times n$ matrices when they are interpreted as vectors of $\mathbb{C}^{n^2}$. The most natural inner product of two such matrices $U = \{u_{ij}\}$, $V = \{v_{ij}\}$ is to define

$$(U, V) = \sum_{i,j=1}^{n} u_{ij} \bar{v}_{ij} = \mathrm{tr}(UV^*).$$

Here tr indicates the trace, and $V^* = \bar{V}^T$ is the adjoint of B.

Lemma 12.4-3. *An $n \times n$ matrix P is orthogonal to the range of the operator L defined by $L(X) = XA(0) - A(0)X$ if and only if P^* is the nullspace of L.*

PROOF. $\mathrm{tr}(L(X)P^*) = \mathrm{tr}(XA(0)P^*) - \mathrm{tr}(A(0)XP^*) = \mathrm{tr}(XA(0)P^* - XP^*A(0)) = -\mathrm{tr}(XL(P^*))$. If P is orthogonal to the range of L, the first member of these equalities is zero for all $X \in \mathbb{C}^{n^2}$. The last member is zero for all X if and only if $L(p^*) = 0$, i.e., if and only if P^* is the nullspace of L.

This lemma proves that the transposes of the d special real matrices which were a basis for the nullspace of L constitute a basis for the orthogonal complement of the range of L. Let these matrices be denoted by $Q_1, Q_2, \ldots, Q_d$ and define the deformation M by

$$M(y) = A(0) + \sum_{j=1}^{d} Q_j y_j. \qquad (12.4\text{-}5)$$

For this deformation, $\partial M / \partial y_j = Q_j$ and the differential in (12.4-3) has indeed rank $(n^2 - d) + d = n^2$. Therefore (12.4-5) is a versal deformation. Moreover, the argument given shows that for no versal deformation of $A(0)$ can m be less than d.

The deformation M in (12.4-5) can be replaced by one that is, in general, considerably sparser. To do this, one observes first that one can add to Q_j any matrix from the range of L, i.e., any matrix of the form $XA(0) - A(0)X$. The resulting new set of matrices, when joined to the range of L will still span all of $\mathbb{C}^{n^2}$. Only the orthogonality property of the Q_j has been sacrificed. To make use of this remark, consider, in particular, matrices X which have only one nonzero entry, which is equal to 1. One verifies directly that the action of L on such matrices can be described as follows: If the nonzero entry is in the ρth row and σth column of X, the $L(X)$ has nonzero entries in the positions $(\rho, \sigma + 1)$ and $(\rho - 1, \sigma)$, at most. If both these positions are in the same block of the partition of X induced by the partition of $A(0)$, then these entries in $L(X)$ have the values 1 and -1, respectively. If one or both of these shifted positions are outside the block containing the position (ρ, σ), the corresponding entry is zero. By adding suitable linear combinations of such special matrices of the form $L(X)$ to the Q_j one can replace all but one of the nonzero entries in Q_j by zero. Finally, the remaining nonzero entry can be changed into 1 by a division of the matrix by a scalar. There is some arbitrariness in the location of this one nonzero entry. Following Arnold it will be taken in the *last row* of a block if the block is diagonal or above the diagonal, and in the *first column* for blocks below the diagonal. Let these matrices be called Γ_j, $j = 1$, $2, \ldots, d$, in some ordering.

The theorem below summarized the results obtained.

Theorem 12.4-2. *Let $A(0)$ be in Jordan form and nilpotent, and let the matrices $\Gamma_1, \Gamma_2, \ldots, \Gamma_d$ be defined as above. If $y = (y_1, y_2, \ldots, y_d)$, then*

$$M(y) = A(0) + \sum_{j=1}^{d} \Gamma_j y_j \qquad (12.4\text{-}6)$$

is a versal deformation of $A(0)$. There is no versal deformation of $A(0)$ with fewer than d parameters.

An equivalent but more explicit formulation of the same result is the following.

Corollary. *Let $A \in H_{nn}(\Omega)$. If $A(0)$ is nilpotent and in Jordan form, then A is, at $x = 0$, holomorphically similar to a matrix of the form*

$$A(0) + \sum_{j=1}^{d} \Gamma_j \rho_j(x), \qquad (12.4\text{-}7)$$

with $\rho_j \in H_{11}(\Omega)$, $j = 1, 2, \ldots, d$.

PROOF. Since M in (12.4-6) is versal, the matrix $A(x)$ is similar to $M(y)$ near $y = 0$ for some vector y whose components are holomorphic in the components of $A(x)$ near $x = 0$ and, therefore, holomorphic functions of x near $x = 0$.

The fact that the matrices of the form $A(0)X - XA(0)$ together with the matrices Γ_j span all the $\mathbb{C}^{n \times n}$ deserves to be stated explicitly.

Lemma 12.4-4. *Every matrix $P \in \mathbb{C}^{n \times n}$ can be represented (not uniquely) in the form*

$$P = A(0)X - XA(0) - \sum_{j=1}^{d} \rho_j \Gamma_j \qquad (12.4\text{-}8)$$

by a suitable choice of the matrix $X \in \mathbb{C}^{n \times n}$ and of the d complex scalars ρ_j.

That the representation (12.4-8) is not unique is almost obvious, since, for given P, it can be interpreted as a system of n^2 linear equations in the $N^2 + d$ unknown represented by the entries of X and the ρ_j, $j = 1$, $2, \ldots, d$.

Arnold's matrices of the form (12.4-7) are not canonical in the sense that every holomorphic matrix is holomorphically similar to only one Arnold matrix. Two different such Arnold matrices may be similar to each other. For instance, the Arnold matrices

$$A = \begin{bmatrix} 0 & 1 & 0 \\ 0 & 0 & x \\ 2x & 0 & 0 \end{bmatrix}, \qquad B = \begin{bmatrix} 0 & 1 & 0 \\ 0 & 0 & 2x \\ x & 0 & 0 \end{bmatrix}$$

satisfy the relation $T^{-1}AT = B$ with

$$T = \begin{bmatrix} 1 & 0 & 0 \\ 0 & 1 & 0 \\ 0 & 0 & 2 \end{bmatrix}.$$

Whether a reduction to Arnold's form is advantageous or not depends on the form of $A(0)$. For matrices $A(0)$ with only one distinct eigenvalue the matrix M is quite sparse if d is small, for instance if $A(0)$ has only one Jordan block. At the other extreme, if $A(0)$ is diagonal then every deformation M is versal.

References

[1] Abramowitz, M. and I. A. Stegun, *Handbook of Mathematical Functions*, National Bureau of Standards Applied Math. Series 55, U.S. Gov't Printing Office, 9th Printing, Washington, D. C. 1964.

[2] Ackerberg, R. C. and R. E. O'Malley, *Boundary layer problems exhibiting resonance*, Studies in Appl. Math. 49 (1970), 277–295.

[3] Arnol'd, V. E., *On matrices depending on parameters*, Uspehi Mat Nauk, No. 2 (1971), 101–114. English translation in Russian Math. Surveys 26, No. 2 (1971) 29–43.

[4] Besicovitch, A. S., *Über analytische Funktionen mit vorgeschriebenen Werten ihrer Ableitungen*, Math. Zeitschr. 21 (1924), 111–118.

[5] Borel, E. *Leçons sur les séries divergentes*, Gauthiers-Villers, Paris 1901.

[6] Brillouin, L., *Remarques sur la mécanique ondulatoire*, J. Phys. Radium [6] 7 (1926), 353–368.

[7] Carlini, F. (Jacobi), *Untersuchungen über die Konvergenz der Reihe durch welche das Keplersche Problem gelöst wird*, Astr. Nachr. 30 (1850) 198–211. Revised translation by Jacobi of *Ricerche sulla convergenza della serie che serve alla soluzione del problema di Keplero*, Memoria di F. Carlini, Milano, 1817.

[8] de Groen, P. P. N., *The singularly perturbed turning point problem: a spectral approach*, Singular Perturbations and Asymptotics, edited by R. E. Meyer and S. V. Parter, Acad. Press, New York, 1980, pp. 149–172.

[9] Drazin, P. G. and W. H. Reid, *Hydrodynamic Stability*, Cambridge Univ. Press, New York, 1981.

[10] Erdélyi, A., *Asymptotic Expansions*, Dover Publications, New York 1956.

[11] Evgrafov, M. A. and M. V. Fedoryuk, *Asymptotic behavior as $\lambda \to \infty$ of the solution of the equation $w''(z) - p(z,\lambda)w(z) = 0$ in the complex z-plane*, Uspehi Mat. Nauk 21, No. 1 (127) (1966) 3–50 (Russian). English translation in Russian Math. Survey 21 (1966), 1–48.

[12] Friedland, S., *On pointwise and analytic similarity of matrices*, Israel J. Math. 35 (1980), 89–108.

[13] Fröman, N. and P. O. Fröman, *JWKB Approximation*, Contributions to the theory, North-Holland, Amsterdam, 1965.

[14] Gans, R., *Fortpflanzung des Lichts durch ein inhomogenes Medium*, Ann. der Phys. (4) 47, (1915), 709–736.

[15] Gantmacher, F. R., *The Theory of Matrices*, Chelsea Publ. Co., New York, 1959.

[16] Gingold, H., *A method of global block diagonalization for matrix-valued functions*, SIAM J. Math. Anal. 9 (1978), 1076–1082.

[17] Green, G., *On the motion of waves in a variable canal of small depth and width*, Trans. Cambridge Philos. Soc. 6 (1837), 457–462.

[18] Hanson, R. J., *Reduction theorems for systems of ordinary differential equations with a turning point*, J. Math. Anal. Appl. 16 (1966), 280–301.

[19] Hanson, R. J., *Simplification of second order systems of ordinary differential equations at a turning point*, SIAM J. Appl. Math. 16 (1968), 1059–1080.

[20] Hanson, R. J. and D. L. Russell, *Classification and reduction of second-order systems at a turning point*, J. Math. and Phys. 46 (1967), 74–92.

[21] Harris, W. A. Jr., *Singular perturbations of two-point boundary problems for systems of ordinary differential equations*, Arch. Rat. Mech. Anal. 5 (1960), 212–225.

[22] Heisenberg, W., *Über Stabilität und Turbulenz von Flüssigkeitsströmen*, Ann. Phys. Leipzig (4) (1924) 74, 577–627.

[23] Hille, Einar, *Analytic Function Theory*, Vol. II., Ginn and Company, New York, 1962.

[24] Hsieh, P.-F., *On an analytic simplification of a system of linear ordinary differential equations containing a parameter*, Proc. Amer. Math. Soc. 19 (1968), 1201–1206.

[25] Hsieh, P.-F. and Y. Sibuya, *On the asymptotic integration of second order linear differential equations with polynomial coefficients*, J. Math. Anal. Appl. 16 (1966), 84–103.

[26] Iwano, M., *Asymptotic solutions of a system of linear ordinary differential equations containing a small parameter, I*, Funk. Ekvac. 5 (1963), 71–133.

[27] Iwano, M., *Asymptotic solutions of a system of linear ordinary differential equations containing a small parameter, II*, Funk. Ekvac. 6 (1964), 89–139.

[28] Iwano, M., *On the study of asymptotic solutions of a system of linear ordinary differential equations containing a parameter with a singular point*, Jap. J. of Math. 35 (1965), 1–30.

[29] Iwano, M. and Y. Sibuya, *Reduction of the order of a linear ordinary differential equation with a turning point*, Kōdai Math. Sem. Rep. 15 (1963), 1–28.

[30] Jeffreys, H., *On approximate solutions of linear differential equations*, Proc. Cambridge Philos. Soc. 49 (1953), 601–611.

[31] Kazarinoff, N., *Asymptotic theory of second order differential equations with two simple turning points*, Arch. Rat. Mech. Anal. 2 (1958), 129–150.

[32] Kazarinoff, N. and R. McKelvey, *Asymptotic solutions of differential equations in a domain containing a regular singular point*, Can. J. Math. (1956), 97–104.

[33] Kopell, N., *The singularly perturbed turning point problem: A geometric approach*, Singular Perturbations and Asymptotics, edited by R. E. Meyer and S. V. Parter, Academic Press, New York, 1980, pp. 173–189.

[34] Kramers, H. A., *Wellenmechanik und halbzahlige Quantisierung*, Z. Physik 39 (1926), 828–840.

[35] Kreiss, H.-O., *Resonance for singular perturbation problems*, SIAM J. Appl. Math. 41 (1981), 331–344.

[36] Langer, R. E., *On the asymptotic solutions of differential equations with an application to the Bessel functions of large complex order*, Trans. Amer. Math. Soc. 34 (1932), 447–480.

[37] Langer, R. E., *On the asymptotic solutions of ordinary linear differential equations of the second order, with special reference to the Stokes phenomenon*, Bull. Amer. Math. Soc. 40 (1934), 545–582.

[38] Langer, R. E., *The asymptotic solutions of ordinary linear differential equations*

of the second-order, with special reference to a turning point, Trans. Amer. Math. Soc., 67 (1949), 461–490.

[39] Langer, R. E., *Turning points in linear asymptotic theory*, Bol. Soc. Mat. Mexicana (2) 5 (1960), 1–12.

[40] Lee, Roy Y., *On uniform simplification of linear differential equations in a full neighborhood of a turning point*, J. Math. Anal. Appl. 27 (1969), 501–510.

[41] Leung, A., *Studies on doubly asymptotic series solutions for differential equations in unbounded domains*, J. Math. Anal. Appl. 44 (1973), 238–263.

[42] Leung, A., *Doubly asymptotic series of nth order differential equations in unbounded domains*, SIAM J. Math. Anal. 5 (1974), 187–201.

[43] Leung, A., *Lateral connections for asymptotic solutions for higher order turning points in unbounded domains*, J. Math. Anal. Appl. 50 (1975), 560–578.

[44] Leung, A., *A doubly asymptotic existence theorem and application to order reduction*, Proc. London Math. Soc. (3) 33 (1976), 151–176.

[45] Lin, C. C., *The Theory of Hydrodynamic Stability*, Cambridge Univ. Press, 1966.

[46] Liouville, J., *Sur le développement des fonctions ou parties de fonctions en séries*, J. Math. Pures et Appl. [1], 2 (1937), 16–35.

[47] McKelvey, R. W., *The solutions of second order ordinary differential equations about a turning point of order two*, Trans. Amer. Math. Soc. 79 (1955), 103–123.

[48] Matkowsky, B. F., *On boundary layer problems exhibiting resonance*, SIAM Rev. 17 (1975), 82–100.

[49] Nakano, M., *On a system of linear ordinary differential equations with a turning point*, Kōdai Math. Sem. Rep. 21 (1969), 1–15.

[50] Nakano, M., *On a system of linear ordinary differential equations related to a turning point problem*, Kōdai Math. Sem. Rep. 21 (1969), 472–490.

[51] Nakano, M., *Second order linear ordinary differential equations with turning points and singularities, I*, Kōdai Math. Sem. Rep. 29 (1977), 88–102.

[52] Nakano, M. and T. Nishimoto, *On a secondary turning point problem*, Kōdai Math. Sem. Rep. 22 (1970), 355–384.

[53] Nishimoto, T., *On matching methods for a linear ordinary differential equation containing a parameter I, II, III*, Kōdai Math. Sem. Rep. 17 (1965), 307–328; 18 (1966) 61–86; 19 (1967), 80–94.

[54] Nishimoto, T., *On matching methods in turning point problems*, Kōdai Math. Sem. Rep. 17 (1965), 198–221.

[55] Nishimoto, T., *A turning point problem of an n-th order differential equation of hydrodynamic type*, Kōdai Math. Sem. Rep. 20 (1968), 218–256.

[56] Nishimoto, T., *On the central connection problem at a turning point*, Kōdai Math. Sem. Rep. 22 (1970), 30–44.

[57] Nishimoto, T., *On an extension theorem and its application for turning point problems of large order*, Kōdai Math. Sem. Rep. 25 (1973), 458–489.

[58] Nishimoto, T., *Global solutions of certain fourth order differential equations*, Kōdai Math. Sem. Rep. 27 (1976), 128–146.

[59] Nishimoto, T., *On the Orr-Sommerfeld type equations, I W.K.B. Approximations; II connection formulas*, Kōdai Math. Sem. Rep. 24 (1972), 281–307; 29 (1978), 233–249.

[60] Olver, F. W. J., *Error analysis of phase-integral methods I; II*, J. Res. Nat. Bur. Standards, Sec. B, 69B (1965), 271–290; ibid., 291–300.

[61] Olver, F. W. J., *Asymptotics and Special Functions*, Academic Press, N.Y. 1974.

[62] Olver, F. W. J., *Connection formulas for second-order differential equations with multiple turning points*, SIAM J. Math. Anal. 8 (1977), 127–154.

[63] Olver, F. W. J., *Connection formulas for second-order differential equations*

having an arbitrary number of turning points of arbitrary multiplicites, SIAM J. Math. Anal. 8 (1977), 673–700.

[64]   Olver, F. W. J., *Second-order differential equations with fractional transition points*, Trans. Amer. Math. Soc. 226 (1977), 227–241.

[65]   Olver, F. W. J., *General connection formulae for Liouville-Green approximations in the complex plane*, Phil. Trans. R. Soc. London, 289 (1978), 501–548.

[66]   Pittnauer, F., *Vorlesungen über asymptotische Reihen*, Lecture Notes in Math., No. 301, Springer-Verlag, New York, 1972.

[67]   Ramis, J. P., *Dévissage Gevrey*, Astérisque, 59–60 (1978), 173–204.

[68]   Ritt, J. F., *On the derivatives of a function at a point*, Ann. Math. 18 (1916), 18–23.

[69]   Rudin, W., *Real and Complex Analysis*, McGraw-Hill Publ. Co., New York, N.Y. 1966.

[70]   Schlissel, A., *The development of asymptotic solutions of linear ordinary differential equations, 1817-1920*, Arch. Hist. of Exact Sci. 16 (1977), 307–378.

[71]   Schlissel, A., *The initial development of the WKB solutions of linear second order ordinary differential equations and their use in the turning point problem*, Historic Math. 4 (1977), 183–204.

[72]   Sibuya, Y., *Sur réduction analytique d'un système d'équations différentielles ordinairès linéaires contenant un paramètre*, J. Fac. Sci. Univ. Tokyo, Sec. I, 7 (1958), 527–540.

[73]   Sibuya, Y., *Simplification of a system of linear ordinary differential equations about a singular point*, Funkc. Ekvac. 4 (1962), 29–56.

[74]   Sibuya, Y., *Asymptotic solutions of a system of linear ordinary differential equations containing a parameter*, Funkc. Ekvac. 4 (1962), 115–139.

[75]   Sibuya, Y., *Formal solutions of a linear ordinary differential equation of the n-th order at a turning point*, Funkc. Ekvac. 4 (1962), 115–139.

[76]   Sibuya, Y., *Simplification of a linear ordinary differential equation of the nth order at a turning point*, Arch. Rat. Mech. Anal. 13 (1963), 206–221.

[77]   Sibuya, Y., *Uniform Simplification in a Full Neighborhood of a Transition Point*, Memoirs of the Am. Math. Soc., No. 149, 1974.

[78]   Sibuya, Y., *Global Theory of a Second Order Linear Ordinary Differential Equation with a Polynomial Coefficient*, North-Holland Math. Studies 18, North-Holland–American Elsevier Publishing Company, Amsterdam–New York, 1975.

[79]   Sibuya, Y., *A theorem concerning uniform simplification at a transition point and the problem of resonance*, SIAM J. Math. Anal. 12 (1981), 653–668.

[80]   Tollmien, W., *Über die Entstehung der Turbulenz*, 1. Mitteilung, Nachr. Gesellschaft der Wiss. Göttingen (1929).

[81]   Tollmien, W., *Asymptotische Integration der Störungsdifferentialgleichung ebener laminarer Strömungen by hohen Reynoldsschen Zahlen*, Zeitschr Angew. Math. Mech. 25/27 (1947), 33–50, 70–83.

[82]   Turrittin, H. L., *Stokes multipliers for asymptotic solutions of a certain differential equation*, Trans. Am. Math. Soc. 68 (1950), 304–329.

[83]   Turrittin, H. L., *Asymptotic expansions of solutions of systems of ordinary differential equations*, Contributions to the Theory of Nonlinear Oscillations II, Am. of Math Studies, No. 29, Princeton 1952, pp. 81–116.

[84]   Wasow, W., *On the asymptotic solution of boundary value problems for ordinary differential equations containing a parameter*, J. Math. Phys. 23 (1944), 173–183.

[85]   Wasow, W., *The complex asymptotic theory of a fourth order differential equation of hydrodynamics*, Ann. of Math., 49 (1948), 852–871.

[86]   Wasow, W., *On small disturbances of plane Couette flow*, J. Res. Nat. Bur. Stand. 51 (1953), 195–202.

[87] Wasow, W., *On holomorphically similar matrices*, J. Math. Anal. Appl. 4 (1962), 202–206.

[88] Wasow, W., *Turning points for systems of linear equations*, I. The formal theory, Comm. Pure Appl. Math. 14 (1961), 657–673, II. The analytic theory, Ibid. 15 (1962), 173–187.

[89] Wasow, W., *Simplification of turning point problems for systems of linear differential equations*, Trans. Amer. Math. Soc. 106 (1963), 100–114.

[90] Wasow, W., *Asymptotic Expansions for Ordinary Differential Equations*, Interscience Publishers, New York 1965.

[91] Wasow, W., *On turning point problems for systems with almost diagonal coefficient matrix*, Funk. Ekvac. 8 (1966), 143–171.

[92] Wasow, W., *On the analytic validity of formal simplifications of linear differential equations*, Funk. Ekvacioj 9 (1966), 83–91 and 10 (1967), 107–122.

[93] Wasow, W., *Simple turning point problems in unbounded domains*, SIAM J. Math. Anal. 1 (1970), 153–170.

[94] Wasow, W., *The central connection problem at turning points of linear differential equations*, Comm. Math. Helvetici 46 (1971), 65–86.

[95] Wasow, W., *Arnol'd's canonical matrices and the asymptotic simplification of ordinary differential equations*, Linear Alg. and Appl. 18 (1977), 163–170.

[96] Wasow, W., *Topics in the Theory of Linear Ordinary Differential Equations having Singularities with Respect to a Parameter*, Inst. Rech. Math. Avancée, Univ. Louis Pasteur, Strasbourg, 1978.

[97] Wentzel, G., *Eine Verallgemeinerung der Quantenbedingungen für die Zwecke der Wellenmechanik*, Z. Physik 38 (1926), 518–529.

[98] Weyl, H. *Mathematische Analyse des Raumproblems*, Springer-Verlag, Berlin 1923. (Reprinted by Wissenschaftliche Buchgesellschaft Darmstadt, 1977.)

Index